Beef Production
and Management

Beef Production and Management

Gary L. Minish
Danny G. Fox

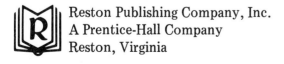
Reston Publishing Company, Inc.
A Prentice-Hall Company
Reston, Virginia

Library of Congress Cataloging in Publication Data

Minish, Gary L.
   Beef production and management.

   1. Beef Cattle.   I. Fox, Danny G., Joint author
II. Title.
SF207.M56   636.2'1'3   79-4617
ISBN 0-8359-0445-8

© 1979 by Reston Publishing Company, Inc.
      A Prentice-Hall Company
      Reston, Virginia 22090

All rights reserved. No part of this book may be reproduced in any way, or by any means, without permission in writing from the publisher.

10 9 8 7 6 5 4 3 2 1

Printed in the United States of America

Contents

Preface **xi**

1 Beef Production: What It Costs and Opportunities for Improving It **1**

   1-1 Economic Costs in Producing Beef, 3
   1-2 Reducing the Cost of Producing Beef, 13

2 Breeding Principles **19**

   2-1 Animal Breeding Fundamentals, 19
   2-2 Selection Response, 22
   2-3 Systems of Selection, 26
   2-4 Selection to Improve Overall Merit, 27
   2-5 Estimated Breeding Values, 29
   2-6 Tools for Beef Cattle Improvement, 29
   2-7 Performance Records, 30

3 Breeds and Breeding Systems **37**

   3-1 Selecting a Breed of Beef Cattle, 37
   3-2 Classification of Breeds, 57
   3-3 Choosing a Breeding System, 59

4 Sire Selection **67**

   4-1 Factors to Consider in Sire Selection, 68

viii   Contents

    4-2 The Mating Game, 83
    4-3 Official Ranking of Five AI Sires, 84
    4-4 Sire Selection Precautions, 91
    4-5 Criteria to Consider in Selecting AI Sires, 92
    4-6 Summary on Sire Selection, 92

5 **Selecting the Productive Female**   95

    5-1 Factors to Consider in Selecting Females for the Herd, 96
    5-2 Replacement Heifer Selection, 104
    5-3 Criteria for Culling Cows, 105
    5-4 Milk Production, 105

6 **Cow Herd Management**   109

    6-1 Beef Cattle Identification, 110
    6-2 Management Procedures for Beef AI, 116
    6-3 Beef Herd Health Program, 122
    6-4 Resource Requirements for a Cow Herd, 128
    6-5 Forage Management Systems, 131
    6-6 Beef Cow Herd Management Calendar, 136

7 **Selling, Buying, and Managing Feeder Cattle**   145

    7-1 Developing a Marketing Strategy, 145
    7-2 Selecting the Most Desirable Type of Feeder Cattle, 149
    7-3 Marketing Alternatives and Sources for Feeder Calves, 154
    7-4 New Feeder Cattle: Nutrition, Health, and Treatments, 162

8 **Nutrient Requirements of Beef Cattle**   177

    8-1 Impact of Frame Size on Nutrient Requirements, 178
    8-2 Water Requirements and Water Quality for Beef Cattle, 180
    8-3 Mineral Requirements of Beef Cattle, 187
    8-4 Growth Stimulants and Feed Additives, 200
    8-5 Energy Utilization by Cattle and the Use of Energy Values in Ration Formulation, 203
    8-6 Protein Requirements and Feed Protein Utilization, 218
    8-7 Summary of Nutrient Requirements for Growing and Finishing Cattle, 226
    8-8 Nutrient Requirements of Breeding Cattle, 233

9 Evaluation of Feedstuffs and Ration Formulation — 245
- 9-1 The Use of Feed Analysis, 245
- 9-2 Feed Composition Values, 256
- 9-3 The Impact of Feed Processing on Feedstuff Utilization, 260
- 9-4 Determining Daily Dry Matter Intakes of Growing and Finishing Cattle, 265
- 9-5 Correcting for Moisture Content of Feeds, 274
- 9-6 Ration Formulation, 280

10 Simple Guidelines for Feeding Beef Cattle — 289
- 10-1 Guideline Rations for the Beef Herd, 289
- 10-2 Guidelines for Feeding Bull Calves, 299
- 10-3 Creep Feeding Calves, 304
- 10-4 Feeding Grain on Pasture, 305
- 10-5 Guideline Rations for Growing and Finishing Beef, 306
- 10-6 Treating Corn Silage with NPN for Growing and Finishing Beef Cattle, 311
- 10-7 Expected Performance with Various Levels of Silage Feeding, 321

11 Systems Analysis: Developing the Most Profitable Management System — 323
- 11-1 Computerized Ration Balancing and Performance Stimulation for Growing and Finishing Cattle, 323
- 11-2 Long-Range Feeding System Planning, 340
- 11-3 Beef Herd Feeding System, 343

12 Marketing Finished Cattle — 353
- 12-1 When to Market Slaughter Cattle and Methods of Marketing, 353
- 12-2 Factors Affecting Carcass Grade, 361
- 12-3 Placing Slaughter Cattle in a Show, 376

13 Facilities and Feed Storage for Beef Cattle — 381
- 13-1 Beef Cow Calf, 382
- 13-2 Facilities for Growing and Finishing Cattle, 386
- 13-3 Making and Storing High Quality Silage and High Moisture Corn, 389

Preface

The beef industry is highly competitive, with over 1,400,000 cow-calf producers and 138,000 cattle feeders. In order to be successful, a beef producer must utilize the best information on breeding, feeding, buying and selling cattle, as well as good general management practices to develop the best operation.

This textbook is designed to provide both the student and the beef producer with specific guidelines for breeding, feeding, and managing a successful beef cattle enterprise. The introductory chapter describes the beef cattle industry in the United States and its efficiency, and it identifies the practices that have the greatest impact on the costs of production. A producer must set priorities for the use of his time and management skills in order to concentrate on those factors that mean the most in terms of profits. In this and subsequent chapters, the management priorities for optimizing returns are emphasized.

The first portion of the book deals with breeding and managing the cow-calf herd. These chapters review the fundamentals needed for selecting breeding stock. The guidelines offered are based on the latest information available on beef cattle breeding and on author Gary Minish's experience from working within the purebred cattle industry. Various mating systems and their limitations are outlined. A description of over 30 breeds of beef cattle is presented, as well as a discussion of the best utilization of each breed under various conditions.

This portion of the text also contains many practical guidelines for managing the beef herd, such as identification systems, herd health programs, and an overall management system for the complete reproductive cycle of the beef herd.

The last half of the book deals with buying and selling, feeding, and managing growing and finishing cattle. Three chapters specifically cover the nutrition of all classes of beef cattle. Author Dan Fox summarizes the latest information on energy and protein requirements, feed evaluation, and ration formulation. Adjustments are included for various factors, such as frame size and cattle type, feed additives, feed quality, and environment, which are based on the author's own research and his experience working with cattle feeders. Simple guidelines for feeding cattle are presented, including guideline rations for the breeding herd and for growing and finishing cattle.

A complete discussion of treating silage with NPN is presented in this part of the book, as well as an entire chapter on the subject of developing the most profitable feeding system. Shelter systems and harvesting and storing high quality silage are discussed in the final chapter.

The authors wish to acknowledge faculty associates who have contributed to their programs and to the data included in this text:

T. L. Bibb, College of Veterinary Medicine
Virginia Polytechnic Institute

J. R. Black, Department of Agricultural Economics
Michigan State University

A. L. Eller, Jr., Animal Science Department
Virginia Polytechnic Institute

T. N. Meacham, Animal Science Department
Virginia Polytechnic Institute

R. H. Nelson, Head, Department of Animal Husbandry
Michigan State University

L. H. Newman, College of Veterinary Medicine
Michigan State University

H. D. Ritchie, Department of Animal Husbandry
Michigan State University

The authors wish to express appreciation for the valuable suggestions and encouragement given by the following reviewers of this text:

E. H. Cash, Animal Science Department
Pennsylvania State University

R. W. Harvey, Animal Science Department
   North Carolina State University

R. H. Nelson, Animal Husbandry Department
   Michigan State University

G. R. Wilson, Animal Science Department
   Ohio State University

Special gratitude is offered to those who field tested the authors' recommendations—the many beef cattle breeders and cattle feeders. Their willingness to share their experiences have made possible the development of recommendations and systems outlined in this text for utilizing the most sophisticated scientific information in managing a practical beef production operation.

1

Beef Production: What It Costs and Opportunities For Improving Efficiency

Beef is one of America's most nutritious and favorite foods. The average American consumes about 129 lbs of edible beef per year. It is consumed for its nutritional value and flavor. The average person in the United States consumes about 245 calories and about 24 grams of high quality protein from beef each day. That is about 10 percent of our daily energy and 50 percent of our daily protein requirement per person.

The role of the beef industry in the United States is extremely challenging—to provide all the beef Americans would like to consume at a price they can afford and of a quality they prefer. At the same time the beef industry must develop breeding and production systems that will allow the beef industry to convert available sources of nutrients such as forages, grain by-products, and wastes into high quality protein that is highly palatable and in an available form for human consumption, which will make the ruminant less competitive with humans for feed grains.

Beef animals produce many other needed by-products such as are illustrated in Figure 1-1. Leather goods, insulin, and the many other valuable products go unnoticed by the feed industry and consumer. Grain is fed to cattle to reduce costs and to improve the quality of beef. Nutrients not utilized are returned to the land to provide nutrients for plant growth.

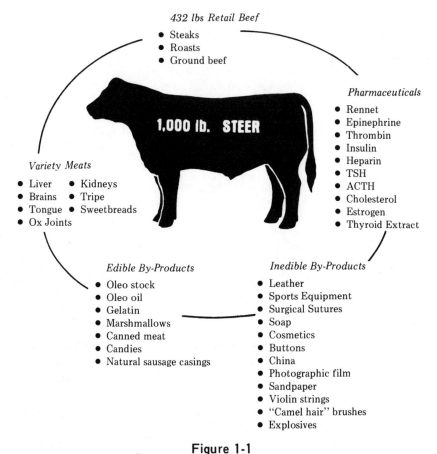

Figure 1-1

Good Things from Cattle
(Courtesy of the Beef Industry Council of the Meat Board)

There are obvious reasons for and benefits of the development and growth of the beef industry in the United States. The purpose of this introductory chapter is to present facts on the development of the beef industry, the efficiency of converting plant and grain nutrients into beef nutrients, the feasibility of various management systems to reduce grain feeding, the opportunity to improve total efficiency through breeding, and opportunities for reducing the cost of producing beef.

1-1 ECONOMIC COSTS IN PRODUCING BEEF

Cost of Producing Calves

The beef cow-calf industry (those that produce the calves) is very diverse. Beef cow herds are concentrated in areas containing surplus forages and crop residues to utilize these sources of nutrients that have little alternative use. Since these are harvested primarily by grazing, the beef herds are typically small; more than 95 percent of those who own beef cows have less than 100 cows, and more than 60 percent of the beef cows are in herds of less than 100 head. It is difficult to determine an average cost due to differences in length of grazing season and winter feeding, fencing and land costs, type of farming operation, equipment, and labor costs.

Budgets for beef herds developed at land grant universities located in the central, plains, and western states were examined to develop estimates of average production costs. The values in Table 1-1 appear to be representative of those of typical beef herds in the United States. Total annual costs appear to be $180 to $250/beef cow unit, which includes a 1,000 lb beef cow, 0.2 of a replacement heifer, 0.05 of a bull, and the calf produced to weaning at about 7 months of age.

Part I of Table 1-1 gives representative costs for those items that everyone would likely incur; i.e., feed and operating costs. Feed costs are more than 75 percent of these costs. Even though much of the feed utilized is forage and crop residues that have no alternative use, costs such as grazing fees, fencing, brush and weed control, fertilizer, labor, costs of harvesting and handling winter feed, and sale value of harvested hay must be considered.

The costs listed in Part II may or may not be real costs. Where surplus family labor is used, it may not be included, especially if the beef herd is a small and supplementary enterprise. Machinery and equipment costs vary considerably, depending on whether or not much of it is already available for use in a cropping operation. In small, part-time operations, the machinery is low cost; used equipment or winter feed is purchased. The importance of interest costs varies, depending on the value of converting other income into capital gains or the value of the beef herd as a forced savings account.

Land charges are not usually included, as the alternative use value of the land is reflected in feed costs to be offset by appreciation in land value. This may not seem like a very sound approach to operating a business involving as much investment as that in a

Table 1-1
Annual Costs for a Beef Herd Unit[1]

I. Total Cash Cost $

Hay - 1.6 tons	56
Pasture - nonsaleable (1.4 tons hay eq)	4
Pasture - improved (2.0 tons hay eq)	30
Grain - (150 lb)	6
Minerals (70 lb)	5
Protein supplement (50 lb)	4
Vet and medicine	3
Machinery & equipment, fuel & repairs	12
Interest on operating capital	7
Bull costs	5
Marketing	7
Total cash costs	139

II. Other Costs (labor, depreciation, taxes, interest, ins.)

Labor	20
Machinery and equipment	28
Livestock	28
Land charge	?
Management charge	?
Total other costs	76

III. Production[2] lb.

Steer calf .41 x 450 lb =	184.5
Heifer calf .21 x 420 lb =	88.2
Cull replacement heifer .05 x 700 =	35
Cull cow .15 x 900 =	135
Total weight/beef cow unit	442.2

IV. Sale Price Needed to Break Even[3]

To cover out of pocket costs ($139)	$/cwt
Steers	38.50
Heifers	32.70
Cull heifers and cows	23.10
To cover total costs ($139 + $76)	
Steers	59.50
Heifers	50.60
Cull heifers and cows	35.70

---

[1] Beef herd unit includes a 1,000 lb. cow, 20% replacement heifers and 5% bulls.
[2] Based on estimated U.S. averages, considering usual levels of fertility, culling, death losses, and weaning weights. Cull cows averaged 20.1% of total cattle slaughter from 1970-1975.
[3] Assumes heifers sell for 85% and cull heifers and cows sell for 60% of steer price, based on average commercial cow and steer calf prices from 1970-1975.

beef herd (typically $1,000 to $2,000 or more is invested/beef cow unit). However, it partially explains why beef cow numbers have increased over the last 25 years with a small apparent return on the capital invested. Many producers will continue to own beef cows as long as returns are adequate to recover direct feed and operating costs. Currently, feeder steers weighing 450 lbs would have to return an average of $54.93/cwt to recover these costs. This assumes that heifers normally sell for 85 percent of steer price and cull cows for 60 percent of steer price.

The price of beef includes a "subsidy" from the beef herd owners in the United States. This subsidy results from their desire to keep beef cows to obtain some return for feed resources that have no alternative use, to maintain the value of land, and as a hobby.

Costs of Feeding from Weaning to Slaughter

Cattle feeding operations tend to be larger and located in areas where feed grains are available at a reasonable cost. Thus, feeder cattle must be assembled from the small beef herds located over a wide area of the United States into larger groups and shipped to feedlots that may be as close as a few miles to as far away as 2,000 miles.

Considerable shipping and handling costs are involved in this transfer of ownership, as are costs due to sickness that result from this translocation. The calf may be sold at weaning and either moved to a feedlot or placed in wheat pasture, crop residues, or other pasture prior to being placed on harvested feeds in a feedlot.

Traditionally, most cattle were fed for slaughter primarily in the corn belt by farmer-feeders in groups of 50 to 150 head. They feed cattle to market grain and roughage grown at a higher price and to utilize their labor in the winter. Thus, many would feed cattle without a return for their labor and interest on investment.

During the 60s, however, a growth of large commercial feedlots occurred, primarily in the plains states, as a result of large scale development of irrigation and increases in the size of operation, forcing many small feeders to retire or quit feeding cattle. As a result, 64.5 percent of our fed cattle are now produced in lots of 1,000 head and larger. There are now 426 feedlots that feed over 8,000 cattle/year; nearly all of these are commercial feedlots, and they now feed 49.1 percent of the cattle. Also there are 1,338 lots, which would almost all be farmer-feedlots, who feed 1,000-8,000 cattle/year and 15.1 percent of the total cattle fed.

Only 35.5 percent of the cattle are fed by 136,262 cattle feeders whose lots hold fewer than 1,000 head.

Thus, cattle feeding has become a large scale industry, and nearly all costs are real and the inputs used such as feed, labor, management, and capital must be priced at their alternative use or market value. As a result, costs of feeding cattle in feedlots are easier to determine than those of the beef herd. However, the average days grazed between weaning and placement in a feedlot are difficult to assess. Also, data on this are not available. It is important to consider effects of differences in environment and overall use of growth stimulants and use of best practices in ration formulation and in developing feeding systems. Estimates of days grazing and feedlot performance were made, based on discussions with other extension specialists at universities located in the corn belt and plains states. Amount of grain fed was based on USDA estimates.

Table 1-2 gives a budget that appears to be representative of the performance and costs of feeding the calf crop in the United States—from weaning to slaughter at low choice grade. Over a long period of time, all of these costs must be recovered for cattle feeding operations to survive, including returns to investment, labor, and management. Feed represents about 70 percent of these costs, primarily due to the cost of harvested forages and grains and protein-mineral supplements. Nonfeed costs, including interest, death loss, medical facilities and equipment, labor, transportation, and marketing, account for approximately 30 percent of the total cost.

A return to management of 5 to 10 percent of costs is necessary to encourage cattle feeding. Below this level, cattle producers tend to reduce cattle feeding. When profits exceed this level, more cattle are fed, resulting in increased demand for feeder cattle, a surplus of fed beef, depressed cattle prices, reduced profit margins, and substantial losses. Thus the price of beef often includes a "subsidy" from cattle feeding as a result of overproduction. Beef production is a good example of free enterprise, with many producers (approximately 1.4 million cow-calf owners and 138,026 cattle feeders) competing with each other.

The long-term return to management in beef cattle production (about 10 percent of costs) represents only about 5 to 10 cents/lb of retail beef. Present production costs without profits are about 80 cents/lb for the carcass and $1.16/lb for retail beef. These costs are based on values in Tables 1-1 and 1-2 with the assumption that each beef cow unit annually produces 0.4 of a 1,050 lb steer, 0.2 of a 850 lb heifer, 0.05 of a 700 lb cull heifer, and 0.15 of a

Table 1-2
Average Costs of Feeding Calves from Weaning to Slaughter

		Heifers	Steers
I.	Grazing Phase		
	Initial weight, lb.	420	450
	Avg. daily gain, lb.	0.90	1.00
	Days grazing	90	90
	Pasture, lb. DM[1]	1071	1120
II.	Growing and Finishing Phase		
	Initial weight, lb.	501	540
	Final weight, lb.	850	1050
	Avg. daily gain, lb.	1.87	2.13
	Days	186	240
	Total feed, lb. DM		
	Silage	497	700
	Hay	443	620
	Grain	1854	2705
	Supplement	112	162
III.	Summary of Feedlot Costs		
	Costs grazing		
	Feed[2]	$ 12.60	$ 13.50
	Non-feed[3]	10.64	11.49
	Costs growing, finishing		
	Feed[4]	130.98	189.43
	Non-feed[5]	54.76	63.40
	Total cost of gain	208.98	277.82
	Cost/cwt gain	48.60	46.30
	Cost/beef cow unit		
	(.2 heifer & .4 steer)	41.80	111.13

IV. Overall Production Costs, Beef Herd Unit

	Weight produced, lb.	Break even cost, $/lb.
Live	760	0.48
Carcass	455	0.81
Retail	318	1.16
Edible	293	1.26

---

[1] Based on expected dry matter intake.
[2] Assumes $1.25/cwt/month for pasture.
[3] Includes per annum interest at 10% on heifers at 40 cents, and steers at 45 cents/lb.
[4] Assumes $3.00/cwt DM for hay and silage; $5.00/cwt DM for grain and $9.00/cwt DM for protein-mineral supplement.
[5] Non-feed costs include $25/head plus 15 cents/head/day for labor, interest, facilities.

1,000 lb cull cow. These estimates are based on USDA values for annual cow, steer, and heifer slaughter, and the number of beef cows in the United States.

The USDA estimates that farmers have actually received about 63.5 percent of retail prices since 1970.

Energetic Efficiency of Beef Production and the Economics of Grain Feeding

Energetic and economic costs of feeding the calves increase as they increase in weight due to an increase in the proportion of fat deposited in the body tissue. Many have concluded that we could reduce the cost of grain by slaughtering at an earlier stage of growth. Figure 1-2 shows the relationship of total energy cost/lb edible beef produced at various steer slaughter weights with various sizes of cows. It is clear that there is a slaughter weight at which energy costs are minimized for each type of cattle. Below this weight, not enough dilution of the fixed energy costs of the beef herd has been obtained. Above this weight, excess fat is produced that is not consumed.

Beef from carcasses can be expected to contain a slight to small amount of marbling and a yield grade of 2-3 at this point of

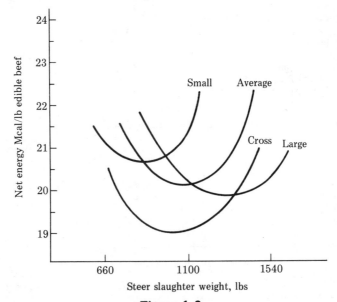

Figure 1-2

Impact of Steer Slaughter Weight on Energy Efficiency

maximum energy efficiency. Steers will be about 90-95 percent and heifers about 70-75 percent of their dam's weight at this stage of growth. At present, steers and heifers are slaughtered at 50-100 lbs above this weight. These weights are likely to be reduced to near the optimum slaughter weight for maximum efficiency (least total cost/lb edible beef) as a result of the recent change in USDA grade standards, with the meat containing the amount of intramuscular fat preferred (low choice).

The efficiency of various types of livestock in converting nutrients fed into food has been determined. On the surface, livestock production appears to be an inefficient means of producing food. However, the portion of nutrients fed that could be used directly for human consumption, the value of by-products, and the recycling of undigested nutrients to replenish soil nutrient and organic matter levels must be considered. Estimates of current forage and grain energy utilized, grain fed and edible beef produced by beef cattle in the United States are presented in Table 1-3. In these calculations, nutrients used were based on present levels of feed usage/beef cow unit, which includes feeding the calves. Our best estimates suggest that overall the cow-calf and feedlot segments operate at 75 to 80 percent efficiency in energy utilization. The factors that cause the energy losses will be discussed in the next section.

Edible beef is assumed to be that portion actually consumed. Other assumptions are given in the footnotes to Table 1-3. Table 1-3 also gives a comparison of energetic efficiency and beef that could be produced with various systems using different amounts of grain. System 1 assumes no grain would be fed, and system 6 assumes high grain levels would be fed to the calves after weaning with the other systems representing intermediate levels of grain feeding.

None of the levels of grain fed in systems 1 through 6 are as high as that fed at present. They are based on systems that approach 100 percent efficiency; no grain is fed to the cow herd; all rations are properly balanced; grain is fed at the time when it would be most efficiently utilized; grazing phases are managed so that overall gain on grass is 1 lb/head/day; growth stimulants are used at 100 percent efficiency; good environmental conditions are assumed in the feedlot; and the steer calves are slaughtered at about 90 percent of the dam's weight and heifers are slaughtered at 80 percent of steer weight. As less grain is fed, less beef is produced because more of the forage energy has to be used postweaning.

Table 1-3

Annual Energy Requirements of Beef Cow Unit on Various Feeding Systems

	Average for present system[1]		Production Systems at 100% efficiency[2]					
			1	2	3 Ration 1st half of postweaning gain	4 Ration 1st half of postweaning gain	5	6
					All forage Ration 2nd half of postweaning gain	All forage Ration 2nd half of postweaning gain		
	40% grain	All forage	All forage	All forage 42% grain	All forage 82% grain	40% grain 40% grain	40% grain 72% grain	77% grain 82% grain
	80% grain							
Total NE, Mcal	7135	7395	7395	6305	6732	6567	6385	6285
Forage NE, Mcal[3]	5549	7395	7395	6305	5993	5438	5147	4855
Cow units maintained, million head[4]	43.7	31.8	31.8	47.6	47.6	49.4	49.4	49.4
Lb. grain/lb edible beef[5]	5.68	0	0	2.08	2.89	3.79	4.33	5.35
Lb. edible beef per capita[6,7]	68.2	50.7	50.7	75.9	75.9	78.8	78.8	78.8
Daily consumption/capita, gm.	84.8	63.1	63.1	94.4	94.4	98.0	98.0	98.0
Energy, calories	212	202	202	236	236	245	245	245
Protein, gm	20.9	15.5	15.5	23.2	23.2	24.1	24.1	24.1
Fat, gm	14.3	8.7	8.7	13.0	13.0	13.5	13.5	13.5

[1] Feed usage based on USDA estimates. Estimated efficiency $\left(\frac{\text{predicted requirements}}{\text{actual energy used}}\right)$ values are: beef herd, 90.0%; grazing, 100%; growing and finishing, 83.3%.

[2] All rations properly balanced so that no energy is wasted. Factors involved are proper feeding of beef herd, good environmental conditions, use of growth stimulants and good nutritional management to make most efficient use of grain, and feeding to low choice grade.

[3] Includes forage portion of silage fed. In systems 3, 5 and 6, there is less silage forage available as more of the grain crop is harvested and fed as grain.

[4] Based on estimate of pasture and hay presently fed.

[5] Includes grain in silage. In systems 1 through 6, only grain fed to beef herd is that in silage that has no alternative use due to drought, etc. In systems 2 through 6, the forage fed is from silage, except in the first phase of systems 2 and 3.

[6] Based on the following assumptions: An average beef cow produces 2 steers, 1 heifer and 1 cull cow for slaughter and 1 replacement heifer over a breeding life of 5 years, based on average beef cow inventory and cow slaughter and cow, calf and feedlot mortality rates. Assuming that the edible portion contains all of the carcass protein and enough fat to result in a 13.8% fat in the edible portion, this results in an average of 312 lb. edible beef/cow unit/yr in the present system and 319 lb. in systems 2 through 6.

[7] Does not include the beef from cull dairy cows or that imported. The beef from the calves produced in system 1 would be considerably older, and likely less fat than those in other systems and therefore would be of a lower quality (standard and good grade).

As a result, the American consumer would have to be satisfied with consuming less beef, and this beef would be of a lower acceptability due to increased age at slaughter and less intramuscular fat. However, the amount of beef produced does not increase directly as more grain is fed. Less forage is available from silage as more of the crop is harvested as grain.

To clearly evaluate the efficiency of grain feeding, adjustments must be made for the feed grain that is fed that is not consumable by humans. Estimates by grain science and extension specialists indicate that 20 to 40 percent of the "feed grain" fed is not consumable by humans, due to by-products from milling, mold, contamination, damaged kernels, and that portion that must be harvested as high moisture grain or silage due to weather conditions. Further, adjustments must be made for the differences between grain and meat in the biological value of the protein. Consideration should also be given to the value of by-products produced and undigested organic matter and nitrogen returned to the soil, which is recycled through new plant growth.

Table 1-4 presents comparisons in efficiency of food production from grain or beef in which adjustments were made for the amount of grain not directly consumable by humans, and differences in protein quality. These calculations assume supplemented nitrogen needs could be met with nonprotein nitrogen (NPN), grain by-products, and plant protein. They also suggest that we obtain somewhat less energy but equivalent amounts of protein when grain is fed during the last part of the growth period.

Some of the energy is recovered in the form of manure applied to the soil to maintain soil organic matter. Further, a high proportion of the nitrogen not deposited as tissue is returned to the soil and recycled, thus reappearing in plant and grain protein. The manure from 2 steers or 1 dairy cow provides the nitrogen needed to grow an acre of corn. Essential by-products include leather and pharmaceutical products. They are worth about $40/beef cow unit/year at slaughter, and this value is multiplied many times during manufacture.

Other livestock species are more efficient than beef cattle in forage utilization. The use of sheep rather than cattle would improve the efficiency of energy and protein utilization from forages, due to the opportunity to produce more offspring relative to the number of breeding animals maintained. However, some of the major problems of sheep include predators, fencing, parasites, intensity of management and housing required, and low value of cull breeding stock. For these reasons, more farmers and ranchers prefer to raise beef cattle. The major problem, however, is lack of

Table 1-4

Efficiency of Feeding Grain to Beef Cattle

	All forage	All forage 42% grain	Ration 1st half of postweaning All forage 82% grain	Ration 2nd half of postweaning 40% grain 40% grain	40% grain 72% grain	77% grain 82% grain
			Amount available/capita/day			
Calories[1]						
Grain	1568	1312	1006	611	400	–
Beef[2]	159	236	236	245	245	245
Total	1727	1548	1242	855	645	245
Protein						
Grain, gm	23.5	18.3	14.1	8.6	5.6	–
Beef, gm	12.4	18.6	18.6	19.3	19.3	19.3
Total	35.9	36.9	32.7	27.9	24.9	19.3
N returned to soil, gm	25.3	33.3	32.2	32.1	32.8	28.9

1,2 Assumes 20% of grain inedible due to byproducts from milling, mold, contamination, damaged kernels and that salvageable only as high moisture grain or silage. All calculations in system 1 are based on the grain used in system 6. In systems 2-5, the grain fed in the system includes inedible grain plus as much of the edible grain as necessary for the feeding system used. The grain is assumed to contain 4 calories/gm, and the beef is assumed to contain 2.5 calories/gm, physiological fuel basis.
3 Assumes 10% protein in grain with a biological value of 56% and 24.6% protein in edible beef with a biological cal value of 80%. Assumes NPN, byproducts and forage protein used to provide supplemental N.
4 Assumes 50% of N excreted is lost as volatiles, in runoff, or is not readily available for plant growth.

consumer acceptance of lamb and mutton compared to beef. Goats are another possibility but also have the problem of acceptability by producers and consumers.

The production of milk is more efficient than the production of meat. However, the problems with total use of dairy cattle to utilize forage are housing, equipment and labor requirements, and reproductive and management problems under range conditions.

On the surface, it appears that swine and poultry are more efficient in utilization of grain, and any grain fed should be utilized through them. As one-stomach animals, they have the same problem as humans in use of contaminated feeds and proteins of a lower quality. Also, cattle can utilize the plant portion of the grain crop. A combination of livestock products appears desirable. Grain should be fed to cattle as a means of reducing maintenance costs and meeting consumer demands for the quantity and quality of feed and milk desired. This also provides a means of salvaging a crop as silage due to climatic conditions, and utilizing contaminated, moldy, or damaged grain.

The ultimate determinant of level of grain feeding is the market price of grain, total demand for beef, and demand for different qualities of beef. Table 1-5 presents estimated postweaning costs of producing a 630 lb carcass at different prices of grain. It was assumed that if the beef industry shifted to all forage or only a small amount of grain just prior to slaughter, the carcasses would grade standard or good and thus would be worth $4 to $6/cwt less than choice grade. No credit was given for a higher yield grade, as all trim fat would be available and the carcasses were thus discounted $2/cwt. These estimates indicate that the industry would switch to less grain (all silage or grazing followed by a high grain ration) at $2.50–$3.00/bushel, depending on overhead costs, use of NPN and continuous feedlot operation vs one group fed/yr. All forage systems, however, would not be least cost until grain approached $4.50/bushel. Under these conditions, abnormally high prices for choice carcasses might further increase the price of grain needed to force a shift to all forage systems.

1-2 REDUCING THE COST OF PRODUCING BEEF

The cost of producing beef has not increased directly proportional to the rate of inflation over the last 25 years. The major reason for this has been an increase in the efficiency of producing beef. More efficient crop production and increases in yields, artificial insemination, crossbreeding, improved genetic selection, development of

Table 1-5
Economics of Feeding Grain—Total Industry Basis

	1	2	3	4	5	6
			\multicolumn{4}{c}{Production systems at 100% efficiency}			
			Ration 1st half of postweaning gain			
	All forage	All forage	All forage	40% grain	40% grain	77% grain
			Ration 2nd half of postweaning gain			
	All forage	42% grain	82% grain	40% grain	72% grain	82% grain
	\multicolumn{6}{c}{Amount/steer, 450–1,050 lb.}					
Expected performance						
Daily gain, lb.	1.00	1.32	1.96	1.85	2.16	2.57
Lb. feed/gain, lb. DM	18.91	12.92	11.55	8.77	7.98	6.35
Turnover rate	.58	.77	1.13	1.07	1.24	1.48
Feed/cost/600 lb. gain[1], $						
Corn at 1.50	223.36	149.94	149.13	141.30	106.74	112.87
Corn at 3.00	223.36	193.41	204.35	199.52	211.40	222.04
Corn at 4.50	223.36	237.60	259.46	288.18	298.30	330.95
Non-feed cost/600 lb. gain[2], $	82.29	85.59	76.75	94.62	82.62	72.30
Feed and non-feed cost/600 lb. gain, $						
Corn at 1.50	306.65	235.53	225.88	205.92	189.36	185.17
Corn at 3.00	306.65	279.00	281.10	294.14	294.02	294.34
Corn at 4.50	306.65	313.19	336.21	382.80	380.92	403.25
Relative value of 630 lb. carcass[3]	-37.80	-25.20	--	--	--	-12.60

growth stimulants, antibiotics, and immunizing agents to control many diseases; identification of nutrient requirements and more efficient ration formulation; improved methods of grain and forage handling, storage and processing methods; and improvements in plant breeding and pasture management techniques are some of the developments that have contributed to increased efficiency. These improvements are the result of combined efforts of university research and extension, industry, and cattlemen themselves. However, there are many opportunities for even further improvements in efficiency as follows.

Adaptation of Known Practices

Table 1-6 lists a number of practices that affect cost of production. The first column gives the level of efficiency of the most efficient producer compared to the average producer. At present,

Table 1-6
Impact of Careful Management on Efficiency of Nutrient Utilization

	Level of efficiency	
	Average producer	Most efficient producer
Fertility[1]	80%	95%
Efficiency of growth[2]	80-90%	100%
Survival rate	97-98%	99-99.5%
Control of diseases[3]	95%	100%
Ration[4]	80-90%	100%
Environment	80-95%	100%
Use of growth stimulant[5]	90-95%	100%
Use of metabolic stimulant[6]	90-95%	100%
Marketing at optimum stage of growth[7]	95%	100%

[1] Primarily an effect of nutrition.
[2] Reflects selection for most energetically efficient cattle.
[3] Reflects impact of high disease incidence on efficiency.
[4] Includes impact of nutrient balance and feeding system used.
[5] Impact of use of DES, Ralgro, Synovex, MGA. The range reflects variation in levels and consistency of usage.
[6] Impact of Rumensin. The range reflects variation in levels used.
[7] Impact of feeding beyond low choice grade.

Table 1-7
Beef Herd Management Errors that Cost Cash

	Most efficient producer	Average producer	Difference per 50 cows	$ value per 50 cow herd
Calf crop/12 mos.	95%	80%	3000#	$1350
Weaning weight	500#	400#	4000#	1800
Yearly feed costs/cow	$60-80	$80-150	$1000	1000
Investment/cow unit	$1,000	$2,000	$20,000	2000
Marketing - % of market top	100%	85%		600
Total/50 cows				$6750
Total/cow unit				135

the beef industry as a whole operates at 75 to 85 percent of efficiency, due to not fully using known techniques. Table 1-7 and 1-8 give estimates of the economic impact on an individual operator who does not use all available information. Differences in costs between poor and excellent managed operations can easily be $100.00 or more/beef cow or calf fed to slaughter.

Need for Further Research

A number of problems remain to be solved, and more research is needed to keep the price of beef at a level that will encourage profitable production at a price which will keep beef competitive at the retail counter.

1. *Improved reproduction.* Crossbreeding alone has shown as much as 20 to 50 percent improved output in the breeding herd. More research is needed to identify nutrition and management needed to obtain early conception after calving. Methods of preventing infertility must be identified. The problems associated with twinning, such as the high incidence of retained placentas, need to be solved. Further work is also needed on the possibilities of multiple births.

2. *Selection for more efficient cattle.* Means of selecting for cattle that are the most efficient in energy and protein utilization

Table 1-8
Extra Costs of Careless Feedlot Management

	Differences in cost/cwt gain	
	Careful feeder	Careless feeder
Feeder quality, health	Buys high performing cattle, has good health program	+ $3
Lot conditions	Most economical; minimum stress	+ $7
Feed intake	normal or above	+ $2
Ration formulation, feeding system	Balanced; least cost; optimum combination grain and roughage feeding	+ $3
Digestive stimulant	Uses	+ $3
Metabolic stimulant	Uses	+ $3
Sale condition	Low choice, yield grade 2 1/2-3 1/2	+ $2
Total cost/cwt gain	Least cost	$ 23
Difference per steer (600 lb. gain)		$138

without increasing or decreasing mature weight must be identified. There are over 30 breeds of beef cattle available and the potential of blending the blood of these breeds could effectively increase the overall efficiency of beef production significantly. These breeds vary from 10 to 50 percent in their genetic potential among breeds for reproduction growth and carcass merit, suggesting that complementarity or the use of breeds that complement each other will affect much improvement in the industry. Individual cattle vary from 30 to 70 percent in efficiency of energy and protein utilization, suggesting that there is opportunity for selection for these traits within breeds. The use of superior progeny-tested sires through artificial insemination (AI) could greatly enhance these programs. Synchronization of estrus will allow AI to expand.

3. Reducing sickness. Means of reducing sickness and death loss in newborn calves and feeder cattle must be identified. Scours and respiratory infections are two of the major problems in this area that need further study.

4. Improvement in utilization of forages and grains. Areas needing basic research are identification of nutrient requirements and nutritive values of feeds under various conditions, so that nutrients are not overfed or underfed. Methods of improving digestion and metabolism are also needed to minimize feed energy and protein losses. Applied research is needed to determine: (a) stages of growth for effective utilization of various combinations of feeds; (b) types of cattle that are most suitable for various types of diets; (c) practical environmental systems that minimize nutrient costs maintenance; and (d) methods of recycling the nutrients in undigested feed residues. Research is needed on forage production and range management and to identify practical breeding, feed production, and management systems that are most efficient in small beef herds. Much of our beef herd management information is usable only in larger herds, as discussed previously, and a large number of our beef cows are in small herds.

5. Systems analysis research. Computers now make it possible to develop programs in which the impact of a large number of variables can be considered simultaneously. With these types of studies, the variables having the greatest impact on costs can be identified, allowing emphasis in research and extension programs to be concentrated on the most important factors.

2

Breeding Principles

Beef cattle breeding is experiencing the use of new breeds, crossbreeding, artificial insemination, performance and progeny testing, computerization, and many other innovative means of progress. Breeding systems are improving significantly and more objective measures of progress and predictability are being attained. The seedstock phase is taking on a new role in the beef cattle industry. This division can best be described as the research and engineering branch of a large and complex manufacturing plant. The division of research and engineering is where the models are designed to provide the commercial beef industry with the genetic improvement for economically important traits.

Against this background, it becomes clear that we have an extremely complex system from which to design future breeding programs. The ultimate goal must be determined for each new generation of cattle breeding based on net merit and maximum profit. This chapter will explore the factors and breeding systems that will affect the rate of improvement in beef cattle.

2-1 ANIMAL BREEDING FUNDAMENTALS

Only two mechanisms are available to the breeder to control the inheritance of his cattle: (1) selection and (2) determining which animals shall mate among those selected. Selection or choice of

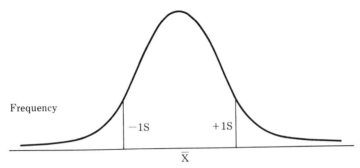

Figure 2-1

Normal Distribution of Weaning Weight

parents is the common element of all breeding programs, and it is the major force which we have to make continued improvement.

Our main goal in beef production is to improve the average performance of the offspring. From the genetic standpoint, this means improving the genotype, and selection is our best tool to accomplish this goal. Through selection, we increase the average performance by changing gene frequency or, more specifically, by increasing the frequency of the desired genes. For most all production traits, we have a normal distribution of genetic material from which to begin.

Figure 2-1 gives an example of the normal curve for weaning weight. Assume that $\bar{X}$ represents the mean for weaning weight of bull calves in your herd, and S signifies the standard deviation or the value that represents the average deviation from the mean. In normal populations, $X \pm S$ contains about 2/3 of the individuals in the population. To further illustrate the mean and standard deviation, let's use bull weaning weight as an example and look at the normal distribution of a herd with 10 animals (Table 2-1).

Table 2-1

Typical Herd Weaning Weights

i	x
1	390
2	410
3	418
4	340
5	390
6	460
7	360
8	392
9	430
10	410
	4000

2-1 Animal Breeding Fundamentals

1. Specifically, we want to know the average weaning weight.
2. Total the weaning weight using the following notation:

$$\sum_{i}^{n} = X_i$$

where Σ = "the sum of"
n = the number of observations (10 in this case)
i = the weaning weight for the ith calf

3. Calculate the average using

$$\overline{X} = \sum_{i=1}^{n} X_i/n$$

In this case, $\overline{X}$ = 4000/10 = 400 pounds

Now suppose we want to measure the amount of variation of weaning weight of bull calves in your herd. If the weaning weights were identical, there would be no variation. As we see in our sample, the weaning weights are not identical. Therefore, some amount of variation is present. The first approach is to see how much each weaning weight differs from the mean.

Table 2-2
Weaning Weight Variation

X	$\overline{X}$	$(X-\overline{X})$	$(X-\overline{X})^2$
340	400	-60	3600
340	400	-60	3600
395	400	-5	25
397	400	-8	64
400	400	-0	0
405	400	5	25
410	400	10	100
410	400	10	100
412	400	12	144
491	400	91	8281
			15,939

The sum of the differences squared is generally referred to as the "sum of squares" (SS).

Variance is

$$S^2 = \sum_{i=1}^{n} (X_i - \overline{X})^2 / n - 1$$

We will shorten this to SS/n–1.
Applying this to the weaning weights, we get:

$$S^2 = SS/n\text{-}1 = 15{,}939/9 = 177.10$$

To convert the variance into the value we are searching for, we simply take the square root, which equals the standard deviation, symbolized by S.

$$S = \sqrt{177.10} = 13.3$$

$\overline{X} \pm S$ contains about 2/3 of the population in the sample of weaning weights.

$$\overline{X} \pm S = 400 \pm 13.3$$
$$= 387 \text{ lbs and } 413 \text{ lbs}$$

Seven of the weaning weights should be between 387 and 413 pounds. In fact, seven of them are. This is an example of normally distributed data. Most of the parameters in beef cattle selection we are concerned with follow the normal distribution.

Diagrammatically, we see our herd situation in Figure 2-2.

If we were to want bulls with weaning weights over 413 lbs in this herd, we would find only 1/6 of the bulls above this value. The objective of beef improvement for the production traits is to change this entire cattle population. Through selecting animals far superior to keep and mate and also by culling the lower producing animals, we can move the mean forward.

2-2 SELECTION RESPONSE

Five major factors determine selection response. These are as follows: heritability, accuracy, selection differential, generation interval, and number of traits selected.

1. Heritability. Cattle vary in their phenotypes for a given trait for either or both of two reasons: (1) they possess different genes which affect the trait, or (2) environments have varied their phenotypes. Heritability gives us a good estimate of how much progress can be made from selection. Table 2-3 notes current heritability estimates for economically important traits in beef cattle, and the emphasis that should be placed on them in selection.

If heritability of a trait is reasonably high (30 percent), a breeder should consider simple phenotypic selection as the most

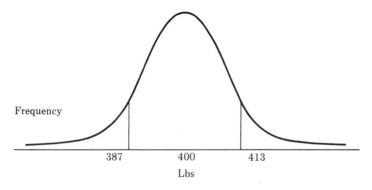

Figure 2-2
Sample Distribution of Weaning Weight

practical means of increasing the average of his herd for the trait. The higher the heritability the faster the progress which can be made. Fortunately, all traits in beef cattle are medium to highly heritable with the exception of fertility. This table suggests that little selection should be placed on reproductive traits because of low heritability. However, even though the heritability is not high for reproductive traits, the economic importance probably justifies high emphasis for selection as indicated in Table 2-3.

2. *Selection differential.* This is also referred to as reach because it reflects how much the selected animals average exceeds the average of the group from which they were selected. Figure 2-3 diagrammatically depicts the entire group from which to select with $\overline{X}$ as the mean and also the group selected for breeding with an average of $\overline{X}_2$. Selection differential (reach) is the difference between the average of the selected group, and the average of the

Table 2-3
Economically Important Beef Traits

Trait	Heritability	Emphasis in Selection
Fertility	10	High
Birth weight	40	High
Weaning weight	40	High
Post-weaning weight	50	Low
Feed efficiency	40	Low
Yearling weight	60	High
Conformation traits	25-40	Med
Carcass traits	30-70	Med

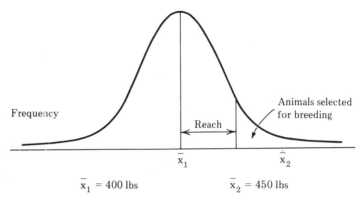

Figure 2-3
Selection for Weaning Weight

group from which they were selected. The greater the reach, the faster the improvement. The selection differential for bulls is generally much greater than for females because fewer are selected for breeding. This is another reason sire selection is so important.

The principle involved with heritability and reach says:

$$\text{Progress} = \text{heritability} \times \text{reach}$$

Consider weaning weights of calves as the trait to be improved. If heritability for this trait is 0.4, the selected replacement breeding stock averaged 450 pounds at a standard weaning age, and the average weaning weight of the calf crop from which they were selected was 400 pounds.

$$\begin{aligned}\text{Progress} &= 0.4 \times (450-400) \\ &= 0.4 \times 50 \\ &= 20 \text{ lbs}\end{aligned}$$

So we can predict that the progeny of these selected animals will average 420 pounds in weaning weight.

3. Generation interval. Generation interval is the average age of the parents when the replacement offspring are born, and it averages between 4.5 and 5.0 years in beef cattle. In order to speed selection response, it is important to shorten the generation interval by reducing the average age of the herd. This is also advantageous from the standpoint of utilizing young, more genetically superior herd sires and putting more daughters, by these superior bulls, back into the herd. Two advantages are thus utilized: a higher frequency of superior genes is introduced, and the generation interval is shortened. This system does dictate superior management and selection techniques. It is impossible to shorten the generation interval below 2.0 to 2.5 years with young parents even if they are allowed to produce only one offspring. If you replace 40 percent of the females per year and use only yearling bulls, you could reach 2.4 years.

Generation interval has a major effect on the progress we can make per year, and continuing the previous example, let's assume we have a four-year generation interval.

The principle now says:

$$\text{Annual Progress} = \frac{\text{heritability} \times \text{reach}}{\text{generation interval}}$$

$$\text{Annual Progress} = \frac{0.4 \times 50}{4}$$

$$= 5 \text{ lbs per year}$$

4. Accuracy. This value is defined as the correlation between the true breeding value of the selected parents and the estimated value. It varies from 0 to 1.0 and is determined by taking the square root of heritability. For a trait with a heritability of 0.4, the accuracy of the animals' own performance is 0.63. Again, this reflects the reason for selection of traits that are highly heritable when we use phenotypic selection.

Table 2-4 gives the relative accuracy values for a 40 percent heritable trait. It is significant to note the relative value of an animal's own performance in comparison to 1/2 sibs, parents, and progeny. The higher the heritability, the more meaningful is the animal's own breeding value estimate; however, this is why with lowly heritable traits, progeny testing and information on ancestors becomes more valuable in estimating the breeding values for selected parents.

Table 2-4

Comparative Accuracy of Selection for a 40% Heritable Trait

Animals Measured	Relative Accuracy
40 progeny	.90
10 progeny	.72
Individual	.63
40 1/2 sibs	.45
10 1/2 sibs	.36
1 parent	.31
2 1/2 sibs	.22

5. *Number of traits selected.* In beef cattle selection, it is normal practice to select for more than one trait. It is important to keep our selection to those traits that are economically important and highly heritable because the greater the number of traits for which we select, the slower the improvement made for each trait. The selection response for the number of traits selected can be estimated by $1/\sqrt{n}$ where n equals the number of traits for which selection is practiced. For example, if we select for four traits, $1/\sqrt{4} = 1/2$; or if we select nine traits, $1/\sqrt{9} = 1/3$ (1/2 and 1/3 are the comparative selection responses we would expect when selecting more than one trait).

2-3 SYSTEMS OF SELECTION

All breeding programs have a choice of many systems from which to choose. Systems of selection may be classified as follows: mass selection, which is selection on the basis of the animal's own phenotype or performance; progeny selection; pedigree selection; family selection; and use of correlated trait selection.

1. Mass selection. Mass selection has been the primary system for improvement of beef cattle, and it will continue to be the main procedure to make genetic progress for highly heritable and economically important traits. With the use of computers and added technology, we can add information from other systems and develop estimated breeding values that improve the accuracy and selection response of phenotypic selection.

2. Progeny testing. Chapter 5 is totally devoted to the design and operation of progeny testing programs. However, its role as an additive means of estimating the breeding value and also improving the accuracy of selection is pertinent to this discussion. Progeny testing is evaluating an animal's transmitting ability for a given trait by studying the performance of its offspring with respect to that trait. Most progeny testing with beef cattle involves the testing of sires because of the number of offspring possible and also because the male contributes more than 85 percent of the selection response. Progeny testing is most useful in selecting for traits which are sex limited such as milking ability and calving ease, or terminally expressed traits like carcass data. Progeny data will play a much broader role in beef improvement with the increased use of artificial insemination. It should be noted that it is of significance only when used along with or following an individual's own performance testing, and is most valuable when incorporated into an animal's estimated breeding value. It should be noted that progeny testing does increase the generation interval.

3. Pedigree estimation. Pedigree information as noted in Table 2-4 when used alone provides low accuracy in estimating the breeding value, and is of almost no consequence if we go beyond the parents for information. However, sire and dam data when incorporated with an animal's own records provide an additive effect in predicting breeding value and increase the accuracy of this estimate. There are other isolated situations when pedigree information should receive some attention: (1) when we have to compare animals with no individual information; (2) when we find it necessary to cull young animals for traits which are not expressed until maturity; or (3) when we have to decide between animals who are equally good in their individual merit. Pedigree information will play more of a role in beef cattle improvement as performance pedigrees are developed.

4. Family selection. Information on a group of ½ sibs can be valuable in calculating the estimated breeding value of an animal, and is used to increase the accuracy of this estimate. Similarly, it is of little consequence unless used in combination with the animal's own performance. In general, family selection is most useful when the genotypic correlation (average relationship) among family members is high such as in inbred herds.

5. Use of correlated traits. Fortunately for beef cattle improvement, most production traits are positively correlated, meaning that selection for one trait results in the improvement of most others. If a certain economic trait is difficult or expensive to measure such as feed efficiency, it is possible to base selection on some easily measured trait such as postweaning gain. Most genetic correlations are not high enough to be of much value; in fact, the sample just mentioned is no doubt overestimated for present-day standards.

2-4 SELECTION TO IMPROVE OVERALL MERIT

This is the most common goal for selection of beef cattle. It provides three choices: (1) tandem culling, (2) independent culling levels, and (3) selection indexes.

Tandem culling. This is the simplest system, and consists of selection solely on the basis of one trait for a few generations, then ignoring that trait for a few generations, and selecting solely on the basis of another trait and so on. This method is least efficient of the three in improving overall merit with respect to all traits of interest, but makes greater improvement in a single trait while that trait is receiving attention.

Independent culling levels. A minimum level of excellence is set for each trait of interest, and those animals which do not meet the minimum are not kept for breeding. The important consideration is where you set this culling level for each trait. The two factors which may serve as a guide are the economic importance of the trait and its heritability. The greater the economic importance and the lower the heritability, the higher should be the culling level. However, if the herd is considerably inferior in one trait, then the culling level for that trait should be set higher. In efficiency of improving overall merit, the use of independent culling levels is intermediate between tandem culling and the use of a selection index.

Selection index. A selection index is a single numerical measure of overall net merit with respect to all the traits being considered in a selection program. Statistically speaking, it is a multiple regression equation, that is, a regression equation which has more than one independent variable.

$$I = a X_1 + b X_2 \ldots + n X_n$$

The values $X_1, X_2 \ldots X_n$ are the individual's phenotype for the n traits under consideration. The constants a, b, ... n are coefficients determined by multiple regression methods from herd data on all traits. Each trait is given a weight (coefficient) in accord to economic value, heritability, and genetic correlation with other traits. Those animals with the highest total index are retained for breeding. It has the advantage over the independent culling level system that might cull an animal that was very superior for all traits except one. However, in practice, a skillful breeder using the method of independent culling levels can do as well as he could with an index, especially if extensive data in a herd is not available. Most breeders will not find a selection index feasible because they do not have enough information to form a basis for estimating accurately the various values needed in the construction of an index. However, it may be worthwhile to obtain the assistance of an animal breeding specialist in the construction of an index.

2-5 ESTIMATED BREEDING VALUES

These are the most valuable parameters we have to make livestock improvement through selection. Most beef cattle improvement associations and major breed associations are computing estimated breeding values along with their accuracy estimates. These data can be utilized by breeders to help make decisions for selection of genetically superior beef cattle. These values combine the animal's own performance and the additive information from ½ sibs, sire, and dam, and progeny into one breeding value for each trait considered. It requires the use of complicated regression formulation and because of the complexity and extensive data involved, computers are used to formulate the information so it can be forwarded back to the breeder for his use in selection programs. Its use will be demonstrated in Chapter 4 on sire selection.

2-6 TOOLS FOR BEEF CATTLE IMPROVEMENT

In order to make rapid genetic improvement, seedstock producers must utilize every tool in the bag to make a maximum selection response. Listed below are the many tools we have available in selection to measure the specified traits.

1. Scales to measure weight at various ages
    a. birth
    b. weaning
    c. yearling
    d. slaughter
2. Son-o-ray
    a. backfat
    b. loin eye
3. K-40 whole body counter to measure lean growth
4. Backfat probe
5. Linear measures of skeletal growth
    a. height
    b. length

6. Testicular circumference measurements as indicators of bull fertility
7. Visual appraisal of:
   a. fat and muscle (cutability)
   b. frame size
   c. skeletal soundness and bone quality
   d. udder, mouth, and eye soundness
   e. sex character (femininity and masculinity)
   f. health

All of these tools will be elaborated upon in future chapters. But it is easy to see that our resources for improving the net merit of our cattle population are extensive.

2-7 PERFORMANCE RECORDS

One of the most valuable management tools we have available to us as cattle producers is performance records. It would be safe to say that the owners of less than 5 percent of our herds keep and utilize records as a benchmark for progress. Performance testing procedures have been available for more than 20 years, and recommended guidelines were developed with the formation of the Beef Improvement Federation (BIF) on February 1, 1968. Since that time more than 50 organizations now provide beef cattle improvement programs for the beef industry. The principal features of an effective record of performance program (outlined by BIF) are as follows:

1. All animals of a given sex and age are given equal opportunity to perform through uniform feeding and management.
2. Systematic records of economically important traits on all animals are maintained.
3. Records are adjusted for known sources of variation, such as age of dam, age of calf, and sex.
4. Records are used in selecting replacements (bulls and heifers) and in eliminating poor producers.
5. The nutritional regime and management practices are practical and comparable to those where the progeny of the herd are expected to perform.

Records of performance are useful primarily to provide a basis for comparing cattle handled alike on your farm or ranch. The performance data you collect not only reflects the genetic potential of your herd but even more the management.

Information Needed for Performance Test

1. Enroll in a performance program. Most states have a Beef Cattle Improvement Association (BCIA) program conducted through the Cooperative Extension Service. Contacting your county agent or land grant university extension service should provide you with the necessary information to enroll. Almost every major breed association provides a viable performance testing program in which you can easily enroll your cattle.
2. Identification. All calves, their sires, and dams must have positive identification.
3. Accurate birth dates must be recorded on all calves.
4. Birth weights should be recorded.
5. Weaning weights should be taken when the calves are between 160-250 days of age; however, it is suggested that weights be taken as close to 205 days of age as possible.
6. Type scores such as frame, muscle, condition (fat), structure, etc should be recorded when weaning weights are taken.
7. Age of dam should be recorded for each calf at birth.
8. Sex of calf should be noted.
9. Creep information should be noted. It can affect the interpretation of your records.
10. Postweaning weights should be recorded for 365 days and should be taken between 350 and 440 days of age. Frame, structure, muscle, and fat scores should again be recorded at the yearling stage.

Measurement of Weaning Weight (205 days)

Weaning weights are obtained in order to evaluate the differences in mothering ability and to measure differences in the growth potential of calves. BIF recommends that 205-day weight be computed on the basis of average daily gain from birth to weaning. This is accomplished by:

1. Subtracting actual birth weight from actual weight at weaning (if actual weight is not available, substitute the appropriate standard birth weight as designated by the respective breed association for the sire breed of calf).

2. Divide the age in days at weaning to obtain average daily gain.
3. Multiply the average daily gain by 205 and,
4. Add back the birth weight that was subtracted initially.

This is summarized in the following formula:

$$\text{Computed 205-day weight (lbs)} = \frac{\text{actual wt} - \text{birth wt}}{\text{age in days}} \times 205 + \text{birth wt}$$

This provides an estimated 205-day weight, unadjusted for age of dam or sex of calf. To establish a uniform procedure for age of dam, BIF recommends the following adjustments:

Age of dam	Additive factors	
	Male calves	Female calves
2-year-old cows	60 lbs	54 lbs
3-year-old cows	40 lbs	36 lbs
4-year-old cows	20 lbs	18 lbs
5–10-year-old cows	0 lbs	0 lbs
11-year-old cows	20 lbs	18 lbs

Weaning Weight Ratio

Records on 205-day weight and 205-day weight ratio, adjusted for age of dam on individual animals, should be evaluated on the basis of each sex (within sex basis without sex adjustment). Weaning weight ratios within sex groups are calculated by dividing each individual's 205-day weaning weight adjusted for age of dam by the average of its sex group and expressing it as a percentage of its sex group average. For weight ratios to be meaningful, contemporaries should be herd mates and similar in age and should all have been exposed to the same environmental influences.

Produce of Dam Summary

A record of lifetime productivity (cow summary) is recommended. It can be valuable for within-herd comparisons. It can be most helpful for identifying both the lowest-producing cows to be culled and consistently high-producing cows.

It is recommended that the cow summary include the following information:

- Measures relating to reproductive efficiency
    1. Age at first calving (days)
    2. Current age

3. Number of calves born (lifetime)
4. Number of calves weaned (lifetime)
5. Average age of calves when weaned
- Measures relating to productivity
  1. Average birth weight
  2. Average weaning weight ratio of all calves weaned
  3. Average adjusted 365-day weight ratio and number of contemporaries
  4. MPPA

Most Probable Producing Ability (MPPA)

It is recommended that MPPA be included on produce of dam summaries and that ranking of dams be based on MPPA for 205-day weaning weight ratio. This is needed to compare dams which do not have the same number of calf records in their averages. For example, suppose six cows have the following records of production:

Cow	No. calves	Avg. wn. wt. ratio	MPPA
A	1	85	94.0
B	2	88	93.2
C	4	90	92.7
D	3	110	106.7
E	4	112	108.8
F	1	115	106.0

In the example, cow A has the lowest lifetime average. However, this is for only a single calf for which environmental conditions or the calf's genetic potential for growth might have been below the average of what the cow would normally produce. One or more calves from cows B or C could also have had a record of 85 or less. All three cows are probably low producers, but the use of MPPA enables more accurate culling and, in this example, indicates that cows B and C are slightly lower producing cows than A.

MPPA for weaning weight ratio is computed by the following formula:

$$\text{MPPA} = \overline{H} + \frac{NR}{1 + (N-1)R} (\overline{C} - \overline{H})$$

where $\overline{H}$ = 100, the herd average weaning weight ratio

N = the number of calves included in the cow's average

R = 0.4, the repeatability factor for weaning weight ratio
$\bar{C}$ = average for weaning weight ratio for all calves the cow has produced

An example of the MPPA for Cow D follows:

$$\text{MPPA Cow D} = 100 + \frac{3 \times .4}{1 + (3-1).4}(110-100) = 106.7$$

Postweaning Phase

Measurement of yearling weight (365 days). Yearling weight should be computed and reported separately for each sex. The postweaning period should start on the date weaning weights are obtained (i.e., actual weaning weight is used as initial weight on test).

Age-of-dam effects on 365-day weight are of the same magnitude as age-of-dam effects at weaning. For this reason, it is desirable to add postweaning gains in a 160-day postweaning period to 205-day weaning weight, adjusted for age of dam to arrive at adjusted 365-day weight. The following formula is recommended (BIF):

$$\text{Adjusted 365-day wt} = \frac{\text{actual final wt} - \text{act. wn. wt}}{\text{number of days between wts.}} \times 160$$

$$+ \text{ 205 day wn. wt adj. for age of dam}$$

The period between weaning and final weight should be at least 160 days. Final weight should not be taken at less than 330 days of age for any animal, and the average age for each sex management group should be at least 365 days.

The procedure of using adjusted 365-day weights as a measure of yearling weight will apply primarily to herds that develop bulls on a rather high level of concentrate feeding starting at weaning time. For herds that prefer to develop bulls more slowly, a long yearling weight (452 or 550 days) may be used as an alternative to adjusted 365-day weights. To compute 452-day and 550-day weights, 247 and 345 would be substituted respectively for 160 in the 365-day adjusted weight formula.

For example,

$$\text{Adjusted 550-day wt.} = \frac{\text{actual final wt} - \text{act. wn. wt}}{\text{number of days between wts}} \times 345$$

$$+ \text{ wn. wt (205 days) adj. for age of dam}$$

2-7 Performance Records

Weight ratios for either adjusted 365-day weight (yearlings), adjusted 452-day weight, or adjusted 550-day weight (long yearlings) should be computed separately for each sex-management group.

Reproduction

Beef producers are encouraged to record reproductive performance in both the female and the male and to build this data into their herd records. They are urged to use this data in culling and selection. The following recommendations (BIF) are made for scoring and recording traits associated with the female.

- General reproductive performance:
  1. Open or pregnant—score 0 for open and 1 for pregnant.
  2. Calving date—record in conventional manner but store and carry in dam summaries in Julian calendar form.
  3. Calf born—score 0 for no calf born and 1 for calf born.
  4. Calf weaned—score 0 for no calf weaned and 1 for calf weaned.
  5. Age at first calving—should be carried in dam records in days.
- Birth weight of calf: The weight recorded in pounds and may be expressed as a ratio within like sex, age of dam, and management groups.
- Calving difficulty:
  1. No difficulty and no assistance (score 1).
  2. Minor difficulty, some assistance (score 2).
  3. Major difficulty, mechanical assistance with jack or puller (score 3).
  4. Ceasarean section, very difficult, or other surgery (score 4).
  5. Abnormal presentation (score 5). *Note*: Scores 1 through 4 may be averaged but 5 should not be included.
  6. Mortality (Score 0 for live and 1 for dead. This should be scored in a separate column.)

The following recommendations (BIF) are made for scoring and recording traits associated with the male.

- Physical exam:
  1. Palpation of scrotum and its contents. (Score 0 for unacceptable and 1 for acceptable.)
  2. Measure and record scrotal circumference in centimeters.
  3. Examine extended penis and prepuce for injury or abnormalities. (Score 0 for unacceptable and 1 for acceptable.)
  4. Palpate internal accessory glands rectally. (Score 0 for unacceptable and 1 for acceptable.)
- Semen evaluation:
  1. volume
  2. concentration
  3. % mortality
  4. morphology

The physical and semen examination should be performed by an experienced veterinarian.

3

Breeds and Breeding Systems

Selecting a breed or combination of breeds to use in your beef herd should be based on the following criteria: (1) marketability in your area; (2) cost and availability of good seedstock; (3) climate; (4) quantity and quality of feedstuffs on your farm; (5) how the breeds used in a crossing program complement one another; and (6) personal preference. As an example of climatic adaptability, British breeds are well adapted to cold climates but do not fare as well in subtropical regions. Conversely, Brahman blood is needed for optimum performance in certain Gulf Coastal areas but is not required in the northern states.

This chapter will first examine the selection characteristics of the breeds with summary tables for easy comparison. Breeding systems will be discussed in the section following breed selection illustrating ways to blend the blood of the breeds to enhance the productivity and efficiency of the entire industry.

3-1 SELECTING A BREED OF BEEF CATTLE*

British Breeds

Compared to breeds that originated on the continent of Europe, those that were developed in the British Isles generally exhibit the

---

*Dr. Harlan Ritchie, Professor of Animal Husbandry at Michigan State University, contributed significantly to this section.

following characteristics: (1) mature and fatten earlier; (2) grow less rapidly and are smaller at maturity; (3) are less muscular; (4) tend to be more fertile; (5) have less difficulty calving; and (6) live and reproduce longer. As a result of these characteristics, the British breeds are thought of as maternal breeds in a crossing program, as they tend to contribute those traits that are deemed important in a productive beef cow.

Angus. Angus cattle are second to Herefords in commercial numbers in the United States; however, they lead all breeds in numbers of purebred registered cattle (264,621 in 1977). They tend to be more popular in the Eastern two-thirds of the country than in the far West. Angus cattle are black and polled. Mature cows weigh about 1,000 lbs and, for a strictly beef-type breed, are considered good milkers. Angus females excel other beef breeds in fertility and calving ease. The breed is nearly pure for the polled trait and Angus bulls can be expected to sire calf crops that are 100 percent hornless. The dark skin pigment provides some resistance against cancer eye and sunburned udders.

Angus calves fatten quickly and grade Choice at a relatively light weight (1,000 lbs). They possess more marbling in the meat than any other breed of beef cattle, which means their quality grade (Prime, Choice, Good, etc) is often higher than that of other cattle. For this reason, many packers pay a premium for Angus or Angus-cross steers. However, feedlot operators sometimes pay less for Angus feeder calves because they tend to mature too quickly, stop gaining efficiently, and become fat at too light a weight. Nevertheless, Angus breeders are working hard at selecting larger cattle and are making considerable progress in changing the genetic ability of their cattle for growth rate.

The disposition of Angus cattle is considered as good or better than that of most European exotic breeds but not quite as docile as that of the Shorthorn or Hereford. Generally speaking, the breed is relatively free of defects such as pendulous udders, balloon teats, and uterine prolapse. The breed carries a red recessive gene at a low frequency (less than 10 percent). When two Angus red carriers are mated, there is a 25 percent chance that their calf will be red. Dwarfism was a problem in the 1940s and '50s but has practically been eliminated from the breed. The frequency of other undesirable recessive genes seems to be very low.

In crossbreeding programs, the Angus contributes polledness, pigment, fertility, early maturity, small cow size, and carcass quality (marbling).

Devon. The Devon, sometimes referred to as the "North" Devon or "Red" Devon, should not be confused with the South Devon, which is a larger-framed, lighter-colored, heavier-milking dual-purpose (meat and milk) breed. The Devon is dark cherry red in color and is horned. Although the cows are excellent milkers, the Devon is considered primarily a beef breed rather than dual-purpose. Mature cows weigh about 1,100 lbs. Devon calves mature early and fatten rather easily. Devons tend to be somewhat angular and appear to be lighter-muscled than some other British beef breeds.

Galloway. Like the Angus, the Galloway is black and polled, although a few are dun-colored. Size of frame is similar to the Angus, but conformation differs in that the Galloway is later-maturing, does not fatten as readily, and is more angular in its shape. Also, the hair coat of the Galloway is much longer and curlier. Some tend to have a nervous disposition. The breed is well-adapted to harsh Northern climates; however, it has not attained widespread popularity in the United States. In its native land, Scotland, the Galloway is a popular crossing breed and the crosses have ranked high in interbreed carcass competition. Mature cows weigh about 950 lbs.

Belted Galloway. Although it is considered a separate breed, the characteristics of the Belted Galloway are similar to those of the Galloway. The obvious difference is the belt of white hair that encircles the body of the Belted Galloway. The latter is also reported to be slightly larger and heavier milking than the Galloway. Mature cows weigh about 1,050 lbs.

Hereford. There are more commercial Herefords in the United States than any other breed of cattle. There were 219,681 purebred Herefords registered in 1977, placing them second to Angus in this respect. The Hereford's white face and underline are a prepotent trademark that can be transmitted through one or more generations of crossbreeding. However, the red body color is recessive to black; hence, the popular Hereford X Angus cross calf usually has a white face and underline with a black body.

The Hereford is particularly noted for its ability to thrive and reproduce under range conditions. Its heavy hide and hair coat adapt it to harsh winter weather and it is able to hold its condition well during extremes in climate and scarcity of feed. Commercial ranchers believe the fertility of the Hereford under range conditions is superior to that of any other breed. Hereford range bulls have a reputation for spreading out and covering the cow herd during breeding season more completely than bulls of other breeds. Hereford breeders are striving to select for more pigment around the eyes and teats as a means of lowering the incidence of cancer eye and sunburned udder in areas where these afflictions can be a problem. Hereford breeders are also selecting for greater milk production, a trait in which the breed needs improvement; however, an overabundance of milk in range country can place too much stress on the lactating cow causing her to lose condition and fail to cycle and conceive on time with the rest of the herd. Hereford cows weigh about 1,100 lbs.

Dwarfism hit the breed hard during the 1940s and '50s but has been virtually eliminated in recent years. Hereford feeder calves are in strong demand by feedlot operators, who know that they can purchase large uniform groups of calves nearly any time of the year. Herefords fatten easily at a young age and can be expected to grade Choice at about 1,100 lbs liveweight; however, their carcasses are not quite as heavily marbled as Angus at a comparable age.

In crossbreeding programs, the Hereford contributes range adaptability, winter hardiness, fertility under limited feed conditions, and moderate mature size with adequate growth rate.

Lincoln Red. The Lincoln Red is a dark red, horned, dual-purpose breed that descended from the local Shorthorn cattle in northeastern England. In some respects, it resembles the South Devon. However, the Lincoln Red is darker colored and does not appear to be quite as large nor as trim and muscular as the South Devon. Lincoln Red cows are excellent milkers, averaging about 7,500 lbs per lactation with 3.5 percent fat in milk-recorded herds in England. Mature females weigh about 1,300 lbs. Growth rate of the calves is similar to the South Devon but they are smaller at birth, averaging 81 and 74 lbs for males and females, respectively, whereas South Devon calves average 94 and 88 lbs, respectively, according to British data. In crossbreeding, the Lincoln Red could add milk production and some improvement in growth rate without an increase in calving difficulty.

Luing. The Luing (pronounced "Ling") is a recently synthesized beef breed that originated on the island of Luing off the west coast of Scotland, where the climate is cold and wet. The breed was started in 1947 by crossing the Shorthorn and Scotch Highland breeds. No other breeds have been introduced. Selection has been for hardy, easy-fleshing cattle with the ability to survive the harsh climate of western Scotland. They are not milked. Average size of mature cows is about 1,100 lbs. Color pattern varies due to the differences present in the two parent breeds. It is considered a maternal breed.

Milking Shorthorn. The Milking Shorthorn and beef Shorthorn originated from the same foundation in England. Selection for beef-type Shorthorns was carried out primarily in Scotland, while English Shorthorns continued to be bred as dual-purpose cattle. Mature Milking Shorthorn cows weigh about 1,250 lbs and produce approximately 10,000 lbs milk with 3.7 percent fat. In recent years, the American beef Shorthorn herd book has been opened to Milking Shorthorn cattle in an effort to improve size and milk production. Depending upon the country or region, they are also referred to as "Dual-Purpose Shorthorns," "Dairy Shorthorns," or "Durhams." In Australia, there is a heavier-milking strain of the Dairy Shorthorn known as the "Illawarra." It is very refined in its make-up and is considered strictly a dairy breed.

Murray Grey. This is an Australian breed that descended from a light roan Shorthorn cow mated in the early 1900s to black Angus bulls. She lived to be very old and produced many progeny from these matings, all of which carried the gray color to some degree. From that time on, these gray cattle were used in Angus herds and the gray color persisted, resulting in the Murray Grey breed. For all practical purposes, the Murray Grey is, in fact, a gray Angus, although Australians suspect that Charolais blood has crept into the breed in recent years. In Australia, the Murray Grey is noted for low birth weight, high calf livability, and excellent carcass quality. In crossing programs, it would contribute the same traits as the Angus.

Polled Hereford. There is very little difference between Herefords and Polled Herefords except for the absence of horns in the latter. However, commercial ranchers on the western ranges feel that Polled Hereford bulls do not spread out and trail cows as well as horned bulls, although there is no scientific evidence to confirm this belief. They also object to the fact that some Polled Hereford bulls allow their penis and sheath lining to protrude, exposing them to the risk of injury in rough country. They register the same complaints about Angus and all other naturally polled bulls, which for some reason seem to share these characteristics. However, more polled bulls are being used in the far West all the time. The

Polled Hereford breed has grown rapidly during the past 25 years and now ranks third in purebred beef registrations (173,010 in 1977).

Red Angus. As noted before, the Red Angus breed originated from the black Angus as a homozygous recessive trait. There were 8,512 Red Angus registered in 1977.

Red Poll. The Red Poll is a deep red, naturally polled breed that was once classed as dual-purpose but in recent years has moved in the direction of being strictly a beef breed. During its dual-purpose era, milk-recorded Red Poll cows averaged about 7,500 lbs milk with 4.2 percent fat. Mature cow size is about 1,150 lbs. According to research at the U.S. Meat Animal Research Center, Red Poll-sired calves are similar to Angus and Hereford-sired calves in calving ease, growth rate, and carcass characteristics. Steers sired by Red Poll bulls reached Choice grade at about 1,050 lbs, which is the same weight at which Angus and Hereford steers graded Choice. Red Poll-sired first-calf heifers weaned slightly heavier calves, but raised a slightly lower percent of calves than Angus or Hereford heifers.

Scotch Highland. This is an environmental breed that, like the Galloway, was developed to withstand the adverse climate of mountainous western Scotland. Most Scotch Highland cattle are dun-colored but some are black, brindle, red, yellow, or silver. Long, widespread horns and a long, dense, shaggy hair coat are characteristic of the breed. They are small in size and lack the natural thickness and muscling of other beef breeds. Mature cows weigh about 900 lbs. Their main area of activity in the United States lies in the Northern Plains.

Shorthorn. The Shorthorn was the first improved breed of beef cattle that achieved prominence in the United States and was the most numerous breed in this country until the early 1900s. Since then, Herefords and Angus have taken over as the two leading beef breeds. There were 25,648 Shorthorns registered in 1977.

Shorthorns may be characterized by their variable color pattern which ranges from red-to-roan-to-white. As the name implies, the horns are relatively small. There is a polled gene in the breed and Polled Shorthorns have increased in popularity so that they now rival horned cattle in numbers.

Shorthorns are noted for their maternal ability. The Shorthorn cow is an excellent milker and weans a heavy calf. When Shorthorn bulls are mated to Hereford cows, the female progeny are superior to their dams in milking ability, and this trait seems to carry through for one or two or more generations of back-crossing with Hereford sires.

In addition to their maternal ability, there is no breed of cattle that is more docile than the Shorthorn. Their disposition is unexcelled and cattlemen appreciate the ease with which they can be handled.

At one time, Shorthorns were recognized as the largest British beef breed. But, when the trend to extremely small-framed, early maturing cattle occurred in all breeds during the 1930s, '40s and '50s, they seemed to lose even more size and growth than Herefords or Angus. This fact, together with their lack of natural muscling, contributed to their decline in popularity. However, with the infusion of large-type Canadian, Australian, and Milking Shorthorn blood into the breed in recent years, it has made a comeback. Once again, some of the largest specimens among British cattle may be found in the Shorthorn breed. If Shorthorn breeders can now improve muscling and cutability without losing size, their future position in the beef industry should improve significantly.

Because of the introduction of new blood and the resultant variation in type, it is difficult to characterize the size of the breed today. Mature cows may range from 900 to 1,600 lbs, but the average is probably close to 1,100 lbs. Shorthorn steers fatten easily and can be expected to grade Choice by the time they weigh approximately 1,050 lbs.

In crossbreeding programs, the Shorthorn contributes several maternal traits—milk production, ease of calving, disposition, early maturity, and moderate cow size.

South Devon. South Devons are light red, horned, dual-purpose cattle. They are probably the largest breed of cattle in Great Britain. Mature cows weigh about 1,450 lbs; in milk-recorded herds, they average about 7,500 lbs milk with 4.0 percent fat. According to data from the U.S. Meat Animal Research Center, half-blood South Devon steers grade Choice when they weigh 1,100 to 1,150 lbs. In most growth and carcass traits, they lie somewhere between the larger exotic (Charolais and Simmental) and traditional British beef breeds (Angus and Hereford). Fertility, birth weight, ease of calving, and weaning percentage are also intermediate, but are closer to the larger exotics than the British beef breeds. In crossbreeding programs, the South Devon could improve milk production and growth rate.

Sussex. The Sussex is a short-horned, dark red breed with a white switch on the tail. It is strictly a beef breed. Mature cows weigh about 1,150 lbs. Compared to the South Devon and Lincoln Red, it is a thicker, smaller-framed, earlier maturing breed. But in British tests, half-blood Sussex calves have been comparable in 200-day weights to those sired by South Devon and Lincoln Red bulls. In England, it has a reputation as an easy calving breed. The Sussex has not yet been promoted in the United States.

Welsh Black. The Welsh Black is a horned, black, large-framed, long-haired breed that was developed under the harsh conditions of the Welsh mountains. It is known for its ability to thrive under sparse feed conditions and for its longevity. The Welsh Black is a dual-purpose breed; milk-recorded cows produce an average of about 5,500 lbs of milk with 3.8 percent fat. Mature cows weigh approximately 1,200 lbs. The Welsh Black cow is highly valued for crossing with terminal sire breeds such as the Charolais. The purebred Welsh Black is later maturing and must be fed for a longer period of time before it accumulates desirable finish.

Continental European Breeds

Compared to British breeds, the Continental exotic breeds are generally larger, later-maturing, heavier-muscled, "growthier," less fertile, and have more difficulty calving (Figure 3-1). Because of their leanness, they exhibit a higher yield grade (cutability), less marbling, and lower quality grade. As a result of these characteristics, they are ordinarily considered terminal sire breeds for crossbreeding programs, although a few of the heavier-milking, more fertile, moderately sized Continental breeds may be classified as either dam or two-way (sire and dam) breeds.

The first of these new breeds came to North America in the late 1960s, and a few have arrived just recently. There is still much to be learned about these breeds under the varying conditions that exist on our continent. Research data on half-blood progeny from several of the new breeds have been collected since 1971 at the U.S. Meat Animal Research Center (U.S. MARC), Clay Center, Nebraska, and at several Canadian Experiment Stations.

Aubrac. The Aubrac is a relatively small yellowish-brown breed that originated in mountainous, south central France. Mature cows weigh about 1,100 lbs. It has been used as a dual-purpose breed, but is not noted for high milk production. Milk-

Figure 3-1
Continental European Breeds

Charolais Bull

Chianina Bull

Gelbvieh Bull

Limousin Cow

Maine-Anjou Bull

Marchigiana Bull

Pinzgauer Bull

Simmental Bull

recorded cows average approximately 5,000 lbs with 4.0 percent fat. The Aubrac ranks at or near the top of the French breeds in fertility and ease of calving. It is one of the few Continental breeds that would be considered maternal. To date, the breed has not been introduced to the United States.

European Friesian (Beef Friesian). This breed is the European version of the Holstein-Friesian. The first European Friesians exported to North America came from Ireland, although they are found all over Europe. Compared to the North American Holstein, they are thicker, heavier-muscled, earlier-maturing, and give less milk. The Friesian and the Simmental are the two most popular breeds in Europe. Compared to the Simmental, the Friesian produces more milk but is not as growthy or muscular. Mature European Friesian cows weigh about 1,350 lbs and half-blood steers should reach Choice grade at about 1,200 lbs. The Beef Friesian is considered to be two-way, but tends to be more of a maternal breed than a sire breed.

Blonde d'Aquitaine. This is a large, blond breed from France that resembles the Limousin in shape and color except it is larger-framed and appears to be even more extreme in its musculature. They are not noted for their milk production and are classified as strictly a beef breed. Mature females in France weigh about 1,500 lbs. It is estimated that half-blood Blonde steers will probably grade Choice at about 1,250 lbs. Their growth rate and cutability are comparable to the Charolais, Simmental, and Maine-Anjou. They are classified as a terminal breed.

European Brown Swiss (Braunvieh). This breed is smaller, beefier, and produces less milk than its North American counterpart. It yields about the same amount of milk as the Simmental but is not as growthy or muscular. Mature cows weigh about 1,300 lbs and half-blood steers should grade Choice at 1,150 lbs. It is considered a two-way breed.

Charolais. In France, the typical Charolais is slightly thicker, heavier-muscled, shorter-legged, and heavier-boned than the American version. As a result, they would appear to have more calving difficulty and a higher percent of Caesarean births than domestic Charolais. The Charolais is strictly a beef breed and is not milked in France. There seems to be little difference in growth rate and carcass composition between the Charolais, Maine-Anjou, Simmental, and Blonde d'Aquitaine. Mature Charolais cows in France weigh about 1,550 lbs. Half-blood Charolais steers reach Choice grade at about 1,250 lbs. Like other large Continental breeds, calving difficulty is a major problem. Other French breeds may excel the Charolais slightly in female fertility; nevertheless, the Charolais is the third leading breed in France in total numbers and has contributed greatly to beef production in North America. Charolais bulls have added growth and cutability to the United States cattle population. A few of the new breeds can equal the Charolais in these two traits, but none can consistently excel them. In the United States, 34,332 Charolais were registered in 1977.

Chianina. This Italian white breed is the largest-framed breed of cattle in the world. Mature cows in Italy weigh about 1,700 lbs. It appears that half-blood Chianina steers need to weigh over 1,300 lbs to safely grade Choice. For this reason, there is growing interest in using 1/2 and 3/4 blood Chianina bulls to produce 1/4 and 3/8 calves, which can grade Choice at 1,100 to 1,200 lbs. Chianina × Angus cross bulls are especially popular for this purpose. The Chianina is not milked and is strictly a beef breed. Growth rate of Chianina half-bloods is similar to other large Continental breeds. Because they are later-maturing, they produce leaner, higher cutability carcasses at a given age or weight than other exotic breeds. At Choice grade and heavier weights, their cutability is comparable to other large exotics. Data suggest that the Chianina is an easier calving breed than some of the other large exotic breeds. Chianina half-blood heifers seem to reach puberty later, and their conception rate is lower. The Chianina tends to have a nervous disposition, which is objectionable to some cattlemen. It is considered a terminal sire breed. Chianina bulls cross especially well with smaller, thicker Angus cows; steers bred this way have ranked very high in interbreed, on-hoof and carcass competition. In 1977, 14,377 Chianina were registered in the United States.

Gasconne. This gray French breed is moderate in size and muscling. It is not milked. Mature cows weigh approximately 1,300 lbs. The Gasconne was developed in mountainous, southern France primarily as a draft breed. It has a reputation for being one of the most fertile French breeds with minimum calving difficulty. Although it is too early to be certain, the Gasconne may emerge as a two-way breed.

Gelbvieh (German Yellow). The Gelbvieh is a reddish-yellow dual-purpose German breed. It is a bit smaller and slightly more refined in its bone than the German Simmental (Fleckvieh) but appears to have slightly more muscle expression, suggesting higher cutability. The Germans do, in fact, claim that the Gelbvieh excels all other German breeds in cutability; however, the difference seems to be slight. In color and shape, the Gelbvieh somewhat resembles the Limousin, and some authorities believe these two breeds share common ancestry. However, the Gelbvieh is larger and is dual-purpose, whereas the Limousin is smaller and has been developed solely for beef production. The mature Gelbvieh cow weighs about 1,450 lbs and in Germany milk-recorded herds average about 8,500 lbs with 4.0 percent fat, which places them behind the Friesian, Fleckvieh, and Braunvieh in milk yield. Data on half-blood steers at U.S. MARC indicate that growth rate and carcass composition are similar to the other large Continental breeds. However, birth weight and calving ease were slightly better than the Maine-Anjou and Chianina. Half-blood steers grade choice at about 1,200 lbs. The half-blood heifers reached puberty at a young age but their conception rate was not as high as British heifers. The Gelbvieh shows promise as a two-way breed.

Limousin. The Limousin is a reddish gold colored, strictly beef-type breed from central France. Among the Continental breeds, it is moderate in size, mature cows averaging about 1,300 lbs. According to U.S. MARC data, Limousin half-blood steers are intermediate in growth rate between the British and the larger exotic breeds. They mature slightly earlier than the Charolais and Simmental, grading Choice at about 1,150 lbs. In cutability, they were the highest of all breeds evaluated, including the Charolais and Simmental. The very muscular appearance of the Limousin is indicative of its high cutability. Limousin-sired calves weigh less at birth and are easier to deliver than calves sired by Charolais or Simmental bulls. Although Limousin cows are not thought of as heavy milkers, they do an adequate job of raising their calves. In U.S. MARC research, age at puberty and subsequent fertility of half-blood Limousin females were similar to that of half-blood Charolais females. Generally speaking, the Limousin is considered a terminal sire breed that works especially well in situations where rapid improvement in muscle is needed but calving difficulty is to be avoided as much as possible. There were 26,450 Limousin registered in the United States in 1977.

Maine-Anjou. The Maine-Anjou is a red and white dual-purpose breed found in northwestern France. It is the largest

French breed in terms of body weight and length; the Blonde D'Aquitaine is probably taller. Mature Maine-Anjou cows in France average approximately 1,700 lbs. Data from U.S. MARC suggest that it may be as rapid in growth rate as any breed evaluated to date. Although Chianina and Gelbvieh-sired calves weighed slightly more at weaning, Maine-Anjou half-blood steers gained faster in the feedlot. Their rate of maturity resembles that of the Charolais and Simmental with half-blood steers reaching Choice grade at about 1,250 lbs. Carcass composition is similar to the Charolais, Simmental, and Gelbvieh. Half-blood Maine-Anjou heifers reached puberty later than Gelbvieh but earlier than Chianina heifers. Calving difficulty and mortality were higher for Maine-Anjou-sired calves than for those sired by other breeds. Milk production is good, but Maine-Anjou cows do not milk as heavily as some of the other dual-purpose breeds; their average in France is about 5,500 lbs with 3.7 percent fat. Because of its milk production, growth, and cutability, it is sometimes thought of as a two-way breed. However, many authorities argue that the mature size of the Maine-Anjou is too large to classify it in this manner and that it should be considered a terminal sire breed. In 1977, 9,000 were registered in the United States.

Marchigiana. This Italian white breed closely resembles the Chianina except for the fact that it is smaller-framed, thicker and more bulging in its musculature. Mature cows weigh about 1,500 lbs. Like the Chianina and Romagnola, it is strictly a beef breed and is not used for milk production. All three of these breeds have white hair but are black-skinned, which presumably gives them some resistance to cancer eye, pink eye, sunburn and other skin problems. They are also heat-tolerant, which would make them well-adapted to the southern United States. The Marchigiana is a terminal sire breed.

Meuse-Rhine-Issel (MRI). The MRI is a red and white dual-purpose breed found in southeastern Holland. It probably originated from the same foundation stock as the Dutch Friesian but underwent selection for red instead of black color and for a shorter-legged, heavier-boned, more muscular body type that tends to be more typically dual-purpose when compared to the conformation of the Friesian. Mature cows weigh around 1,450 lbs. According to research in the Netherlands, MRI bull calves averaged 6 lbs heavier at birth and gained faster (2.53 vs. 2.28 lbs per day) up to a slaughter weight of 935 lbs than Friesian bull calves. Their carcasses were also heavier-muscled and carried less finish than the Friesian carcasses. MRI cows average about 10,500 lbs milk with

3.7 percent fat, compared to 11,500 lbs milk with 3.9 percent fat for Dutch Friesian cows. About 24 percent of the cattle in the Netherlands are MRI's, 74 percent Friesians, and 2 percent other breeds. Because of its dual-purpose characteristics, the MRI is considered a two-way breed in crossing programs.

Other red and white breeds in northwestern Europe. Several red and white breeds in northwestern Europe resemble the MRI and probably originated from a somewhat common ancestry. The most prominent of these are the Rotbunte (German Red and White) and three Belgian breeds: the Red and White Campine, the Red and White East Flemish, and the Red West Flemish. These breeds tend to be spotted except for the Red West Flemish, which is nearly solid red. Cows in these breeds weigh from 1,350 to 1,500 lbs at maturity and average about 10,000 lbs milk with 3.7 percent fat. Compared to the Friesian, they produce less milk and are more muscular in their conformation. These differences are becoming greater as more North American Holstein semen is used in European Friesian herds to further improve their milk production. In North America, the MRI herd book will be open to these other red and white breeds.

Montbeliard. The Montbeliard is a heavier-milking strain of the French Simmental. Compared to the beefier Pie Rouge, the Montbeliard is flatter-muscled, more refined, and generally more dairylike in its conformation. Most authorities feel that the Montbeliard has less calving difficulty than the heavier-muscled, heavier-boned Pie Rouge, but there is no research information yet to support this belief. Montbeliard cows weigh about 1,400 lbs and produce approximately 10,000 lbs milk with 3.7 percent fat. They account for about 6 percent of the cattle in France compared to 3.5 percent for the Pie Rouge. The Montbeliard is considered a two-way breed.

Normande. The Normande is a medium-sized dual-purpose breed that originated in the Normandy region of northwestern France. It ranks a close second to the Friesian in total numbers, accounting for about 19 percent of the French cattle population. Mature cows weigh about 1,350 lbs and produce an average of 8,500 lbs milk with 4.0 percent fat. The Normande bears an interesting and variable color pattern. Most specimens are mahogany and white or black and white with rings around the eyes. However, a few are blond, others are white, and some are nearly all black.

Although little research data are available, the Normande does not appear to be as growthy or as muscular as the larger exotics such as the Charolais, Simmental, etc. However, they are reported to calve more easily and may be more fertile than the larger breeds. They are being publicized as an all-around, relatively trouble-free two-way breed.

Norwegian Red. This red dual-purpose breed accounts for about 60 percent of the cattle population in Norway. Because the breed has developed since World War II from the amalgamation of three older breeds, it is difficult to characterize the Norwegian Red. Cows range from 1,000 to 1,400 lbs with an average weighing about 1,250 lbs. Milk-recorded cows average about 10,500 lbs of milk with 4.2 percent fat. In color, size, and conformation, the Norwegian Red resembles the American Milking Shorthorn, but appears to be slightly trimmer and more muscular. Calving ease is reported to be similar to the British breeds. In crossbreeding programs, the Norwegian Red is considered a maternal breed that could increase milk production and perhaps cause a slight improvement in growth with no increase in calving difficulty.

Parthenaise. The Parthenaise is a fawn-colored dual-purpose breed that originated in western France. Mature cows weigh about 1,400 lbs and produce approximately 6,000 lbs milk with 4.4 percent fat. Emphasis has been on milk production for buttermaking. The Parthenaise has a good reputation for fertility and ease of calving. In this respect, it ranks near the top of the French breeds, along with the Aubrac, Gasconne, Salers, and Tarentaise. As of yet, the Parthenaise has not been introduced to the United States.

Pinzgauer. The Pinzgauer is a dual-purpose mountainous breed that originated in the Alpine regions of Austria, southeast Bavaria, and northeast Italy. It makes up about 14 percent of the cattle population of Austria. The Pinzgauer has a very distinctive color pattern—chestnut brown with white topline, buttocks, and underline; the skin is pigmented even in the areas covered by white hair. It somewhat resembles the Simmental and Gelbvieh in conformation but is not quite as large; mature cows weigh about 1,350 lbs. Average production of milk-recorded Pinzgauer cows is about 8,000 lbs with milk fat test averaging about 4.0 percent. The Pinzgauer is noted for its sturdiness and hardiness under rugged mountainous conditions. It is reported to be more fertile, less prone to calving difficulty, and generally more trouble-free than some of the larger, heavier muscled Continental breeds. The Pinzgauer is considered a two-way breed.

Piedmont (Piemontese). The Piedmont differs from the other Italian white breeds in that it is smaller and nearly all of the bulls in AI service are of the double-muscled type. In this area of Italy (northwest), strong selection pressure has been placed on muscling. As a result, milk production and size have declined and calving difficulty has increased, although not as much as one might expect. Mature cows weigh about 1,300 lbs. In addition to being extremely heavy-muscled, Piedmont cattle are characterized by thin, relatively fine bone. Carcasses from young (15-month) Piedmont bulls are practically devoid of fat and excel Piedmont × Friesian crossbred bulls in percent muscle by 7 percent (54 vs. 47 percent). To date, the Piedmont has not been promoted in the United States.

Romagnola. The Romagnola is another of the Italian breeds with white hair and black skin. Compared to the Chianina and Marchigiana, it is a deeper-bodied, bulkier breed that is not quite as trim or upstanding. Most of the other traits are comparable to those exhibited by the Chianina and Marchigiana. Mature cows weigh about 1,500 lbs. It is strictly a beef breed that is considered a terminal sire breed in crossing programs.

Salers. The Salers is a dark-red, dual-purpose breed that originated in the mountainous region of south central France, where the soil and climatic conditions are relatively severe. Due to these conditions the cattle seem to be adapted to less than optimum feeding and management programs. The Salers is reported to be a relatively trouble-free breed that is above the average of the French breeds in fertility and calving ease. Size and muscling are moderate; double muscling does not appear to be a problem. Mature cows weigh about 1,300 lbs. They are tall cattle, characterized by upright "lyre" shaped horns. Milk-recorded cows average about 6,200 lbs of milk with 3.7 percent fat. The Salers is considered a two-way breed.

Simmental (Fleckvieh, Pie Rouge). The Simmental, along with the Friesian, is one of the two most numerous breeds throughout Europe. It originated in Switzerland but is also very popular in Germany, Austria, and France as well as several eastern European countries. In Germany and Austria, it is known as the "Fleckvieh" and in France as the "Pie Rouge." Color is red and white with the red varying from light yellow to dark red. Mature cows weigh about 1,500 lbs with milk-recorded cows averaging approximately 9,000 lbs milk with 3.9 percent fat.

In France, there are two other Simmental strains besides the Pie Rouge; namely, the Montbeliard and the Abondance. The Montbeliard, as noted previously, is a more dairylike strain of the

French Simmental. The Abondance is somewhat smaller than either the Pie Rouge or Montbeliard, with mature cows weighing about 1200 lbs. It is also less numerous than the other two strains.

In Europe, the various Simmental strains are recorded in separate herd books, but in the United States they are recorded by one registry association. In 1977, 67,271 cattle were registered by the American Simmental Association.

Tarentaise. The Tarentaise is one of the smaller-framed Continental breeds imported to date. Mature cows average about 1,200 lbs. Because birth weight, growth rate, and mature size tend to be highly correlated, Tarentaise calves are usually smaller at birth and grow less rapidly than those of the larger breeds. In general, their growth rate is reported to be similar to that of the traditional British breeds. Although they are promoted as an easy-calving breed, their assistance rate is reported to be no better than that of the British breeds. The Tarentaise originated in the Alpine region of southeastern France and has been developed as a dual-purpose breed. Muscling and cutability appear to lie between the British and larger Continental breeds. The Tarentaise is basically red in color with black pigment around the body orifices—eyes, muzzle, vulva, etc. Because the region of origin is mountainous and not particularly fertile, the Tarentaise has achieved a reputation for vigor and hardiness. It is considered a maternal breed.

Brahman and Brahman Crosses

American Brahman. The American Brahman is primarily a blend of three Zebu (humped) breeds from India: (1) Guzerat; (2) Nellore; and (3) Gyr. In addition to these three breeds, there are many other breeds or strains of Zebu cattle throughout the tropical and subtropical regions of the world. In recent years, an extremely large-framed breed from Brazil, known as the "Indu-Brazil," has been infused into the American Brahman. The Indu-Brazil is also a blend of the Guzerat, Nellore, and Gyr breeds. Guzerat and Nellore cattle are gray to black in color, whereas the Gyr breed is usually red or red with white speckling. It has been roughly estimated that the American Brahman is composed of approximately: 60 percent Guzerat blood; 20 percent Gyr and Indu-Brazil blood; and 20 percent Nellore breeding. In addition, there is a very small, but undetermined, percentage of British blood in the American Brahman. In 1977, the American Brahman Breeders Association registered 31,356 head of cattle.

The American Brahman was developed out of the need for cattle that could tolerate the subtropical environment of the southern United States. It is well known that British breeds are not as well adapted to the rigors of a hot climate as Zebu-type cattle. However, neither straight Brahman nor straight British cattle are as productive as Brahman × British crossbreds, which exhibit a high percent of heterosis over the average of the parent breeds. The degree of heterosis is particularly evident in cow fertility, bull fertility, and growth rate. Straight Brahman cattle are low in fertility and postweaning rate of gain. In addition, straight Brahman beef tends to be low in marbling and palatability. Purebred Brahmans also suffer from a relatively high incidence (3 to 4 percent) of a "weak calf syndrome" that is not generally found in Brahman × British crosses. Compared to British breeds, Brahman cattle exhibit a more nervous disposition.

A number of breeds have evolved from the crossing of Brahman and British breeds. The most notable are the Santa Gertrudis, Brangus, Beefmaster, and Braford. In addition, there are several other less numerous Zebu-cross breeds.

Santa Gertrudis. The Santa Gertrudis is approximately 5/8 Shorthorn and 3/8 Brahman blood. Its color is a deep red and it is large in size, mature cows averaging about 1,400 to 1,500 lbs. It was developed by the King Ranch, starting in about 1920. The Santa Gertrudis exhibits an extremely rapid growth rate that is comparable to the largest of the European breeds. Fertility, however, does not seem to be as high as that of the other synthetic Brahman × British breeds. There were 26,410 Santa Gertrudis cattle registered in 1977.

Brangus. The Brangus is composed of 5/8 Angus and 3/8 Brahman blood. Initial development of the breed occurred during the 1940s and '50s. Brangus cattle are black and polled. Like Angus, the recessive red gene is present in the Brangus breed, and led to the foundation of a Red Brangus Association. Compared to other Brahman × British hybrids, the Brangus tends to be slightly smaller in size and smoother in conformation. Mature cows weigh about 1,250 lbs. In 1977, 16,316 Brangus cattle were registered.

Braford. Braford cattle are approximately 5/8 Hereford and 3/8 Brahman. Although an "International Braford Association" was founded in Fort Pierce, Florida in 1969, the Association is relatively inactive and serves primarily to provide standards for selecting Braford cattle. Research has shown that the Brahman × Hereford cross cow may very well be the most productive beef female for the Gulf Coast region of the United States.

Beefmaster. The Beefmaster is estimated to be slightly less than 50 percent Brahman, with the balance about equally divided between Hereford and Shorthorn blood. It was developed through the efforts of Tom Lasater, who assumed management of his father's Texas ranch after the latter's death in 1930. Lasater stressed six criteria in his selection program: disposition, fertility, weight, conformation, hardiness, and milk production. Cows that were poor milkers or failed to wean a calf for any reason were culled. Color was never a consideration in Lasater's program. Consequently, color is variable, but red predominates. Beefmasters are generally horned, but the polled condition does exist.

Charbray. The Charbray is a Charolais x Brahman cross. There is no longer an association for the registration of Charbray cattle.

Sahiwal. The Sahiwal (pronounced Si-Wall) is a relatively small, heavy-milking, pure Zebu breed from Pakistan. It is a fawn-colored animal. According to U.S. Meat Research Center data, Sahiwal-sired calves were heavier at birth but weighed less at weaning time than calves sired by Hereford and Angus bulls.

Barzona. The Barzona is a combination of four breeds: Africander (a Zebu type), Hereford, Santa Gertrudis, and Angus. It was developed near Kirkland, Arizona by F. N. Bard, who started in 1942 with Africander x Hereford Cross females. Two-thirds of these F_1 females were mated to Santa Gertrudis bulls, and the remaining third to Angus bulls. The resultant progeny formed the basis of the Barzona breed. Their color is a dark reddish brown. In both color and shape, they somewhat resemble the Santa Gertrudis.

Brahmental. A Brahmental is a Brahman x Simmental cross. It is just in the developmental stages. Brahmentals will be recorded by the American Simmental Association. A Brahmental must have a minimum of 25 percent Brahman blood and a minimum of 37½ percent Simmental blood.

American Dairy Breeds

Holstein-Friesian. The Holstein-Friesian is by far the most numerous breed of dairy cattle in North America. There were 279,146 Holsteins registered in the United States in 1975. The American Holstein is unexcelled in milk production by any breed in the world. Average milk production for the breed in the United

States lies somewhere between 14,000 and 15,000 lbs with 3.6–3.7 percent fat. It is a large breed, mature cows averaging about 1,500 lbs in weight, although there are a few cows that weigh 1,700 to 2,000 lbs. Compared to the European Friesian, the American Holstein is larger-framed, more dairylike in its conformation, and produces 30 percent more milk. However, the gap is being narrowed, because American sires are being used to improve the milk production of European Friesians.

Half-blood Holstein × British cross cows milk extremely well and produce a heavy calf at weaning time. However, they require significantly more feed than British beef cows in order to come through the winter in good condition and maintain fertility during the breeding season. In many areas a 1/4 Holstein × 3/4 British cow would make a more ideal beef female.

Brown Swiss. Compared to the Holstein, the Brown Swiss is somewhat more rugged and beefier in its conformation. As a result, Brown Swiss cross cows may tend to come through a cold winter in better condition than Holstein cross beef cows. However, the straightbred Brown Swiss cow is not as inherently fertile as the straightbred Holstein cow. Mature Brown Swiss cows average about 1,500 lbs in body weight and produce about 12,000 lbs of milk with 4.0 percent fat.

Jersey. Because of its small size, its inherent fertility and calving ease, and its high milk production relative to body size, the Jersey breed has recently received considerable attention as a producer of F_1 beef females when crossed with either Angus or Herefords. Research at Texas, Illinois, and the U.S. Meat Animal Research Center has shown very favorable results when Jersey × British cross females are mated to a large muscular bull of a third breed, such as a Charolais. However, the Jersey × British cross male calves are discounted in price, which is an economic disadvantage to the producer of the F_1 cattle.

Mature Jersey cows weigh 900 to 1,000 lbs and average about 9,000 lbs milk, with 5.1 percent fat.

Guernsey. The Guernsey breed is seldom used to produce dairy × beef cross females. Mature Guernsey cows weigh 1,100 to 1,400 lbs and produce about 10,000 lbs of milk with 4.75 percent fat. In recent years, breeders have significantly increased the body size of the breed.

Ayrshire. Because of low numbers, the Ayrshire breed does not contribute much genetic material to the beef cattle industry.

Mature cows weigh about 1,200 lbs and produce an average of 11,000 lbs of milk with 4.0 percent fat.

Other Breeds and Crosses

Hays Converter. The Hays Converter was synthesized by combining the blood of two dairy breeds and one beef breed; namely, the Holstein, Brown Swiss, and Hereford. Color is either black and white or red and white. The cows are large-framed and heavy milking. Their musculature tends to be medium to flat. The breed takes its name from its developer, Senator Harry Hays, former Minister of Agriculture, Calgary, Alberta, Canada.

Longhorn. Longhorns in America are descendants of cattle brought to Mexico by Spanish explorers several centuries ago. They found their way into the United States and formed the basis of Western cow herds during the 1800s. They were late-maturing, slow-growing cattle that were highly fertile and able to withstand the privations of sparse range country. They gradually dwindled in numbers to no more than 1,500 during the early 1960s. In 1964, the Texas Longhorn Breeders Association was founded in San Antonio, Texas. Since then, some interest in breeding Longhorn cattle has developed in certain areas of the West.

Beefalo. The Beefalo is 3/8 Buffalo (Bison) and 5/8 Bovine. It has been developed and promoted in recent years by Bud Basolo of Tracy, California. Its main attributes seem to be increased winter hardiness and ability to subsist on low quality forages.

3-2 CLASSIFICATION OF BREEDS

Classification by Mature Size, Milk Production and Muscle Type

Tables 3-1, 3-2, and 3-3 are an attempt to classify breeds according to mature size, milking ability, and type of muscling (thick, moderate, or flat). It should be remembered that this is a very general classification and there are many exceptions in any breed of cattle. This can be a reasonably useful method of classification because the breeds with greatest mature size and thickest muscling tend to exhibit the most rapid growth rate and the highest incidence of calving difficulty. Conversely, those breeds with the smallest mature size and flattest musculature tend to grow

Table 3-1
Large-Sized Breeds

Muscle type	Milking ability		
	High	Medium	Low
Thick	Simmental	Maine-Anjou	Charolais, Blond d'Aquitaine, Marchigiana
Moderate	North American Brown Swiss	--	Chianina, Romagnola, Santa Gertrudis
Flat	North American Holstein	South Devon	--

Table 3-2
Medium-Sized Breeds

Muscle type	Milking ability		
	High	Medium	Low
Thick	MRI	Gelbvieh, Pinzgauer	Limousin, Piedmont
Moderate	Montbeliard, Braunvieh, Beef Friesian	Normande, Salers, Parthenaise	Gasconne, Brahman, Brangus, Braford, Beefmaster, Barzona
Flat	Milking Shorthorn, Norwegian Red	Lincoln Red, Welsh Black, Hays Converter	--

Table 3-3
Small-Sized Breeds

Muscle type	Milking Ability		
	High	Medium	Low
Thick	--	--	--
Moderate	--	Tarentaise, Abondance, Aubrac	Angus, Hereford, Shorthorn, Murray Grey, Sussex, Beefalo
Flat	Jersey, Guernsey, Ayrshire	Red Poll, Devon, Sahiwal	Galloway, Scotch Highland, Luing, Longhorn

least rapidly and exhibit the lowest incidence of calving difficulty. Furthermore, some knowledge of mature size and milking ability is needed to accurately predict the nutrient requirements of the cow.

Classification as Sire, Dam, or Two-Way Breeds

Another common method of classifying breeds is to divide them into sire or dam breeds, depending upon whether their traits tend toward the paternal or maternal side. An ideal sire breed in a crossbreeding program would have the following characteristics: rapid growth rate (medium to large mature size), high carcass cutability (moderate to thick muscling), adequate calving ease, and adequate carcass quality (marbling). An ideal dam breed in a crossbreeding program would exhibit the following characteristics: high fertility, medium to high milking ability, and small to medium mature size. Of course, no breed can lay claim to all of these characteristics, and one must make compromises when selecting breeds for a crossbreeding program. Some breeds have a near-equal balance between sire and dam traits and may be classified as two-way breeds. Table 3-4 is an attempt to classify cattle breeds as sire, dam, or two-way breeds.

3-3 CHOOSING A BREEDING SYSTEM

Commercial Herds

Selecting a breeding plan to use in your beef herd should be based on the following factors: (1) herd size, (2) resources, (3) breeds, and (4) expertise. As an example, owners of herds with less than 50 cows and minimum resources should first consider a purebreeding system. Preference in selection of a breed should be given to those that best combine reproduction, performance, and carcass merit. The British breeds appear to best meet these criteria. At the most, owners of small herds with minimum to maximum resources and expertise could consider a simple crossbreeding program. Table 3-5 suggests breeding systems based on herd size and resources.

1. Straightbreeding. Many commercial herds will remain straightbred and very productive. Small herds (less than 50 head) represent over 60 percent of the U.S. cattle population and this

Table 3-4

Classification of Cattle Breeds as Sire, Dam, or Two-Way Breeds

Sire Breeds	Dam Breeds	Two-Way (Sire & Dam) Breeds
Blonde d'Aquitaine	Abondance	Beef Friesian
Charolais	Angus	Braunvieh
Chianina	Aubrac	Gelbvieh
Limousin	Ayrshire	Maine-Anjou
Marchigiana	Barzona	MRI
Piedmont	Beefalo	Montbeliard
Romagnola	Beefmaster	Normande
	Braford	North American Brown Swiss
	Brahman	Parthenaise
	Brangus	Pinzgauer
	Devon	Salers
	Galloway	Santa Gertrudis
	Gasconne	Simmental
	Guernsey	South Devon
	Hays Converter	
	Hereford	
	Jersey	
	Lincoln Red	
	Longhorn	
	Luing	
	Milking Shorthorn	
	Murray Grey	
	North American Holstein	
	Norwegian Red	
	Red Poll	
	Sahiwal	
	Scotch Highland	
	Shorthorn	
	Sussex	
	Tarentaise	
	Welsh Black	

Table 3-5

Breeding Systems, Herd Size, and Resources

Less than 50 cows

1. Minimum Resources: straight British or British x British cross
2. Medium Resources: British x Continental European
3. Maximum Resources: British x Dairy

More than 50 cows

1. Minimum Resources: straight British or Brit x Brit x Brit cross
2. Medium Resources: Brit x Brit x Continental
3. Maximum Resources: Brit x Continental x Dairy

alone will maintain straightbred cattle that perform and adapt well to all environments. Cattle with Brahman breeding have a greater heat and insect tolerance. The British breeds have considerable cold tolerance due to their heavy hides and hair coats. The Angus, Hereford, and Shorthorn breeds are also noted for their ability to breed and reproduce where feed is limiting such as in our western range country. Thus, for the seedstock and commercial breeder it is important to select a breed and type of cattle that will perform best in your environment.

Mating unrelated animals within the breed (outcrossing) by using bulls with different bloodlines from outside breeders will maximize performance. Commercial breeders should not inbreed. To maximize profit the straightbred commercial breeder must practice rigid selection, culling, and recordkeeping. He does have access to the whole breed for selection of herd sires and replacement females and should take advantage of this opportunity. Straight breeding is easy as long as the cattle are popular, adaptable to the area, and strong genetically.

2. *Crossbreeding.* Crossbreeding combines the advantages of hybrid vigor and complementarity of breeds. A systematic and organized crossbreeding program can increase total performance up to 20 percent or more above the average of the parents. Figure 3-2 provides a simple example of how to calculate the response from heterosis.

In general, those traits exhibited early in life such as reproduction through weaning can be improved the most by crossbreeding. The most response to crossbreeding (heterosis) occurs in the lowly heritable traits. This indicates the advantage of utilizing a crossbreeding program in your herd for reproduction since progress is extremely slow with selection (heritability 0.10). Table 3-6 also suggests that careful study should be used in selecting the breeds for your crossbreeding program.

Figure 3-2

Example of Heterosis

Assume:

Breed A weaning wt. = 500#
Breed B weaning wt. = 400#   Avg. = 450#

If crossbred progeny actually average 475#

lbs. Heterosis = 475-450 = 25#

% Heterosis = (25 ÷ 450) X 100 = 5.5%

Table 3-6
Examples of Terminal Bulls of F_1 Females

F_1 Female	Terminal Sires
Angux x Hereford	Charolais, Simmental, Limousin, Shorthorn, Maine Anjou, Chianina
Angus x Shorthorn	Hereford, Charolais, Simmental, Limousin, Maine Anjou, Chianina
Shorthorn x Hereford	Angus, Simmental, Limousin Maine Anjou
Hereford x Charolais	Angus, Simmental, Maine Anjou, Limousin
Angus x Charolais	Hereford, Simmental, Chianina, Maine Anjou, Limousin
Holstein x Angus	Hereford, Simmental, Limousin, Charolais, Maine Anjou
Holstein x Hereford	Angus, Simmental, Limousin, Charolais, Maine Anjou

Several types of crossbreeding programs are available for the commercial producer. Examples of recommended systems along with breeds which might be used are presented below.

- Three-breed terminal cross: This system ranks first for most small and large herds. Three-breed terminal cross is a program wherein the producer either buys or raises F_1 females. For the small herds they will most likely choose to buy F_1 females, whereas herds over 100 head may elect to produce their own on a within-herd basis. Selection of maternal breeds with traits such as: fertility, moderate size, longevity, easy calving, good milking, and good disposition for the F_1 female is desirable. The sire breed should possess high fertility, rapid growth rate, cutability and gradability, and calves of adequate shape and birth weights for easy calving. In the terminal three-breed cross all offspring go to market; no replacements are kept. Table 3-6 provides examples of terminal bulls on F_1 females. Logical breed combinations are listed to maximize dam and calf heterosis and carcass merit.

- Three-breed rotational cross: A three-breed rotation is the use of three breeds of sire in a systematic rotation with females kept from these crosses. This system comes very close to maximizing heterosis. Three breeds should be selected and each breed used

Table 3-7

Example of a Simplified Three-Breed Rotational Cross

Assume you have a herd of Angus cows and are going to crossbreed. You want to use Hereford and Charolais with Angus in the cross	
Years	Bull to Use[1]
1978 and 1979	Hereford Bull A
1980 and 1981	Hereford Bull B
1982 and 1983	Charolais Bull A
1984 and 1985	Charolais Bull B
1986 and 1987	Angus Bull A
1988 and 1989	Angus Bull B
Then start breed sequence over	

[1]Each sire breed could be used three years prior to a breed change.

consecutively for two to three years. Table 3-7 gives an example of a simplified three-breed crossing program in which 75 percent of the maximum possible hybrid vigor is obtained.

Examples of other breeds that complement each other well in a three-breed rotational cross program similar to the one just outlined are:

1. Angus x Hereford x Shorthorn
2. Hereford x Angus x Holstein
3. Shorthorn x Hereford x Simmental
4. Shorthorn x Hereford x Maine Anjou
5. Hereford x Angus x Simmental
6. Red Angus x Red Holstein x Charolais
7. Angus x Holstein x Charolais
8. Angus x Limousin x Chianina
9. Hereford x Red Poll x Limousin

• Two-breed crisscross: Basically crisscrossing is a two-breed rotation in systematic backcrossing. Crossbred females are retained and bred to one of the parental breeds. Then always thereafter the daughters are bred to the breed that sired their dams. This plan provides replacement heifers and takes the advantage of the crossbred female. Figure 3-3 illustrates the use of a Hereford and Angus crisscross, two breeds that complement each other extremely well and are used extensively in this system.

Figure 3-3
Example of a Two-Breed Crisscross

Assume equal numbers of Angus and Hereford bulls and females are available to start the cross

          Angus (A)   Hereford (H)
     o = bulls   ♀ = females   ♂ = bull   ♀ = females

1. A ♂ X H ♀ = 1/2 A 1/2 H   H ♂ X A ♀ = 1/2 A 1/2 H
2. A_2 ♂ X 1/2 H 1/2 A ⇐ 3/4 A 1/4 H H_2 ♂ X 1/2 A 1/2 H = 3/4 H 1/4 A
3. A_3 ♂ X 3/4 H 1/4 A ⇐ 5/8 A 3/8 H H_3 ♂ X 3/4 A 1/4 H = 5/8 H 3/8 A

   Equil. 2/3 A 1/3 H   2/3 H 1/3 A

Table 3-8
Impact of Heterosis on Production per Cow

Item	Heterosis %	Heterosis lbs.[1]
2-Way Cross Calves from PB Dams:[1]		
Weaning percentage	4.3%	18#
Weaning weight	4.7%	20#
Gross output per cow	9%	38#
3-Way Cross Calves from X-Bred Dams:[2]		
Weaning percentage	4.7%	21#
Weaning weight	5.0%	23#
Gross output per cow	19%	82#

[1] Assuming mid-parent average of 425 lbs.
[2] Assuming mid-parent average of 450 lbs.

Research has indicated about an 8 percent greater weaning percentage and 5 percent greater calf weaning weight from this crossbreeding system. Other breeds that complement each other well for this plan are:

1. Hereford x Shorthorn
2. Charolais x Angus
3. Chianina x Angus
4. Red Angus x Shorthorn

5. Simmental × Hereford
6. Limousin × Hereford
7. Maine Anjou × Shorthorn

Impact of Heterosis

The total impact of heterosis on beef production is significant. A systematic and organized program such as those just described with a 10-15 year plan can increase performance up to 20 percent or more. Success requires top management, the use of superior tested bulls, and a planned marketing program. Table 3-8 indicates the impact crossbreeding can have on production. This table especially illustrates the additive value of utilizing the crossbred female in a crossbreeding program.

Purebred

The purebred breeder has two choices in designing his breeding plans: closebreeding, mating related animals; or outcrossing, mating unrelated animals within a breed.

1. Outcrossing. This is the best system for most purebred breeders. You can use the entire breed as a source of improvement. Crossing widely, unrelated lines may result in mild hybrid vigor. This is often overemphasized. The primary emphasis in outcrossing decisions should be centered on selecting genetically superior bulls with different pedigrees or bloodlines.

2. Inbreeding. Only the extreme top purebred breeders should inbreed. Line breeding is the most common form of inbreeding; concentrating the blood of a truly outstanding proven animal, most often a sire. Figure 3-4 illustrates the pedigree of a linebred bull A. His inbreeding coefficient is 0.22.

Half-sibs B and C were mated to produce A. Sire D is represented on both sides of the pedigree. Great grandsire G shows up four times in this pedigree. It represents a breeding plan that will concentrate the genes, both good and bad. There is a tendency to sacrifice vigor, livability, fertility, and growth when inbreeding is intense. However, phenotypically inbred animals often do not look as good as they may be genetically. Figure 3-5 illustrates another common form of intense inbreeding practiced by many herds that are closed to outside bloodlines.

A sire daughter mating with an inbreeding coefficient of 0.25 represents a typical plan of intense inbreeding. The owners of purebred herds that have contributed the most predictability and

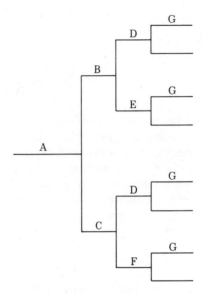

Figure 3-4
Example of Linebreeding

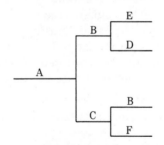

Example 3-5
Example of Intense Inbreeding

genetic superiority to their respective breeds have utilized inbreeding and in many instances maintained a closed herd. Just a handful of purebred breeders are in this category, and they are real engineers and designers of their respective breeds.

The old adage that "In and In will keep you In; and Out and Out will get you Out" in regard to purebred breeding plans may just sort the breeder from the multiplier.

4

Sire Selection

Selection represents the major directional force available to the beef producer for creating genetic change. Herd sire selection will determine more than 85 percent of the improvement made through selection decisions. Fifty percent of the genes in a herd comes from the last bull used, 75 percent from the last two, and more than 85 percent from the last three. Figure 4-1 gives an example of the impact that sire selection can have on genetic improvement in a beef herd for yearling weight.

Herd A, the stud herd, has the selection of bulls with the best yearling weight in the breed to use naturally or artificially. This

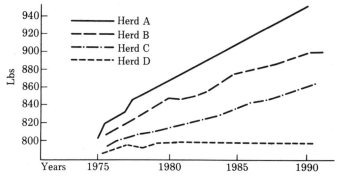

Michigan State University; William Magee.

Figure 4-1
Influence of Sire Selection

herd also practices female selection for yearling weight. Its improvement in 15 years for yearling weight is 140 pounds. Herd B is restricted to selection of performance tested bulls from Herd A and also practices female selection. Its improvement for yearling weight is 120 pounds in 15 years. Herd C could use the top performance tested bulls from stud Herd A, but will not practice selection for females. Its improvement is 100 pounds, indicating that over 83 percent of the improvement for yearling weight in herds B and C is attributed to sire selection. Herd D used only female selection for yearling weight and paid no attention to the yearling weights of the herd sires. This produced approximately 20 pounds improvement in yearling weight in 15 years.

4-1 FACTORS TO CONSIDER IN SIRE SELECTION

The seedstock producer sells genotypes, and the commercial producer phenotypes, so the genetic makeup or breeding value is the major consideration in selling or buying herd sires. The traits that should receive major emphasis are those that are economically important and highly heritable. What then should be considered in sire selection?

Pedigree Data

The best time to use a pedigree is before the animal is born. We can examine his pedigree and get some estimate from his sire, dam, grandsire, grandam, etc, about what he is going to be. A performance pedigree with objective information is of some value in estimating the breeding value of a calf before birth. Distant ancestors are of little importance; for example, a great grandsire has a relationship of only 12.5 percent. The individual performance of the sire and dam and even more valuable the progeny data of the sire and dam (paternal and maternal half-sibs) would be excellent indicators of the genetic potential of an unborn bull.

Performance pedigrees can be of additive value when selecting a potential herd sire following birth when used in conjunction with the bull's own individual breeding value. Figure 4-2 gives an example of a useful performance pedigree of a potential herd sire.

The breeding value for an unborn sire can be estimated based on the performance of the sire and dam and paternal and maternal sibs. This comes under planned mating to engineer and design the models (potential herd sires) that will outperform those presently in use.

Figure 4-2
Sample Performance Pedigree

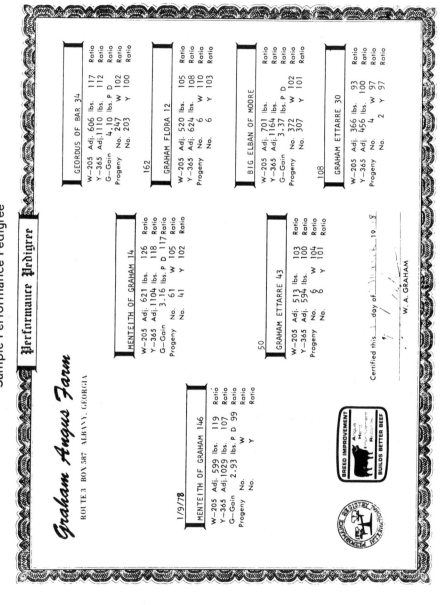

Performance Data

For traits that are highly heritable (above 40 percent) and economically important, the bull's own credentials are the best indicators of his breeding value. After the bull is born, the emphasis should shift to the bull's own phenotype and away from the pedigree. The individual traits that should receive major consideration are:

1. Birth weight. The heritability of birth weight is 0.48, so significant selection pressure can be applied to this trait. Birth weight is positively correlated (0.39) with future growth rate. However, this is one illustration where two things we want are negatively correlated—light birth weight and heavy yearling weight.

Birth weight is highly related to calving difficulty (dystocia). Figure 4-3 relates the relative importance of the factors that have been found to affect calving difficulty. Note that birth weight in both the Angus and Hereford first calf heifers significantly indicates that birth weight provides an objective indicator of a bull's predisposition to calving difficulty. Table 4-1 also gives the increase in calving difficulty with increased birth weights.

2. Weaning weight. The heritability for weaning weight is 30 percent primarily a maternal trait. Recent research indicates that selection for yearling weight will improve the breeding value for weaning weight more rapidly than direct selection for weaning weight itself. For this particular reason and the fact that the environment plays such a significant role in the differences at this period in the growth pattern, it is suggested that weaning weight receive minor consideration in sire selection for growth potential.

3. Yearling weight. This trait is highly heritable, 60 percent, and a bull's own performance record is a good indicator of his breeding value for yearling weight. This is the most valuable parameter for predicting the genetic growth potential of a herd sire. Bulls that excel in growth to this point will sire commercial calves that grow more rapidly and efficiently to desired slaughter weights.

4. Mature weight. This trait is positively related to the previously mentioned growth values. Selection for increased growth in the other categories will ultimately result in increased mature weight. It is suggested that sires can be identified that have moderate birth weight and rapid growth rate to 12–18 months with moderate mature weights. With the high maintenance cost, con-

Figure 4-3
Relative Importance of Factors Affecting Calving Difficulty

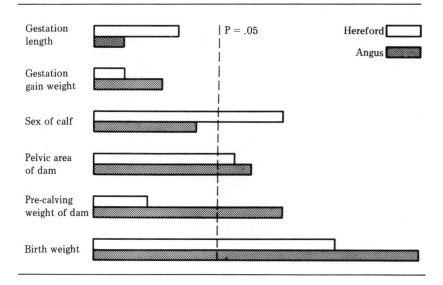

Table 4-1
Birth Weight and Calf Losses in Three-Year-Old Heifers Calving for the First Time

Birth Weight of Calves	No. of Calves	Calves Dead at Birth (%)	Cow Experiencing Calving Difficulty (%)
40-49 lbs.	2	0	0
50-59 lbs.	22	18	5
60-69 lbs.	97	4	12
70-79 lbs.	131	2	18
80-89 lbs.	55	16	40
90-100 lbs.	10	20	30

sideration should be given to the selection for this type of growth pattern with more stress on early gain without increasing mature size. Mature weight has played a far too influential role in selection programs and should only be used if other performance data is unavailable. The recent surge for size has stimulated the breeder to be overwhelmed by big mature cows and bulls, and ultimately this is selecting for a high maintenance cost situation, which goes against the rule of selecting for economically important traits.

Figure 4-4
Modern Herd Sire (by F. C. Murphy, Artist,
Courtesy of the American Angus Assn.)

Physical Traits

The physical traits that should be emphasized in the artist's conception of today's ideal herd sire are fertility, frame, structure, composition (muscle and fat), and body capacity as they relate to maximum production efficiency (Figure 4-4).

1. Fertility traits. The most important rating or score a herd sire can receive is high fertility. Physically, we can examine several traits that reflect a high score for fertility. In order for a bull to cover the country and seek out the cow herd for breeding, he needs an excellent sense of sight, *good eyes.* Another physical trait of importance is sound feet and a skeleton structured for longevity. Corns, prolapsed soles, or even slight founder can affect a bull's ability to breed a sufficient number of cows in a 60-day breeding season. These are all things we can visually appraise and/or score.

4-1 Factors to Consider in Sire Selection 73

The most important physical characteristic we can examine and/or measure is the scrotal circumference and shape (Figure 4-5). Scrotal circumference, size, and shape are closely related to sperm cell production.

To obtain a valid measurement, it is important to first visually assess scrotal shape. Bull #1 has a straight-sided scrotum, often associated with testicles of moderate size. Bull #3 has a tapered or pointed scrotum usually associated with undersized testicles. Bull #2 has a normal shape, with a distinct neck and the scrotum descending down to the hock level.

Most of the bull population one year of age and older would fall into a circumference range of 25–47 cm. The normal for a year-old bull is between 32–34 cm.

Few bulls one year of age would classify as satisfactory with a scrotal circumference of 30 cm or less, and 34 cm for a two-year-old or older.

Overfitted or fat performance tested bulls may average 2 to 3 cm larger than those in good condition. Scrotal circumference is an accurate and highly repeatable measurement when obtained by use of a flexible centimeter tape slipped over the bottom of the scrotum and pulled snugly to the point of greatest diameter of the scrotal sac with the testes fully descended (Fig. 4-6).

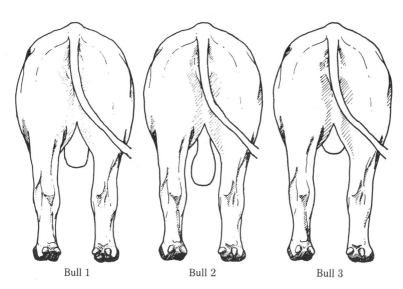

Bull 1   Bull 2   Bull 3

Figure 4-5
Testicle Shape

74   Sire Selection

2. Frame: Skeletal size is rapidly becoming one of the most important traits to evaluate in beef cattle (Figure 4-7). Visual appraisal of frame size is highly heritable (0.6) and the repeatability of scoring cattle for frame runs higher (0.8-0.9) than any other trait we physically appraise.

The reason frame size is so important is its high association in identifying the physiological maturity pattern of cattle on the growth curve. For example, frame size can be effectively used with breed type to predict the optimum slaughter weight to finish feeder calves and the optimum slaughter end point (slaughter weight) for fat cattle to maximize total efficiency and carcass merit. For example, a 3 frame Angus steer should be marketed at 975 lbs to produce maximum pounds of edible beef per pound of energy fed to the steer and his dam. This also approximates a carcass composition of low choice, yield grade 2. This is the optimum end point on the growth curve for a 3 frame Angus steer, whereas 1075 lbs would be the weight to slaughter a 5 frame Angus steer.

So, the frame size of bull selected should relate to frame size of the cow herd and how we want the progeny to serve the commercial industry. Currently, a bull that will sire a frame size feeder and slaughter steer of 4-5 can be marketed in the preferred 1000-1200 lb bracket and produce maximum production efficiency. This sire may need a 4, 5, 6, or 7 frame, depending upon the present herd average which he will serve.

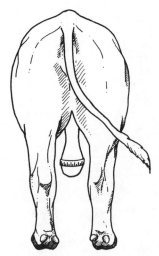

Tape measure correctly applied around scrotum.

Figure 4-6
Measuring Scrotal Circumference

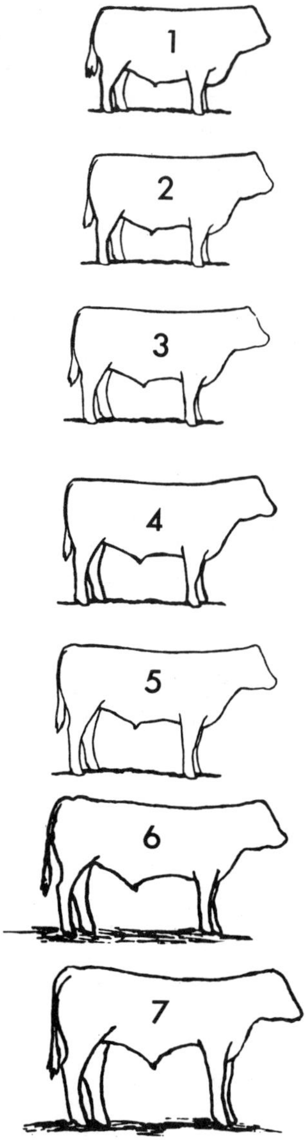

Frame scores	Height measured in inches at hip	
	205 Days	365 Days
1	36–37	41–42
2	38–39	43–44
3	40–41	45–46
4	42–43	47–48
5	44–45	49–50
6	46–47	51–52
7	>47	>53

The base point is 45 inches hip height at 12 months of age for a frame score of 3. Allow two inches for each frame score at the same age. Allow one inch per month from 5 to 12 months of age, 0.50 inch per month from 12 to 18 months and 0.25 inch up to 2 years. Daily adjustment may be made as follows:

$$\frac{\text{number of days}}{365} \times 0.025 + \text{actual height} = \text{adjusted height}$$

Heights for heifers are generally 2 inches less at the same age than those shown above.

Figure 4-7

Frame Evaluation

76   Sire Selection

3. Structure traits. Although geneticists and other researchers have played down the importance of structural traits because they are difficult to measure in a quantitative manner, experience has taught us that a hereditary tendency is clearly evident. Ignoring these traits has resulted in herd sires that physically break down under breeding conditions and daughters that have to be culled from the herd too early.

Specific problems (Figure 4-8) include bulls with straight shoulders, small inside toes, hooves that toe in considerably and result in a short constricted stride. Leg problems of the rear limbs include cocked ankles that give rise to worn off inside toes. Another serious rear limb problem is post legs with straight pasterns leading to many stifled herd bulls.

Structure problems just mentioned are readily passed on to growing and finishing cattle, and these unsound characteristics and poor performance appear to be related. Herd sires that are predisposed to the unsoundness characteristics mentioned and diagrammed below should be eliminated from the test stations, and receive critical scrutiny in the show ring and on the ranch.

4. Muscling. In the beef business we are in the business to produce muscle, and animal breeding specialists will agree that it should be contributed from the sire. Similar to frame, we can

Post-legged and straight shouldered

Toes in—worn off inside toes   Cocked ankles—worn off inside toes

Figure 4-8

Common Abnormalities in Structure

4-1 Factors to Consider in Sire Selection   77

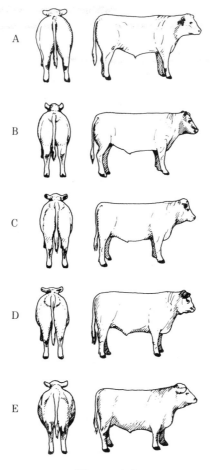

Figure 4-9
Muscle Patterns

definitely move in two extremes—from too little to an excess. The other confounding factor is that cattle can exhibit a high percentage of muscle, and it can be expressed in different shapes.

A herd sire should pass on enough muscle to produce 2 cutability carcasses when the cattle are marketed at the optimum weight for their respective frame pattern. On the chart (Figure 4-9) a sire should be preferred with a muscle pattern example B relating to a cutability of 2.

Category A is approaching an excess and the shape that parallels a cutability 1 market animal. For several reasons I oppose this

78 Sire Selection

extreme thickness of muscle. First, sires within this extreme muscle category appear to be associated with calving difficulty, reduced marbling (too many cattle grading standard and good), and a restriction in the freedom of movement (ability to walk with a long free stride) that a slightly longer, more expandable muscle pattern will allow.

And lastly, but more important in the excess muscle category A we could be moving towards an animal with the characteristics of double muscling, a trait we can't afford in the U.S. beef industry because of its relationship to low fertility and lack of marbling. This may be the reason we see too many of our extremely coarse-muscled steers (cutability 1) void of intramuscular fat and grading standard. Opening the muscle pattern up seems to ease the problem. The characteristics of a full double-muscled animal are:

1. increased development of the hindquarter accentuated in appearance by a groove between the major muscle and a rounded ham appearance
2. thick, open shoulders
3. an unusually wide, stretched stance with the front and hind legs extended, generally creating a sway-back appearance
4. lean, trim appearance, often with a cylindrical middle and tucked up flank
5. general lack of masculinity, other than muscularity in bulls, and lack of femininity in heifers and cows
6. fine bone
7. short tail
8. small testes
9. tail set forward on top of rump

The general overall trim appearance, thicker quarter with a bulging, thicker round, and a higher tailhead setting are the best indicator points to observe (Figure 4-10).

5. Body capacity. Both weight for age and feed efficiency are economically important to most beef enterprises. In reviewing the sires that have excelled in our performance and progeny tests, they tend to have more volume of chest and barrel capacity than has been visually rewarded in our live animal and carcass shows. Our efforts to select carcass cattle that are extremely trim and tight wound appear to be the reason for de-emphasizing cattle with the depth, spring of rib, and body capacity necessary for maximum

Figure 4-10

Example of a Double-Muscled Bull

performance and durability. It would appear that efficient production of market animals will demand selection for an individual with more skeletal size and attendant body capacity than has been true in the past. The swine industry is a good example of what can happen when you select for extremes in carcass merit at the expense of performance and durability. They are now in a rebuilding stage of increasing their frame size, expanding the body cavity, increasing the bone, and opening the muscle pattern of their seedstock. This new seedstock model appears to be improving the reproduction, performance, and durability of the commercial hog with a minimum reduction of carcass merit (lean to fat ratio).

Visually, the change towards more depth of rib and volume will take time to accept. The three profiles in Figure 4-11 suggest the direction we should consider.

Progeny Data

After a bull has sired enough calves to evaluate and test, the selection emphasis should shift from the individual to his progeny to estimate breeding value. This is especially true for lowly heritable traits (reproductive efficiency) or terminal traits (carcass) that cannot be accurately measured or evaluated at all up to this stage. It generally requires data on approximately five progeny to equal the

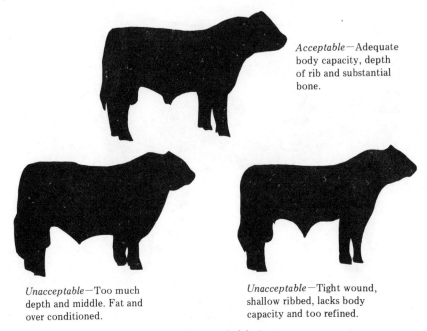

Figure 4-11

Body Capacity

value of the individual's own performance for most of the traits we have discussed. Table 4-2 gives the accuracy values for estimating a bull's breeding value based on his own performance, his progeny, and then a combination of the two. It is assumed that 1.0 represents complete accuracy.

Progeny testing is costly and should be restricted to bulls that have exhibited outstanding individual evaluations and those primarily destined for extensive use artificially. Cost generally dictates this.

National Sire Evaluation

The purpose of the National Sire Evaluation Program is to provide comparative information of important traits on sires across herds throughout the breed. This information is presented as Expected Progeny Differences (EPD) and Possible Change for each economic trait tested. The EPD is an estimate of how future progeny are expected to perform relative to the progeny performance of reference sires when they are mated to comparable cows and the progeny are treated alike. The EPD is the best estimate available from

Table 4-2
Breeding Value Accuracy

Animal's Own Performance

Heritability	Accuracy of Breeding Value
.20	.45
.40	.63
.60	.78

Progeny Performance Only

Heritability	Accuracy of Breeding Value No. of Progeny			
	10	20	40	80
.20	.53	.72	.82	.90
.40	.78	.83	.91	.92
.60	.80	.88	.94	.97

Individual's Own Performance and Progeny Data Combined

Heritability	Accuracy of Breeding Value No. of Offspring			
	10	20	40	80
.20	.66	.75	.84	.90
.40	.80	.86	.92	.95
.60	.88	.91	.95	.97

existing progeny data of the superiority or inferiority of the future performance of progeny of a sire when compared with other reference sires. When reference sires are used in several herds through artificial insemination, comparisons can be made between sires through the tie provided by the reference sires common to each herd (Figure 4-12). This, in essence, is estimated breeding value.

For each EPD, a "Possible Change" (prediction error) is listed which estimates the accuracy with which the number and distribution of progeny available allowed the EPD to predict future progeny performance. It indicates the amount of change, either plus or minus, that is possible in the EPD when additional progeny are included, and it gives a range to expect when using a selected sire when the EPD for a trait between two or more sires is almost identical. This is basically the only time to use possible change. It also is of some value in determining how extensively to use a sire with a superior EPD.

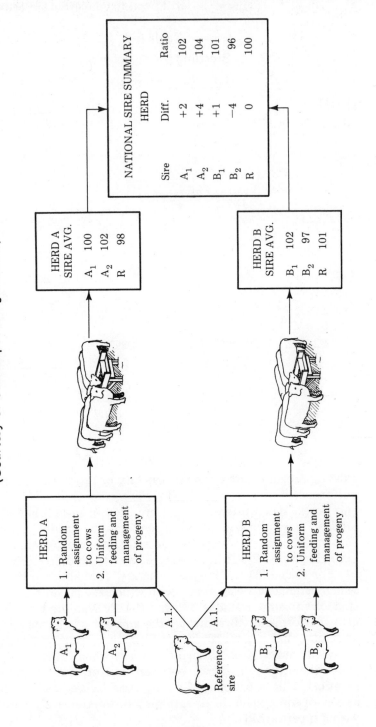

Figure 4-12
National Sire Evaluation
(Courtesy of U.S. Dept. of Agriculture)

4-2   THE MATING GAME

One of the most challenging aspects of seedstock production is the "Mating Game"—that is, the selection of the proper sire to use in a particular herd. Many breeds of beef cattle now have some program for sire evaluation on a national scale. This particular illustration will be limited to the National Angus Sire Evaluation Program. The information contained in this section is simulated and assigned to a given bull for educational purposes only.

As a participant, imagine yourself in the position of a registered breeder who has the option of using one of the five bulls. Rank the bulls 1 through 5 in order of preference.

Materials Available for Evaluation

Pedigree. The pedigree section lists two generations of ancestry behind each individual bull. The breeder code throughout the pedigree tells if the bull is or is not a product of a within-herd breeding program. The breeding value ratio (BVR) found on the pedigree is a mathematical method of measuring an animal's genetic potential based on the performance of related animals, such as parental and maternal half-brothers and sisters and progeny. BVRs are listed as ratios. The breeding value ratio may be found on any or all of the six ancestors in the pedigree. The pedigree will also indicate close or line breeding.

Individual performance. This section for individual performance lists all available data on each bull for weaning and yearling weights. Both weaning and yearling data are listed as adjusted weights at 205 and 365 days and as ratios.

National sire evaluation. The information in this section is listed in seven different columns and is developed from the bull testing program. The first column for *traits* lists the five economically important traits for which each bull was evaluated.

A column for *units* explains the units in which the trait was recorded. This includes scores for calving ease, pounds for weaning and yearling adjusted weights, percent for cutability and one-third grade for quality grade. Figures in the other five columns relate to the units column.

Next is *management groups* or the number of cow herds that each bull was tested in. Most bulls will be tested in less than five management groups.

The column for *number progeny* lists the total number of offspring from each bull that contributed to the results of each trait.

Expected progeny difference is a figure predicting the sire's future progeny performance based on the available results in the test for each trait. This figure is listed as a plus or minus number of pounds, percent cutability, or one-third of the grade.

Possible change is the variation (either plus or minus) that is possible for each expected progeny difference. This figure decreases as the number of offspring increases because large numbers tend to reduce the prediction error. The possible change is a measure of the range you might expect when using such a sire.

The column for *rank* shows where each sire ranks in expected progeny difference for each of the particular traits. A separate ranking appears for each trait.

Individual classification. This area is to allow for subjective visual appraisal and evaluation of the rather highly heritable traits involving growth, composition, and structure. In this illustration, the individual traits are scored on the basis of 1 to 7.

4-3 OFFICIAL RANKING OF 5 AI SIRES

As a registered Angus breeder who has the option of using one of the five bulls on an AI certificate basis, rank the following bulls 1 through 5 in order of preference. The authors have made comments on each individual bull, and our ranking is listed below.

Evaluation of Sire 1

1. Sire 1 has a complete performance pedigree with all ancestors bred and tested by the owner-breeder (breeder code 20). Sire 1 is 12.5 percent inbred—a result of a half brother-half sister mating. The performance information on this bull's ancestors is consistently high.
2. This bull has exceptionally good individual performance records with ratios far above the average of his contemporaries.
3. This bull visually will add frame and structural soundness. He scores slightly above average in composition, and this should be considered.

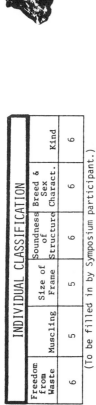

BREEDER CODE: 20

A	Code: 20
Breeding Value Ratio	
Weaning - 103	
Yearling - 103	

I	Code: 20
Breeding Value Ratio	
Weaning - 104	
Yearling - 102	

B	Code: 20
Breeding Value Ratio	
Weaning - 101	
Yearling - 102	

C	Code: 20
Weaning - 104	
Yearling - 106	

B	Code: 20
Breeding Value Ratio	
Weaning - 101	
Yearling - 102	

D	Code: 20
Weaning - 102	
Yearling - 104	

INDIVIDUAL CLASSIFICATION

Freedom from Waste	Muscling	Size of Frame	Soundness of Structure	Breed & Sex Charact.	Kind
6	5	5	6	6	6

(To be filled in by Symposium participant.)

INDIVIDUAL PERFORMANCE

205 Day Adj. Weight		365 Day Adj. Weight	
Wt.	Ratio	Wt.	Ratio
590	118	1095	115

NATIONAL SIRE EVALUATION

Traits	Units	No. Mgt. Grps.	No. Prog.	Exp. Prog. Diff.	Poss. Change	Rank
Calving Ease	Score	10	100	-.10	+.02	A
205 Day Adj. Wt.	Pounds	10	100	+15.2	+5.2	A
365 Day Adj. Wt.	Pounds	20	100	+25.1	+8.5	A
Percent Cutability	Percent	20	100	+.55	+.31	A
USDA Qual. Grade	1/3 Grade	20	100	+.01	+.04	B

4. Sire 1 ranked extremely high (A) in 4 out of 5 major traits evaluated in the National Sire Evaluation Program. His overall breeding potential is superior.

Evaluation of Sire 2

1. Sire 2 has a complete performance pedigree on his ancestors. He is the result of a sire bred back to his dam (25 percent inbreeding), which increases the chances that the bull will sire like-kind progeny. Another positive point for his pedigree is that each ancestor was bred by the same individual breeder (breeder code 30), who has been on a continuous performance testing program. Both sire and dam have superior performance data.
2. The bull has outstanding individual performance data with yearling and weaning ratios far above the herd average, and the actual adjusted weights are equally impressive.
3. Individually, the bull scores well in all categories and would contribute frame, composition, and structural correctness to most herds in which he could be used.
4. Angus bull 2 ranked high (A) in 4 out of 5 major traits on the National Sire Evaluation Program. This should be given major consideration in making the sire selection decision.

Evaluation of Sire 3

1. Sire 3 only has performance information on his sire and dam. There are six different breeders represented in his pedigree. He was designed by a committee.
2. Sire 3 was superior in his individual weaning weight record and dropped considerably in yearling performance. The record is indicative of an early maturing bull.
3. Visual scores on this bull picture a small-framed, early maturing, fat bull that has little to offer in physical traits. His visual scores are representative of his growth pattern.
4. The bull's sire evaluation data leaves much to be desired. He is far below average for two extremely important traits: yearling weight and cutability.

Evaluation of Sire 4

1. The pedigree on Sire 4 is of very little value in predicting his breeding potential. The only information is the sire and dam's

BREEDER CODE: 30

E	Code: 30
Breeding Value Ratio	
Weaning - 109	
Yearling - 110	

F	Code: 30
Breeding Value Ratio	
Weaning - 110	
Yearling - 109	

G	Code: 30
Breeding Value Ratio	
Weaning - 104	
Yearling - 104	

F	Code: 30
Breeding Value Ratio	
Weaning - 110	
Yearling - 105	

H	Code: 30
Breeding Value Ratio	
Weaning - 103	
Yearling - 104	

J	Code: 30
Breeding Value Ratio	
Weaning - 104	
Yearling - 105	

2

INDIVIDUAL CLASSIFICATION

Freedom from Waste	Muscling	Size of Frame	Soundness of Structure	Breed & Sex Charact.	Kind
5	6	6	6	6	6

(To be filled in by Symposium participant.)

INDIVIDUAL PERFORMANCE

205 Day Adj. Weight		365 Day Adj. Weight	
Wt.	Ratio	Wt.	Ratio
600	114	1140	116

NATIONAL SIRE EVALUATION

Traits	Units	No. Mgt. Grps.	No. Prog.	Exp. Prog. Diff.	Poss. Change	Rank
Calving Ease	Score	2	40	-.11	+.04	A
205 Day Adj. Wt.	Pounds	2	40	+18.1	+9.5	A
365 Day Adj. Wt.	Pounds	2	20	+25.0	+15.4	A
Percent Cutability	Percent	2	20	+.60	+.40	A
USDA Qual. Grade	1/3 Grade	2	20	+.02	+.18	B

87

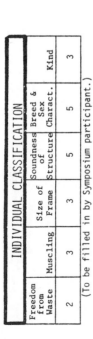

BREEDER CODE: 40   3

K	Code: 44
Breeding Value Ratio	
Weaning – 99	
Yearling – 100	

L	Code: 43
Breeding Value Ratio	
Weaning – 103	
Yearling – 100	

M	Code: 45
Breeding Value Ratio	
Weaning –	
Yearling –	

N	Code: 46
Weaning –	
Yearling –	

O	Code: 42
Breeding Value Ratio	
Weaning –	
Yearling –	

P	Code: 41
Weaning –	
Yearling –	

INDIVIDUAL CLASSIFICATION

Freedom from Waste	Muscling	Size of Frame	Soundness of Structure	Breed & Sex Charact.	Kind
2	3	3	5	5	3

(To be filled in by Symposium participant.)

INDIVIDUAL PERFORMANCE

205 Day Adj. Weight		365 Day Adj. Weight	
Wt.	Ratio	Wt.	Ratio
620	110	925	102

NATIONAL SIRE EVALUATION

Traits	Units	No. Mgt. Grps.	No. Prog.	Exp. Prog. Diff.	Poss. Change	Rank
Calving Ease	Score	2	40	-.00	+.04	B
205 Day Adj. Wt.	Pounds	2	40	+5.1	+10.0	B
365 Day Adj. Wt.	Pounds	1	20	-10.4	+16.1	D
Percent Cutability	Percent	1	20	-1.00	+.42	C
USDA Qual. Grade	1/3 Grade	1	20	+.10	+.15	B

BREEDER CODE: 50

4

Q Code: 52	S Code: 53
Breeding Value Ratio	Breeding Value Ratio
Weaning – 100	Weaning –
Yearling –	Yearling –

	T Code: 52
	Weaning –
	Yearling –

R Code: 50	U Code: 51
Breeding Value Ratio	Breeding Value Ratio
Weaning – 99	Weaning – 100
Yearling –	Yearling –

	V Code: 50
	Weaning – 101

INDIVIDUAL CLASSIFICATION

Freedom from Waste	Muscling	Size of Frame	Soundness of Structure	Breed & Sex Charact.	Kind
6	7	5	3	4	4

(To be filled in by Symposium participant.)

INDIVIDUAL PERFORMANCE

205 Day Adj. Weight		365 Day Adj. Weight	
Wt.	Ratio	Wt.	Ratio
550	106	1240	119

NATIONAL SIRE EVALUATION

Traits	Units	No. Mgt. Grps.	No. Prog.	Exp. Prog. Diff.	Poss. Change	Rank
Calving Ease	Score	2	40	+.21	+.05	D
205 Day Adj. Wt.	Pounds	2	40	+1.0	+11.0	B
365 Day Adj. Wt.	Pounds	1	20	+24.8	+16.0	A
Percent Cutability	Percent	1	20	+1.0	+.41	A
USDA Qual. Grade	1/3 Grade	1	20	-.30	+.17	D

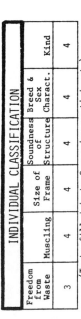

5

BREEDER CODE: 60   Code: 61

Breeding Value Ratio
Weaning -
Yearling -

                  Code: 61   Code: 61

Breeding Value Ratio
Weaning -
Yearling -

                  Code: 60   Code: 64

Breeding Value Ratio
Weaning -
Yearling -

                                                Code: 60

Weaning -
Yearling -

INDIVIDUAL CLASSIFICATION

Freedom from Waste	Muscling	Size of Frame	Soundness of Structure	Breed & Sex Charact.	Kind
3	4	4	4	4	4

(To be filled in by Symposium participant.)

INDIVIDUAL PERFORMANCE

205 Day Adj. Weight		365 Day Adj. Weight	
Wt.	Ratio	Wt.	Ratio
640	NA	1100	NA

NATIONAL SIRE EVALUATION

Traits	Units	No. Mgt. Grps.	No. Prog.	Exp. Prog. Diff.	Poss. Change	Rank
Calving Ease	Score	2	40	-.01	+.05	C
205 Day Adj. Wt.	Pounds	2	40	-.3	+10.0	C
365 Day Adj. Wt.	Pounds	1	20	-3.2	+16.0	C
Percent Cutability	Percent	1	20	-.10	+.43	C
USDA Qual. Grade	1/3 Grade	1	20	+.04	+.16	B

weaning ratios which are just average. In addition, a mixture of breeders are represented in his pedigree.
2. Sire 4 has an exceptional yearling weight and ratio, which is very impressive. We see this often in test stations and many sell extremely high based on this information.
3. The visual scores on this bull indicate an extremely heavy muscled, high cutability bull with good frame and some serious structural problems.
4. The National Sire Evaluation documents the bull's individual performance and visual scores by ranking A in cutability and yearling weight, respectively. His low rating (D) in quality grade and especially in calving ease should cause some serious concern.

Evaluation of Sire 5

1. This bull's pedigree leaves much to be desired. There is no performance information.
2. Sire 5 has good actual weaning and yearling weights, but compared to what—no ratios.
3. This bull is just average visually with the exception of his poor trimness score.
4. The National Sire Evaluation is the only real estimate of this bull's breeding potential. He ranks low (C) in calving ease, 205-day adjusted weight, 365-day adjusted weight, and cutability.

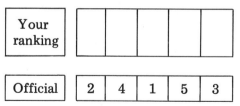

Your ranking					
Official	2	4	1	5	3

4-4 SIRE SELECTION PRECAUTIONS

Sire selection is one of the most important decisions that cattle breeders must make. In the process, several precautions could prevent unnecessary problems.

- Precautions in buying tested sires
  1. Beware of overfatness.
  2. Check feet and legs.
  3. Isolate for at least 30 days.
  4. Let down gradually if fat.
  5. Do not overwork; use common sense.
- Precautions in using crossbred bulls
  1. Progeny may segregate widely and lack uniformity.
  2. Try to buy 3/4 or higher bulls.
  3. Spend top dollar and buy the best bull possible; don't skimp on price.
  4. Select for top performance and composition, but beware of calving ease.
  5. Try to see the dam and her record; she may be a bad cow.

4-5 CRITERIA TO CONSIDER IN SELECTING AI SIRES

1. Conception rate
1. Calving ease (most important in 1st-calf heifers)
3. Weaning weight (more important for those selling calves than for those selling yearlings)
4. Yearling or final feedlot weight (more important for those selling yearlings)
5. Carcass composition—cutability (avoid extremes)
6. Carcass quality grade—marbling
7. Disposition
8. Maternal characteristics of daughters
9. Trait ratios of 100 or higher in sire summaries

4-6 SUMMARY ON SIRE SELECTION

1. Sire selection accounts for 85-90 percent of herd improvement in 15-20 years.

Table 4-3
Estimating the Price for a Commercial Herd Sire

Cows per bull	Slaughter beef price per cwt					
	$ 30	$ 35	$ 40	$ 45	$ 50	$ 55
20	600	700	800	900	1000	1100
25	750	875	1000	1125	1250	1375
30	875	1050	1200	1350	1500	1650
35	1050	1225	1400	1575	1750	1925
40	1200	1400	1600	1800	2000	2200

2. Buy from a reputable herd.
3. Check for unsoundness:
   - undersized testicles (should be 29 cm or more at 12 months)
   - unsound feet and legs
   - bad disposition
4. Stress high performance (weaning and/or yearling weight) ratios over 100.
5. Acceptable conformation:
   - avoid extremes in fatness
   - avoid extremes in muscling
6. Select from proven sires and dams with records.
7. Distant ancestors are of minor importance.

The normal tendency of most bull buyers is to get one as cheap as possible—far too often at the expense of making genetic progress or just plain good sense. Table 4-3 is a good guideline in estimating the price a producer can afford to pay for a commercial herd sire.

5

Selecting the Productive Female

A fast moving evolution is occurring in the beef cattle industry. During the past few years, we have modified our thinking on the kind of cattle that we once considered to be ideal. The changes we are witnessing have been largely prompted by the all too familiar cost-price squeeze that is plaguing the beef industry as well as other phases of American agriculture. Any time production costs rise without an attendant increase in the price received for the product, the natural response of the producer is to strive for improved efficiency. Simply stated, our goal must be to identify cattle that possess the genetic ability to make more money for all phases of the beef industry.

In our effort to develop and engineer more efficient cattle, sire selection has unjustly overshadowed the importance of selecting or owning a highly productive cow herd. The last chapter compared the selection response of sires versus females, and to some extent, this is misleading. If two breeders start a herd with two extremely different cow herds, and both practice the same progressive sire selection program, the herd owner with the superior cow herd will almost always maintain a significant advantage. This is most evident when we find a superior beef herd. It is generally backed up with a highly reproductive, consistent producing, good milking, large-framed, feminine, sound, and uniform cow herd. These herds practice ruthless cow culling and critical selection of replacement heifers out of their tested and proven herd sires. In our effort to

96 Selecting the Productive Female

profit test herds and to develop more efficient females, there are traits that should be considered very fundamental in selecting the productive female for the beef cattle industry. Against this background, if breeders are to make progress in the years ahead, they must focus a great deal of attention to the selection program that will now be discussed.

5-1 FACTORS TO CONSIDER IN SELECTING FEMALES FOR THE HERD

Health

Foundation animals should be healthy and free of disease and parasites. Too many breeders overlook the health aspect when purchasing animals for the breeding herd. The time to be most concerned about herd health is before you start, not after you have become established. It is heartbreaking to have to liquidate a good herd and start over because adequate precautions were not taken at the beginning. Important infectious diseases to be concerned with in breeding cattle are:

1. Tuberculosis (TB): Breeding animals may generally be sold or shipped in from an accredited herd, or animals showing a negative test within 30 days. An accredited herd is one in which all animals over two years have passed a negative test within 12 months.
2. Brucellosis (Bangs): It is good practice to purchase females that are calfhood vaccinated for brucellosis from 2 to 6 months of age. This is also a suggested management procedure for all replacement heifers that will be retained in the herd. This vaccination must be done by a veterinarian. Cattle from a certified Bangs-free herd generally need not be tested for interstate shipment, if from a herd that tests negative within 12 months and the animal shows a negative test within 30 days, or from animals officially vaccinated under 24 months of age from a certified herd or a herd testing negative in a 4- to 6-month period. A certified brucellosis-free herd must pass two consecutive blood tests within a 60-day period and needs to be recertified every year by having a negative blood test.
3. Leptospirosis (animals of all ages affected): This disease may result in abortion at any stage of pregnancy. Diarrhea, anemia,

and mastitis may be other symptoms (blood urine). Testing is not the answer, so the herd must be vaccinated unless it is in an isolated area. All replacement heifers should be injected prior to the breeding season.

4. Vibriosis: This is a technical name for venereal disease caused by bacteria. It can cause sterility and/or abortion. There is an agglutination test for it. In the bull, you can draw semen and isolate it. It is transmitted by the bull from cow to cow. A satisfactory vaccine is available, and it is suggested purchases not be made from infected herds and only noninfected animals be bred. The best rule is to practice a vaccination program for vibriosis.

5. Trichomoniasis: This is also a venereal disease caused by a protozoa. It is characterized by early abortions that may go unobserved. Transmission is nearly always carried out through the act of breeding. There is no simple diagnostic test for it. Avoid buying cattle from herds in which there is a known history of trichomoniasis.

6. Anaplasmosis: This disease, caused by a protozoan parasite, causes the destruction of the red blood cells. It is much more common in warm southern climates. Infected animals that recover become carriers, and a test has been developed to detect carriers. Infected animals can be successfully treated by a veterinarian.

It is wise to isolate all newly-purchased cattle from the rest of the herd for about 30 days and observe them closely for any disease symptoms.

Reproductive Efficiency

Female selection should heavily stress reproductive efficiency. Figure 5-1 gives the relative economic return for reproductive efficiency compared to performance and carcass traits, and thus emphasizes why it is imperative that reproduction receive major attention in female selection and culling programs. In essence, the percentage of calves weaned per cow returns in dollars is twice as much as growth rate and 20 times as much as carcass merit for comparative increments of improvement. A breeder's goal should be to wean more than a 90 percent calf crop per female exposed. In a program limiting the calving season to 45–60 days, one can expect about 5 percent of the cows not to settle. A calf crop weaned may be 5 percent less than the number of cows settled.

The reasons for a cow failing to wean a calf break down approximately as follows:

1. 56.8%—failure of the cow to settle
2. 11.1%—stillbirths
3. 3.9%—deformed
4. 16.4%—abortion
5. 11.8%—accident, disease, or predator

Any good cattleman realizes that getting his cows settled promptly and then delivering live, healthy calves out of these cows is basic to the profitability of his operation. We ordinarily measure this in terms of percent calf crop (number of calves at weaning time divided by the number of cows at breeding time). Research has shown that those traits related to fertility and reproduction are low in heritability. In other words, they cannot be expected to show a significant response to selection pressure. Practical experience and research does reveal a high repeatability for reproductive problems, and the economic return is significant enough to place as much selection pressure on reproduction as on the other major traits.

Longevity is important. Long-lived productive cows mean that a smaller percent of the herd is young replacement heifers; therefore, reproductive efficiency will be higher. You can expect about a 10 percent lower calf crop weaned in the case of first calf heifers.

The average length of a cow's life in the United States is 8.5 years (4-5 calves). In the superior herds, it amounts to 5.5 to 6.5 productive years if they calve first at two years of age. A herd

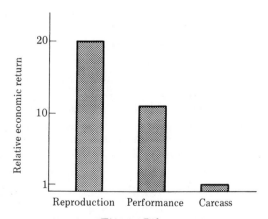

Figure 5-1

Relative Economic Return of Three Major Selection Traits in Beef Cattle owner should set a goal for less than 45 percent of a heifer crop to be retained every year. This allows for about 20 percent of the cow herd being replaced every year, or complete replacement every 4 to 5 years. Most ranchers cull cows at 8 to 9 years of age. The number of cows to be culled and the number of replacements needed can be calculated as follows:

$$\% \text{ cows to be replaced} = \frac{100}{\text{productive life}}$$

$$= \frac{100}{6}$$

$$= 16.6\,\%$$

$$\% \text{ of heifers to be retained} = \frac{\% \text{ of cows to be replaced} \times 2.0 \times 1.6}{\% \text{ of calf drop weaned}}$$

$$= \frac{16.6 \times 2.0 \times 1.2}{90}$$

$$= 44.2\,\%$$

Suggested Procedures to Improve the Reproductive Efficiency of a Cow Herd

1. Save approximately 50 percent more replacement heifers than needed; i.e., if you need 100, save 150.
2. Breed replacement heifers at 14–15 months of age.
3. Breed replacement heifers 20–30 days prior to the beginning of breeding mature cows.
4. Breed replacement heifers and cows for a 45–60 day breeding and calving season.
5. Pregnancy check all females 60 days following breeding season. Cull those that fail to settle.
6. Cull cows that fail to calve every 12 months.

Individuality or Type

Figure 5-2 depicts today's ideal beef female. Excellence in type depends upon your goal. If you want to produce functional, highly productive herd bulls or choice feeder calves, you should buy the best females you can find. Regardless of body conformation, breeding females must be sound in their feet and legs with no evidence of lameness. Females with sound clean bone joints and a good solid foot on each corner will save labor and veterinary

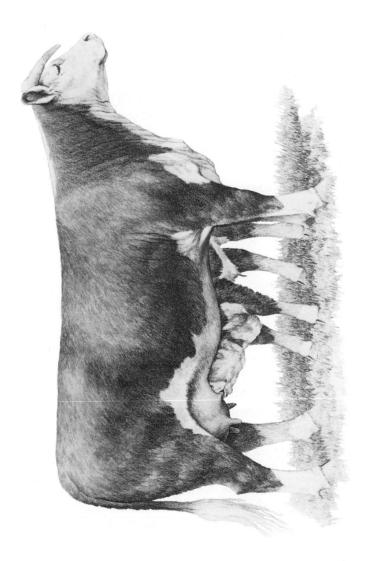

Figure 5-2
A Model of Today's Ideal Beef Female
(Courtesy of the American Hereford Assn., Kansas City, Missouri)

5-1 Factors to Consider in Selecting Females for the Herd 101

bills. They also will have a longer productive life. Their mammary system should exhibit a strongly attached udder and no blind teats. Femininity, udder soundness, and development of extra genitalia are of major emphasis in selecting and culling females. Figure 5-3 gives examples of a feminine, highly fertile appearing cow and a coarse, lowly fertile cow. The highly fertile cow is in beautiful proportion, and presents a graceful, feminine appearance. She is lean and clean in her face, neck, and throat, and is long and smooth in her muscling. She is trim in her brisket, over her shoulders, and in her flanks. She displays good body and rib capacity, and is long in her hip and high and wide at her pins. The lowly fertile cow is a coarse-fronted, unbalanced, extremely deep-fronted cow that exhibits extra fat and is overconditioned. Her udder also appears nonfunctional.

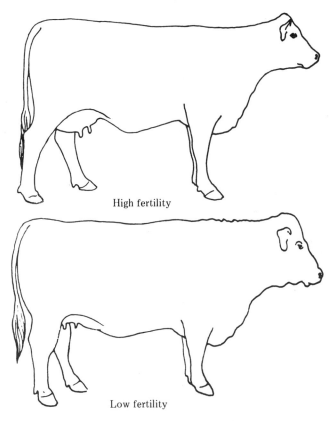

Figure 5-3
Fertility and Udder Soundness

Figure 5-4 gives examples of variously shaped udders and differently formed teats. Today, it is of much more interest to beef cattlemen to cull cows with poor udders and pay attention to using bulls that sire well-formed functional udders.

Size is always an asset as long as it is associated with good conformation. Table 5-1 gives performance standards for female selection.

Avoid cattle that are extremely small in size because larger cows: (1) live longer; (2) have higher calving percentages; and (3) wean heavier calves. However, there may be no advantage in *extremely* large cows because it takes more feed to maintain them (approximately 7 percent more per 100 pounds live weight). We are interested in breeding animals that will produce market steers that perform efficiently up to at least 1,100 pounds and grade choice. Extremes are problems; animals that are too small finish at too light a weight and those that are too large will mature at too heavy a weight.

Select females that are long and smooth in their conformation (muscling). Purchase or keep those that are lean and clean in their head and neck, deep-ribbed, wide-sprung in their chest with an abundance of body capacity. Choose females that have total dimension to their rump—long from their hooks to pins and wide between their pins, accentuating a large pelvic cavity. Heifers that are broody in appearance with a deep jaw, broad muzzle, and a large growthy appearance to their heads are preferable. When selecting replacement females, avoid coarse, heavy-fronted heifers that give the overall impression of masculinity rather than femininity.

Age is an important factor when you purchase or select females for a herd. Heifer calves have their productive life ahead of them. However, 20 percent or more may turn out to be nonbreeders, and others may need culling. You also have to wait 1½ years before they reproduce. You should buy 150 percent as many heifers as you want to end up with. The initial investment is generally lower than older females, and their genetic potential should be superior if they are out of young proven sires.

Yearling heifers have essentially the same advantages or disadvantages as calves when compared to mature cows. Yearlings are one year closer to maturity than heifers, and if they are the right kind, they can be bred to calve at two years of age. The initial investment is higher than calves, and you run the risk of buying females from which breeders have already selected the good replacements. With yearlings, you may make a deal to have them bred to an outstanding bull, if the seller will consent.

5-1 Factors to Consider in Selecting Females for the Herd

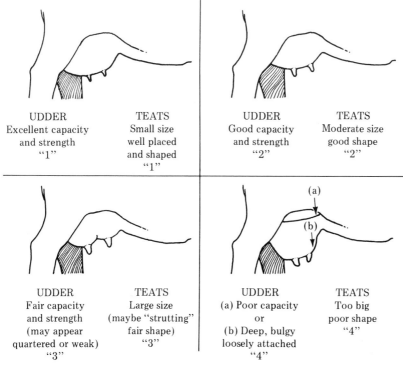

Figure 5-4

Udder and Teat Evaluation

Table 5-1

Performance Standards for Female Selection

Item	British	BREED TYPE Medium Exotic	Large Exotic
205-day adj. wt., lb.	450	500	550
Weaning grade[a]	13+	13+	13+
365-day adj. wt., lb.	600	650	700
Yearling grade[a]	13+	13+	13+
Frame size (1-7)	5	6	6
Mature wt., lb.	1000-1200	1200-1300	1300-1400

[a] 11 = high-good; 12 = low-choice; 13 = average-choice, etc.

Mature cows have an obvious advantage in that you don't have to wait for them to reach the productive age. Be careful that you don't buy someone's problem cows. Mature cows have less time to become acclimated to the new environment before they produce calves. There is a lot to be said for purchasing cows with progeny data and calves at side. It is then possible to predict the genetic ability of the cow. You can observe the milking ability of a cow with a calf at side and often buy them bred back and get a 3 in 1 package. Of course, the cost will be higher. Visual appraisal, except for unsoundness, is not very valuable in selecting mature cows. If you decide to buy mature cows, it is probably safest to select younger cows from 4 to 8 years of age that are in the most productive years of their life. Good sources are dispersal sales, drought areas, and performance and progeny-tested herds that stress a genetically superior young cow herd, and are willing to let their older, but still productive, cows go in order to keep their herd size constant.

5-2 REPLACEMENT HEIFER SELECTION

A young, highly productive cow herd that is making rapid genetic progress depends on progressive selection procedures for replacement heifers. The following is a systematic program for selecting replacement heifers.

1. Rank all heifers on 205-day adjusted weights.
2. Cut off bottom 1/3; consider culling all those under 90 ratio unless circumstances are unusual.
3. From the top 2/3, remove those that are:
   a. structurally unsound
   b. small framed; 3 or under
   c. unfeminine
   d. overfat
4. Rank on 12-month ratios.
5. Remove those that failed to grow adequately.
6. Expose to a bull for 60 days; pregnancy check 60 days after breeding; cull those that are open.
7. Cull again after first calf.

5-3 CRITERIA FOR CULLING COWS

The progressive breeder who adds young, genetically superior replacements will simultaneously be culling the lower end of his mature cow herd. Ruthless cow culling is a must, and the following guidelines should be followed.

1. Cull *open* cows after a 45-60 day breeding season regardless of their records. Pregnancy check to make this decision.
2. Cull cows with poor progeny records:
   a. Poor growth performance ratios under 90 at weaning and/or yearling.
   b. Inconsistent records from year to year. Usually vary with the quality of the bull used, indicating little prepotency for performance.
   c. Low quality calves with poor visual grades and scores.

5-4 MILK PRODUCTION

The primary goal of any breeding program is to produce cows with the genetic capability to make a profit. In commercial cow-calf operations, increasing the weaning weight of calves has been the primary goal to attain this economic return. Because of the high relationship between weaning weight and milk production, new heavy milking breeds such as Simmental, Maine-Anjou and several of the dairy breeds have been infused into the genetic make-up of our beef herds. This appears good on the surface, but recent data indicate that extra high milk yields negatively affect the total performance and economic merit of the cow.

The average daily milk production during lactation for a beef cow is 12 pounds with a conversion of 1 pound of calf gain per 10 pounds of milk. As milk yield increases much above this range, conversion rate is reduced and feed requirements for the cow increase substantially. Cattlemen also complain about cows that give too much milk because they are predisposed to mastitis and spoiled udders. The calves of such cows often contract milk scours. In addition, increased milk yields and weaning weights have indicated a strong relationship to poor reproductive perform-

Table 5-2
Milk Production, Weaning Data and Reproductive Performance[1]

Item	Hereford	Hereford x Holstein	Holstein
Total lactation, lbs.	3360	5040	6720
Daily milk yield, lbs.	14	21	28
Adj. weaning wt., lbs.	604	658	763
Rebreeding conception, %	96.2	89.3	59.0
Days post-partum to apparent conception	75	76.5	94.5

[1] Oklahoma State University.

ance (Table 5-2). The gross return per calf is higher for the heavier milking cows with the larger calves at weaning; however, when this is adjusted for land and supplement requirement and conception rate, net return favors beef cows that are considered good milkers, averaging 14 pounds per day (Table 5-3). The recommended average amount of milk that a beef cow should yield per day during lactation is impossible to estimate. It should be determined in a breeding program by the nutritional requirements, the desired calf performance, and the cow's reproductive efficiency. By combining these factors with present economic costs and returns, net merit is the best criteria to set milk production goals.

Visual appraisal traits such as dairy character (freedom from fatness, a long lean neck, sharp withers, bold prominent ribs, and flat concave muscling), body capacity, and the mammary system are lowly correlated (0.10-0.24) with milk production. However, frame size is positively related to production. Heifers with more skeletal dimension (taller and longer) eat more feed and produce more milk. Larger heifers can be bred earlier, calve easier, and are more efficient converters of forage to milk and calf weight. They also experience less udder problems as a result of being higher off the ground and are less susceptible to dirt, infection, and injury. A cow's first lactation yield has a high positive phenotypic and genetic correlation with lifetime yield. Selection for milk production

Table 5-3

Economic Analysis[1]

Item	Hereford	Hereford x Holstein	Holstein
[a]Land requirement percentage	100	110	137
[b]Total cost/female, $	113.99	123.28	159.02
Return adj. for conception, $	50.35	45.64	9.58

[1]Oklahoma State University.
[a]Expressed as % of Herefords as determined by forage intake in dry lot.
[b]Combination of land and supplement cost.

can be made early, and a cow that gives too little or too much can be eliminated from the herd after her first lactation. Most Probable Producing Ability (MPPA) is the best tool to identify the lowest producing cows. It can be effectively used to compare dams that do not have the same number of calf weaning weights in their averages. The formula for computing MPPA is presented in Chapter 2, Section 2-7. This formula is presently the best procedure for ranking and culling cows in a herd on milking ability.

6

Cow Herd Management

A well-organized comprehensive management program is essential to maximize production and profit with a beef cow herd. Table 6-1 gives an estimate of the variation between herds due to environment, which is primarily differences in management within a climatic area of the country. Cattle vary in their phenotypes for a given trait for either or both of two reasons: (1) they possess different genes which affect the trait; or (2) environments have varied their phenotypes. In most instances the environment can be greatly improved with management. A program to follow should include positive and accurate identification, the use of comprehensive records, a preventive herd health program, and a well-designed cow herd management calendar. Nutrition is a major in-

Table 6-1

Environmental Effects on Economically Important Beef Traits

Trait	% Variation Due to Environment-Management	Emphasis in Management
Fertility	90	High
Weaning Weight	60	High
Post-weaning Weight	50	High
Yearling Weight	40	High
Carcass Traits	30-40	Medium

gredient in superior management, and two chapters have already been assigned to thoroughly relate its significance. Nutrition will be mentioned again from time to time in this chapter where appropriate.

Accordingly, management is the key to a successful beef program. This chapter will suggest principles and guidelines for top cow herd management and pull this together in a sequential production management calendar.

6-1 BEEF CATTLE IDENTIFICATION

Identification is one of the most important points of a workable records system for both performance and parentage. It also serves as a means of legal identity. A satisfactory identification system should provide the following: (1) positive identification, (2) necessary information about the animal, and (3) easy recognition.

Identification Systems

Tattooing, nose printing, freeze branding, and fire branding all represent a permanent method of identification. Eartags represent a less permanent but helpful and attendant system to maintain easy identify. The combination of the eartag along with one of the four permanent techniques is recommended.

Tattoo. Most registered cattle are required to carry a tattoo, and the majority of commercial producers utilize the tattoo as an effective means of permanent identification. Some other visible form of identification should be used in addition because tattoos cannot be read without catching the animal. Good tattoos can be achieved if the following steps are followed:

1. Clean the area of the ear to be tattooed with chlorhexidine (a virucide-fungicide). This is superior to alcohol since it will help prevent the spread of warts (a virus) from one cow to another. Likewise, clean the tattoo equipment after each use.
2. Set or insert tattoo numbers to the desired numerals. Check by tattooing a piece of cardboard prior to tattooing the calf.
3. Firmly clamp the ear with the tattoo pliers between the ribs in the ear.
4. Rub the tattoo punctures full of ink with your finger until the bleeding ceases. Currently, green fluorescent ink appears to

work the best. Calves can be tattooed at any age but preferably at birth with the same tattoo number in each ear.

Nose print. Nose printing appears to have tremendous potential as a means for identification of beef cattle. Research has indicated that no two animals have an identical pattern. Cattle nose prints like human fingerprints are individually unique and unchanging throughout the life of the animal. Figure 6-1 illustrates differences that exist in nose patterns that can be successfully

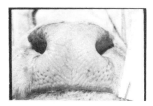
(a) long ridge pattern

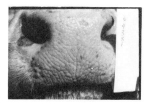
(b) broken ridge pattern

Each print is made up of ridge patterns that are long or broken. Sometimes both are present in the same print.

(c) regular ridges

(d) irregular ridges

Ridges are either regular or irregular. May contain both regular and irregular ridge pattern.

(e) all lines

(f) lines and dots

Nose prints also appear to be structured in lines and dots. Prints may show either one or both characteristics.

Figure 6-1
Identifying Nose Patterns

lifted as prints for positive identification. The following procedure is recommended for nose printing beef cattle:

- Materials and facilities
    a. Squeeze chute
    b. Fast-drying India ink
    c. Good quality bath towels
    d. 5 x 8 rubber stamp pad
    e. 5 x 8 card, preferably yellow
- Procedures
    1. Place the animal securely in the squeeze chute with a rope halter on the animal. Secure the head with a halter and rest its head on your left hip.
    2. Ink stamp pad.
    3. Dry nose with towel.
    4. Ink calf's nose with pad.
    5. Roll card upward over the nose for print. (Use a hard-surface block behind the card).

It is critical that inking and printing be done in less than 10 seconds after drying the nose as perspiration causes ink to smear on the card if you wait too long.

Freeze brand. Freeze branding (cryogenic branding) is a process of using superchilled branding irons, killing the melanocytes (color-producing cells), and thus producing a white hair number against a solid body color. Animals with black or red hair coats are most suitable for freeze branding. The following procedure is recommended for a successful freeze brand:

- Materials and facilities
    1. Squeeze chute
    2. A set of 4-inch copper branding irons (0 through 9) with a 3/8 to 5/8-inch thick face
    3. A pair of small animal clippers
    4. A coolant of either dry ice and alcohol (95% ethyl, methyl, or isopropyl) or liquid nitrogen
    5. An insulated container for coolant and irons. A styrofoam picnic cooler is excellent for dry ice and alcohol, but a special wide-mouth container is preferred for liquid nitrogen.

- Procedures
  1. Prepare coolant of dry ice and alcohol by breaking the ice into small pieces into the container and pour in sufficient alcohol to cover the heavy portion of the irons. Proportions are not critical. Add dry ice as needed. For liquid nitrogen, pour in a sufficient quantity to cover the heavy part of the iron. Keep irons in coolant. Irons are as cold as the coolant, and ready to use or reuse when bubbling over the iron has subsided.
  2. Clip the area to be branded as closely as possible, brush over the area, and finally wet and clean the area with air temperature alcohol.
  3. Apply the supercooled iron with firm pressure for the following recommended time:

	Contact Time Seconds	
Age of Animal	Dry Ice Alcohol	Liquid Nitrogen
Up to weaning (4-8 mo)	25	15
Yearling (9-18 mo)	30	20
Adult (Over 18 mo)	35-40	25-30

  Mature animals with thicker hides should use longer times. Underbranding causes poor results while overbranding caused by too much time kills the hair follicles and produces a hair-free brand. *Use a stop watch to be exact!*
  4. The supercooled iron freezes the skin. As it thaws, swelling and reddening will be observed. When the hair grows back, it will be white. The white number will be complete in about 3 months.
  5. Branding can be done anytime, but the best time is in the fall or very early spring, coinciding with normal hair regrowth pattern.

Hot brand. Fire brands are satisfactory and practical for identification or ownership (holding) branding. Fire brands are quickly applied and inexpensive, producing a permanent, hair-free scar in the shape of the iron used. Hair should be clipped in the fall or winter ahead of calving to make the brands completely legible. Overbranding or underbranding produces poor legibility. The following procedure is recommended for a successful hot brand:

- Materials and facilities
    1. A complete set of numbered irons (0 through 9) with 4 to 6-inch numbers and a 1/4 to 3/8-inch thick face. Commercial electric branders also work well.
    2. A heater of wood or bottled gas for irons.
    3. A pair of cattle or sheep clippers.
    4. Squeeze chute for firm restraint of the animals.
- Procedure
    1. Brands are usually placed on the hip but may be on the rib, shoulder, or thigh.
    2. Normally for identification purposes, only yearling replacement heifers and cows are branded. Holding brands may be put on any age animal including young calves.
    3. Clipping the branding site is not necessary but is recommended for best results.
    4. Apply the iron. It should not be too hot—ash gray in daylight or cherry red if held in the bottom of a 5-gallon bucket. Hold the iron on the skin firmly with a slight rocking motion until the branded area is a rich buckskin color. Avoid overbranding and never brand wet cattle.

Ear tags. *Metal ear tags* are self-piercing aluminum or steel ear tags, which are inexpensive and easy to apply. They work especially well for identifying calves at birth and can be the only identification up to weaning. This type of tag should be placed on the top side of the ear near the head, leaving at least 1/2 inch for growing room. If the tag is placed in the ear too tightly on a young calf, infection may result.

There are several kinds of *large rubber-type ear tags* on the market that are large and easily readable. These tags can be ordered prenumbered or as blanks with ink, allowing the user the option to number the tags. This type of tag should be applied according to manufacturer's instructions and can be expected to last for 3 to 5 years. Most cattle now carry these easy-to-read tags, and they have replaced the neck chain or rope tagging systems.

Numbering System

The numbering system should be simple and prevent duplication. There are two basic systems to consider:

1. Identifying the calves consecutively within a year. This system is preferred by most commercial and registered producers. It is generally preferable to use consecutively numbered ear tags for tagging calves at birth. The last digit of the year of birth is the first number on the tag, designating the year of birth. In herds of up to 100 cows, tags for say 1979 are numbered 901 to 999. In herds of up to 1000 cows, tags are numbered 9001 to 9999. Calves so tagged should be tattooed and/or freeze branded with the same number.

One can prevent any possible duplication of numbers which might occur in replacement females when this system is used longer than 10 years. To guarantee no duplication it may be desirable to assign low numbers to heifer calves for the first 10 years and switch to low numbers for bulls and high numbers for heifers at the start of the second 10-year period. Table 6-2 provides an example of this system.

2. Identifying the calf with its dam. This system is designed for use in small to medium sized herds where cow herd numbers are randomly assigned. The calf number (tattoo and ear tag) is the same as the dam's herd number except that the last digit of the year of birth is added as either the first or last digit of the number (Table 6-3).

3. Use of numbers in a registered name. In naming registered animals, the farm, ranch, breeder name, or prefix could be used first to designate the breeder. Following this, in bulls it is recommended to use the sire line designation and in females the family line designation next in the name. The final part of the name is the tattoo number. For example, the first registered Angus heifer born at Virginia Polytechnic Institute in 1979 from the Blackbird line

Table 6-2

Consecutive Numbering System for Identification

		Heifer calves		Bull calves	
		Under 100 cows	Over 100 cows	Under 100 cows	Over 100 cows
1st 10-year period	Beginning 1979	901 up to 949	9001 up to 9499	950 up to 999	9500 up to 9999
2nd 10-year period	Beginning 1989	950 up to 999	9500 up to 9999	901 up to 949	9001 up to 9499

Table 6-3
Identifying Calf with Dam

For 1982 Calf Crop	
Dam Herd Number	Calf Number
50	250 or 502
251	2251 or 2512

would be named V.P.I. Blackbird 901. This keeps the names simple yet meaningful.

6-2 MANAGEMENT PROCEDURES FOR BEEF AI

Successful beef artificial insemination is critically dependent upon management, semen quality, cow fertility, heat detection, and technician efficiency. Calving results with AI will range from 25 to 100 percent—the difference is management.

Semen Quality

The need for good quality semen is obvious. Semen purchased from reputable AI organizations usually meets the highest standards and providing proper care and handling procedures are followed, should give satisfactory results. It is recommended that stored semen be checked under the microscope from time to time to assure it has maintained its quality.

Cow Fertility

To maximize conception and calving rate the health status and nutrition of the cow herd must be exacting. The prevention and control of diseases that especially affect the reproductive tract are particularly important. These are vibriosis, trichomoniasis, leptospirosis, and infectious bovine rhinotracheitis (IBR). Calfhood vaccination for brucellosis is also highly recommended. More specifics on these diseases and their prevention will be discussed in the following section on herd health.

Nutrition is a critical factor in maintaining the cow fertility necessary for maximum results through AI. Chapter 8, Nutrient Requirements of Beef Cattle, thoroughly covers nutritional recommendations and their effects on cow fertility. It is recommended that nutrition be given special attention in your AI program. Without proper nutrition, you can expect extremely poor results.

Heat Detection

This is the area that is most often discussed as a problem with AI. It is one element in the system that can be done well with some planning and good management.

Heat or estrus. This is the psychological and physiological expression of sexual receptivity to the male. This condition lasts approximately 18 hours and occurs about every 21 days in a cyclic fashion until the female becomes pregnant. Heat is an indication that an egg is about to be released from the ovary and that sperm should be deposited to fertilize the egg. The characteristic behavior of the cow in heat is due to the action of estrogen (female sex hormone) on the brain of the cow.

Factors affecting heat. A number of factors can modify the response of the cow to this hormonal stimulation. *Disease*—the sick cow will not show heat. *Weather*—adverse, changing weather will modify the expression of heat. *Hunger* and *thirst*—will inhibit normal estrus behavior. *Fear and anxiety*—will block the heat symptoms. The happy, contented, comfortable cow in good health is going to show the best heat pattern and be the easiest to detect.

Move the cows to the breeding pastures early so they may settle down and be at home. Ensure adequate feed and water. Move quietly through the herd prior to breeding so the cattle become adjusted to observation. During hot weather, do your heat detection early and late in the day when it is cooler. Avoid checking for estrus when cattle are being fed. They are usually more interested in eating.

The cow in heat. The heart of the detection program is recognizing the cow in heat. There are a number of symptoms or signs to guide us. There is one positive sign that must be observed before we say "she's in." This is standing still when other cows mount her. This is the classic "standing heat." Six to ten hours before standing heat, the cow will display a number of signs. She is restless, bawls and walks excessively. She will attempt to mount other cows. A clear mucus discharge may be seen coming from the vulva or smeared on the buttocks and tail. The cows in heat will usually be bunched together with a lot of activity. There may be mud on the shoulders and roughed up hair on the rump where other cows have been riding. Bull calves will often follow the cow as she comes in heat. The vulva of the cow in heat is usually moist and swollen.

Heat detection aids are listed below:

1. Detector pads—plastic devices that contain a capsule with a colored dye. These pads are glued to the rump of the cow. When she is mounted by another cow sufficiently, the capsule breaks and the dye is released coloring the pad. These pads are lost occasionally in brushy country and may be accidentally activated.

2. Gomer bulls—These are bulls that have been altered so that fertilization cannot occur and equipped with a marking device. There are several methods: the bull may be vasectomized, penectomized, have a rerouted penis, or be equipped with pen-o-block which prevents extension of the penis. These methods which prevent actual copulation are recommended to reduce spread of disease. These bulls should come from within the herd from a disease control standpoint.

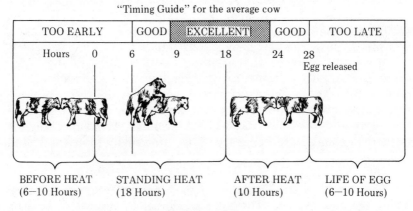

Figure 6-2

When to Breed

The marking device normally used is the chin-ball marker. It works like a ballpoint pen. When the ball makes contact with the cow's back, a stripe is "drawn." The mark should be a stripe; a few spots mean the cow was not standing.

Checking for Heat. How well this important job is done will determine the percentage of your cows found in heat and, therefore, determine the overall success of your AI program.

1. The person assigned this job is obviously a key person in the program. He or she must be enthusiastic and willing to take the time to patiently observe the cattle. They must give the job their undivided attention.
2. The breeding pastures should be relatively free of brush and rough cut-up areas so that the cattle can be seen readily. Areas where the cattle congregate, water holes, etc., should be located and used as observation points. The breeding corral or pen should be located near these areas to minimize the problem of penning cows in heat.
3. The cattle should be checked at least twice a day, early and late, for an hour or so each time. If the cattle are scattered out, it often helps to quietly bunch them up. Bringing the cows together stimulates the estrus behavior.
4. When a cow is spotted in standing heat, her number is written down. Cow identification has been discussed earlier, but we need to emphasize here the need for a system that allows easy identification from a distance. The cows identified in heat should then be moved quietly to the breeding pen.

Insemination—When and How?

The question of when to breed is very important. If we breed too early or too late, conception will not take place. There is, however, several hours leeway in which to inseminate. As a rule of thumb, cows first observed in standing heat in the morning are bred that afternoon. Those in heat in the evening are bred the next morning. The diagram below may help answer the question (Figure 6-2).

The physiological requirements are that the sperm must be in the female several hours before fertilization can occur. The cow will ovulate about 10-14 hours after the end of estrus. If the cow is inseminated late in the standing heat period or shortly after, the chances for conception are optimal.

The actual insemination must be done with a minimum of stress to the cow. She is quietly moved from the holding pen to the chute. Unless the cattle are wild, their heads should not be caught. A bar behind the cow will hold her in place. Do not use metal squeeze chutes. They make too much noise and cattle have bad memories about them (branding, dehorning, vaccinating, etc). If the cow is stressed with rough handling, hot shots, whips, sticks, etc, ovulation may be blocked and no egg released. The inseminator, after properly thawing the semen, will deposit it in the uterine body. This is also done with a minimum of handling and loss of time.

Often it is more practical to bring the cattle in heat to the breeding pens once a day rather than twice. The cattle detected in heat in the evening check are left out overnight and brought in after the morning check. The cattle detected that morning are also brought in and held until evening to breed. If this system is used, calf creep holes should be constructed in the holding pens so the calves can get to their mothers. This reduces the stress on the cow.

Length of Breeding Season. The AI breeding season may be for one, two, or three, or more heat periods. A common system is to breed artificially for two heat periods (45 days) and then utilize clean-up bulls to take care of those failing to settle and the few not detected in heat. You must remember that your clean-up bulls will sire 10-40 percent of the calves, so use good ones.

The artificial breeding should start about 50 days after the first calves are born. It takes 30 to 60 days for the reproductive tract of the cow to return to normal after calving. Normally the cow should be inseminated at the first heat period after 50 days postcalving.

The producer can realistically expect 60-70 percent conception on first service, a total of 80+ percent with two services. With top management all the way through, these percentages can be increased.

Estrous Synchronization

In beef cattle, estrous sychronization holds great promise as a technique for concentrating AI breeding time, which would result in a fairly short period of calving. To date, there are no drugs in the U.S. that have been formally cleared by the FDA for synchronizing (grouping heat periods) so cows can be bred at one time. However, there are several products that have been researched

Figure 6-3
Syncro-Mate-B Treatment
and its Theoretical Response

Implant cows and inject	Remove implant	Breed at 48-54 hr. or at detection	100% in estrus
0	9	10-12	13

Days

and for which the manufacturers are actively seeking FDA clearance. These will most likely be marketed by mid-1979, so we will briefly review estrous synchronization and project the future use of these new products.

Products under study. Cows can be synchronized with either one of two types of products. The first is a progestogen, which is a synthetic form of the hormone progesterone. This product is manufactured under the name of Syncro-Mate-B. The second type is prostaglandin, which will be marketed as Lutalyse (a natural prostaglandin from sea coral) and Estrumate (a synthetic prostaglandin analogue).

Progestogen and beef AI. The Syncro-Mate-B treatment consists of a progestogen implant and an injection of progestogen and estrogen. Nine days later, the ear implant is removed with small forceps, and within 24 to 96 hours, the majority of cows will be in estrus (heat). The technique and theory of the SMB treatment is illustrated in Figure 6-3. Studies with SMB have demonstrated that a single insemination following the implant removal will give satisfactory conception rates without the need for estrus detection.

Prostaglandins and beef AI. Treatment with the prostaglandins is accomplished by a single or double injection schedule. Since only cows with a functional corpus luteum will respond to treatment, a double injection regime of two injections 11 to 12 days apart is normally employed. The use of the double injection procedure allows all cattle to have a mature corpus luteum at the second injection, and they should come into estrus (heat) within 48 to 96 hours. The technique and theory of the prostaglandin treatment is illustrated in Figure 6-4.

Figure 6-4
Lutalyse and Estrumate Treatment and its Theoretical Response

Inject all cows with prostaglandin	60 to 70% in estrus	Inject all cows with prostaglandin	Breed at 70 to 80 hr. or at detection	100% in estrus
0	2-4	11	13	15

Days

Studies with prostaglandins have shown that a single timed insemination after the second injection will give satisfactory conception rates without estrus detection. In cases where semen is high or in limited supply, breeding after heat detection may be advisable since some cows do not respond to treatment. The latter also applies to SMB treatment.

Synchronization will primarily reduce the labor and time requirements of standard AI breeding programs from 25 days to 3 to 5 days. It will also reduce management efforts and the difficulty of detecting estrus. This, together with higher pregnancy rates early in the breeding programs and a reduction in time, labor, and management, could encourage widespread application of beef AI.

6-3 BEEF HERD HEALTH PROGRAM

A herd health program for any phase of the cattle business must be centered around the prevention of diseases rather than treatment. This does not mean that close observation of animals and treatment of those that get sick is not vitally important, but preventative rather than therapeutic medicine is the key.

All health programs should be set up with the advice of the herd owner's veterinarian because of his professional training, and also because he is the person that should be most aware of the problems in the area. Therefore, his advice and consultation should be received when setting up a program. It is also important to realize that no program will be successful without cooperation between the owner, herdsman, veterinarian, and other interested personnel.

Basic Principles of a Herd Health Program

1. Prevent exposure of animals to disease-producing organisms and situations:
   a. Sanitation—cleanliness
   b. Isolation of newly acquired animals for 10 days to 3 weeks
   c. Good environment
   d. Eradication of certain diseases when possible
2. Maintain a high level of resistance in animal population:
   a. Proper nutrition—feed, water, mineral, and vitamins
   b. Vaccination—immunization procedures
   c. Selection of sound animals
3. If disease occurs, keep down the spread:
   a. Isolation of sick animals
   b. Close observation of herd
   c. Obtain a diagnosis early—exam and autopsy
   d. Treatment based on diagnosis
4. Adequate record system

Cow-Calf Operation

- Breeding herd (cows, bulls, replacement heifers)
  1. Fertility test bulls prior to breeding season
  2. Vaccinate for leptospirosis

     (Three strains—pomona, hario, grippotyphosa), IBR, BVD, and PI_3, and vibriosis, prior to the beginning of the breeding season and while females are open (not pregnant)
  3. Treat for internal parasites (worms) at least twice a year— more frequently if necessary
  4. Treat for grubs and lice—follow recommendations on all products available
  5. Practice good external parasite control programs
  6. Pregnancy exam all females after end of breeding season (60 days) and cull open animals

7. Test all new additions to the herd for brucellosis, tuberculosis, and anaplasmosis

- Calving time
    1. Follow close observation of cows at calving time.
    2. Remember the pasture is probably best calving area.
    3. Keep close calvers in the area where handling facilities are available.
    4. Have veterinary assistance available and also have your veterinarian instruct you as to how he wants you to handle maternity cases, i.e., what you should do, when you call him, equipment you may need etc.
    5. First calf heifers usually are going to have more trouble than older animals and will need closer observation and more assistance.
- Calves
    1. *Dip* navels on all newborn calves with a disinfectant such as iodine.
    2. Make sure calves nurse and get colostrum (cow's first milk) within 2 hours. If necessary keep some colostrum frozen for emergencies and also have some form of an esophageal feeder available for use on weak calves.
    3. Inject newborn calves with vitamin A and D during the first few days of life.
    4. Identify calves soon after birth.
    5. Castrate and dehorn calves at an early age if possible—easier and fewer problems at early age—use caustic dehorning material.
    6. Vaccinate all calves against *Clostridium chauvoei* and *C. septicum* (Blackleg and Malignant Edema) during the first 2 months. It is important to repeat this again at 6–8 months.
    7. Check with your veterinarian about use of vaccines for other clostridial diseases. Also consult with the veterinarian about the use of and need for scour vaccines (Reo-Corona Virus) in your herd.
    8. Brucellosis (Bangs) vaccinate all replacement heifers between 2 and 6 months of age.
    9. Routine vaccination for IBR-BVD-PI$_3$ should be done 2–3

weeks prior to weaning or 2-3 weeks past weaning but usually not at weaning because of stress present at this time.
10. Treat for grubs and lice in the fall of the year—follow directions on product labels.
11. Treat for internal parasites on a routine basis—usually before weaning, but more frequently if the need arises.
12. Eye problems—*first obtain a diagnosis* of problem, and then follow veterinarians' advice as to treatments and possible preventative measures. Good fly control and close observation so that proper treatment may be given early are two things that will greatly reduce losses.

- Several Herd Health Practices
  1. Good basic, sound, adequate nutritional program.
  2. Adequate mineral and salt intake including magnesium.
  3. Vitamin A and D supplementation by feed or injection.
  4. Check with your veterinarian on need for selenium—vitamin E preparations in area and on your farm.
  5. Keep feet trimmed and corns removed from animals, especially bulls. Get this work done before the breeding season.
  6. If artificial insemination is used and teaser bulls are a part of the program, make sure the teaser has had his penis removed, blocked, or deviated.

Several diseases were mentioned above and may need to be enlarged upon. This may help to understand the reasons for preventing these conditions.

Brucellosis. This is a bacterial disease commonly referred to as "Bangs" that affects several species of animals including cattle and man. It affects cattle primarily by reproductive losses—abortions. There are federal-state programs for the eradication of this disease because of animal losses and the human health hazard. There is no treatment for animals. Brucellosis can only be prevented by the calfhood vaccination of heifers between 2 and 6 months.

Leptospirosis. "Lepto" is a bacterial disease of cattle caused by the organism Leptospira. There are 3 strains—harjo, pomona, and grippotyphosa—that are primarily involved and animals should be routinely immunized against them. Two other strains—canicola and icterohemorrhagiae—may cause problems in certain herds. These diseases cause abortions, infertility, weak calves, systemic infections, and death in some animals. Treatment is difficult.

Vibriosis. Vibriosis is considered to be a venereal disease spread at the time of breeding from an infected male to the females or from an infected female to the male and then to the rest of the herd. It causes abortions and infertility in the female. Treatment is difficult and prevention is accomplished by annual vaccination of breeding animals or artificial insemination.

IBR. Infectious bovine rhinotracheitis (IBR) is a viral disease that may cause respiratory infections, encephalitis (brain infection), conjunctivitis, abortion, and reproductive tract infections. It is prevented by use of vaccines. There are modified live virus (MLV) products for intranasal (IN) or intramuscular (IM) use. Do not use any of these products unless you have discussed precautions concerning their use with a veterinarian.

PI_3. Parainfluenza type 3 (PI_3) is a viral disease causing primarily respiratory problems in cattle. It is considered to be a secondary factor in a lot of "shipping fever" outbreaks. There are also MLV-IN and IM products and killed virus products for use in immunization programs.

BVD. Bovine virus disease (BVD) affects cattle by causing abortions, diarrhea, chronic digestive disturbances, weak calves at birth, fetal anomalies, conjunctivitis, dermatitis, and nervous signs. The only vaccines available are MLV products for IM use. Do not use these products without discussing their use with a veterinarian.

Blackleg and Malignant Edema. These are diseases caused by the organisms Clostridium chauvoei and C. septicum. These are organisms that live in the ground and may enter a calf through wounds, ingestion, and navel cords. These organisms produce substances (toxins) in the animal's body that are rapidly fatal. The use of vaccine (bacterins—killed bacteria) can prevent their occurrence in the young animal. Repeated doses are indicated.

Grass Tetany. Grass or winter tetany is a condition caused by a deficiency of the mineral magnesium (Mg) in the animal's body. It can cause tetany (convulsions), paralysis, blindness, and sudden death. It is treatable if caught early, but prevention is more successful. The addition of Mg to the animal's diet on a daily basis

Worksheet 6-1

Herd Health Schedule Calendar

		Your herd
Move heifers into calving pasture	Jan. 1	_____
Begin calving heifers - 60 day season	Jan. 15-Mar. 15	_____
Move cows in calving pasture and sort heavy cows weekly	Jan. 15	_____
Begin calving cows - 60 day season	Feb. 15-Apr. 15	_____
Calves: Identify, disinfect navel cord, give Vit A-D, vaccinate for blackleg-malignant edema, castrate and dehorn (may use caustic)	As born	_____
Fertility test bulls (prior to breeding season)	Feb. 15	_____
Vaccinate bulls - Leptospirosis (3 strains), IBR, BVD, PI_3, vibriosis and treat for internal parasites (worm)	Feb. 15	_____
Replacement heifers (those to be bred for first time): Vaccinate against leptospirosis (3 strains), IBR, BVD, PI_3, and vibriosis and treat for internal parasites (worm); do these things at least 2-3 weeks prior to breeding	Mar. 1	_____
Start breeding heifers	Apr. 1	_____
End breeding of heifers - remove bulls	May 30	_____
Cow Herd: Vaccinate open cows and 1st calf heifers for leptospirosis (3 strains), IBR, BVD, PI_3, vibriosis, and treat for internal parasites (worm); do these things at least 2-3 weeks prior to breeding	Apr. 5	_____
Start breeding cows	Apr. 15	_____
End breeding cows - remove bulls	Jun. 15	_____
Calves: Revaccinate for blackleg-malignant edema; castrate and dehorn any missed earlier; Brucellosis vaccinate heifers (2-6 mos of age)	Jun. 1-Jul. 1	_____
Pregnancy exam - heifers	Jul. 30	_____
Pregnancy exam - cows	Sept. 15 to weaning	_____
Treat cows and replacement heifers for internal parasites, grubs, and lice - can do at pregnancy check	Sept. 15 & before Nov. 1	_____
Calves: Worm, treat grubs and lice	Oct. 1	_____
Revaccinate blackleg and malignant edema (prior to sales)	Oct. 1	_____
Vaccinate IBR-PI_3 (nasal vaccine); if using modified live virus (MLV) intramuscular (IM), do after weaning	Nov. 15	_____
Wean calves	Oct. 15-Nov. 1	_____
Start magnesium supplementation for cows (may consider year-round supplementation)	Oct. 15-Nov. 1	_____
Late gestation heifers - Vit A-D and lice treatment	Dec. 15	_____
Late gestation cows - Vit A-D and lice treatment	Jan. 15	_____

will prevent grass tetany. This can be done on a seasonal or year-round basis. Herds in areas where grass tetany is a problem should probably receive a supplement year round.

Any type of health program has certain procedures, vaccinations, and management decisions that must be performed at a given time. An attempt has been made to put a year-round health schedule together incorporating some things that are considered to be important in getting the most number of cows bred, pregnant, and calved successfully and the greatest number and pounds of calf to weaning time (Worksheet 6-1). The schedule included uses the following premises in arriving at the dates:

1. 60-day calving season for cows
   60-day calving season for heifers
2. Spring calving season
3. 285-day gestation length (pregnancy)
4. Heifers start calving 30 days prior to cows

6-4 RESOURCE REQUIREMENTS FOR A COW HERD

Before engaging in any enterprise, consider your available resources, i.e., your land, labor, capital, and management capabilities. A beef cow herd tends to fit on land that is not capable of growing high-value cash crops. The labor requirement per cow may be substantial but tends to be flexible as to the timing of when things have to be done. Facilities and capital items required are quite flexible and vary by geographic area and by individual operator. The amount of cash required per year for nonfeed items varies between $40 and $80 per cow. Remember that the cash inflow or capital turnover is slow with a beef cow enterprise. There may be only one payday each year.

Some estimates of the resources required for a 50-cow herd are presented below.

Land Required for Pasture and Hay

1. Land required for pasture:
   a. Minimum of 1 acre per cow or a total of 50 acres on fertile, highly-productive land.
   b. Maximum of 10 acres per cow or a total of 500 acres on heavily wooded and/or infertile, unproductive land.
   c. Average of 3 to 4 acres per cow or 150 to 200 acres on marginally productive permanent pasture land not suitable for cash crops.
2. Land required for hay:
   a. Depending upon the length and severity of the winter feeding period, hay requirements vary from 2.0 to 3.5 tons per cow unit (including the hay required for replacement heifers and herd sires). Average for the central United States would be about 2.5 tons per cow unit.

6-4 Resource Requirements for a Cow Herd

    b. Hay yields tend to vary from 1 to 5 tons per acre, with an average of about 2.5 tons.

    c. Land required for hay production would, therefore, vary from a low of 0.4 acres to a high of 3.5 acres per cow unit. Average would be about 1 acre of hay land per cow unit.

    d. Total land for hay production for a 50-cow herd would vary from 20 to 175 acres, with an average of about 50 acres.

3. Total land requirements:

    a. Minimum of 1.4 acres per cow unit or 70 acres for a 50-cow herd on the most fertile land.

    b. Maximum of 13.5 acres per cow unit or 675 acres for a 50-cow herd on heavily wooded and/or infertile, unproductive land.

    c. Average of 4 to 5 acres per cow unit or 200 to 250 acres for a 50-cow herd on marginally productive land.

4. In the major areas of the United States, investment in land per cow unit currently ranges from $800 to $2,000 per cow unit, with an average of about $1,500.

Investment in Cattle

50 females @ $250 to $500 per cow...	$12,500 to $25,000
Two bulls @ $400 to $1,000..........	$800 to $2,000
Total........................	$13,300 to $27,000

If artificial insemination were used during the early part of the breeding season, one bull would be adequate.

Investment in Buildings, Equipment, and Machinery (1977 Prices)

All-purpose building for calving, hay and grain storage.....................	$5,000 to $12,500
Corral or handling facility with headgate...	$500 to $1,500
Fencing.............................	$500 to $5,000
Automatic watering system.............	$500 to $1,000
Feeding bunks and racks...............	$250 to $1,000
Self-feeders for minerals..............	$200 to $400
Small equipment (syringes, needles, halters, etc.)......................	$100 to $200

Sprayer for weed control	$300 to	$800
Used tractor with frontend loader	$2,500 to	$5,000
Used 3-bottom plow	$300 to	$400
Used hay wagon	$250 to	$500
Used 10-foot disc harrow	$400 to	$500
Used manure spreader	$200 to	$1,000
Used 10-foot spring tooth harrow	$150 to	$200
Used mower-conditioner	$1,000 to	$2,500
Used drill	$400 to	$600
Used hay racks	$200 to	$500
Used hay baler	$1,000 to	$2,500
Total	$13,750 to	$36,100

For a beef cow herd to be profitable, the above investments must be held to a minimum. They could be further reduced by the following set of conditions:

1. Use of old existing buildings or woods for shelter.
2. Use of old existing buildings for feed storage.
3. Storage of hay outside in form of large stacks or bales.
4. Custom hiring of tillage, seeding, hay harvesting, and other machine work.
5. Use of farm ponds for water.
6. Feeding hay on frozen ground instead of in stacks or bunks.
7. Use of electric fencing as much as possible.

If the above conditions existed, investment in buildings, equipment, and machinery could possibly be reduced to $3,000-$5,000.

Labor Requirements

The labor required for animal management ranges from 7 to 35 hours per cow unit per year, with an average of about 20 hours. This does not include the labor required to produce feed for the cow herd, which will probably average 5 to 10 hours annually per cow unit. Maintenance of machinery, buildings, facilities, and fencing will add another 2 to 5 hours per cow unit per year to the labor requirement. Total labor in a 50-cow herd will average close to 30 hours per cow unit or 1,500 hours total.

A small herd of 25 to 50 cows is a supplementary enterprise which utilizes family labor that might not otherwise be used. Normally, little outside labor is hired except for harvesting of forage.

Table 6-4
Annual Cow Costs (1977)

Item	Cost per Cow Unit	Total Cost for 50 Cows
Winter roughage	$ 90	$ 4,500
Pasture	36	1,800
Salt & mineral	4	200
Grain + supplement	15	750
TOTAL FEED COSTS	$145	$ 7,250
Veterinary & drugs	5	250
Machinery & fuel expenses	16	800
Breeding costs	7	350
Hired labor	4	200
Utilities	3	150
Marketing & transportation	4	200
Repairs, insurance & interest on improvements	12	600
Depreciation on improvements	8	400
Interest on livestock investment	32	1,600
Real estate taxes	4	200
TOTAL NON-FEED COSTS	$ 95	$ 4,750
TOTAL ALL COSTS	$240	$12,000

Annual Production Costs

Listed below are annual costs for maintaining an average 50-cow beef herd (Table 6-4). These figures are based on recent USDA surveys as well as estimates from individual states. A charge for land is not included in this list. Generally, annual land charges vary from $80 to $120 per cow unit.

6-5 FORAGE MANAGEMENT SYSTEMS

The cost of supplying forage represents over one-half of the total cost of maintaining a beef cow unit. Efficient management of pastures and winter feed sources can determine success of the beef cow enterprise. Both the yield and quality of the forage are crucial. Several techniques are available to increase forage production. In evaluating a new forage production practice, consider not only the cost, but the risk of price and yield changes associated with each particular practice.

Several techniques can be used to improve the quantity and quality of forage from your pasture program. You may choose to improve the production of seeded pasture or to renovate the pasture completely. Your decision should be influenced by expected pasture yield, the pasture composition of desired grass-legume mixtures and undesired weeds and brush, and the cost of the various improvement-renovation practices.

Pasture Improvement

On old or presently seeded pasture, possible pasture improvement techniques are:

1. Fertilization. Grass pastures respond to nitrogen fertilization and/or high nitrogen fertilizer in relation to phosphorus and potassium content; e.g., an N-P-K mix of 2-1-1 or 4-1-2. Legume pastures require potassium fertilization.

To assure high utilization of fertilizers, the pH or soil acidity measure should be no less than 6.0 and preferably closer to 7.0. Soil testing is a relatively inexpensive method to obtain information on your soil quality, and application of lime can correct the acidity problem.

Fertilizer may be topdressed in winter or early spring. Apply fertilizer at a time when the field remains structurally solid but you can still minimize the risk of surface runoff and loss of fertilizer. Heavily utilized pastures can benefit from fertilizer directly after being harvested when in a resting-rebuilding stage of development.

2. Weed control. Mowing or clipping pastures at about a 3 in. height is one method of controlling weeds but may not be as effective as spraying with herbicides.

Good weed control will reduce competition against your desired forage species for needed plant nutrients, moisture, and space. Clipping pasture in late summer-early fall before the weeds can reseed can be an especially effective time for weed control. Forage production will improve and pasture carrying capacity will increase.

Spraying may provide more effective weed control but necessitates investment in a sprayer and herbicides, some of which are relatively expensive. Nevertheless, a combination of clipping and spraying is generally a profitable procedure for pasture improvement.

3. Drainage. Some forages, especially alfalfa, cannot tolerate "wet feet." Tile drainage may be the only effective way to rid your pasture of excess water and improve forage production potential. However, tilling is a high investment practice; make certain the increase in forage production justifies the expense. It is usually most feasible on highly fertile alfalfa-producing land.

4. Grazing management. Avoid overgrazing. The various forages have seasonal growth patterns unique to each species. Most species do not tolerate hard continuous grazing in which the plants are clipped close to the ground. Research suggests that forage yields tend to be lower with continuous harvest as opposed to periodic harvest. This suggests that rotation grazing, whereby your pasture may be subdivided into grazing plots sufficient for 20-30 days continuous grazing, then rested for 30-40 days, may increase your forage production potential.

Pasture Renovation

Complete renovation of pasture may require more dollars than improvement of old pasture but also may provide a higher payoff in terms of increased forage and beef production.

There are several items of importance in renovation:

1. Fertilization and liming. Phosphorus is especially important in establishing a new seeding. A soil test will suggest recommended levels of fertilizer and lime. The fertilizer may be applied at seeding time, but lime should be applied in the previous fall or spring.

2. Seedbed preparation. A seedbed is desired in which the old sod is killed and good soil-seed contact is provided for new seed. The amount of tillage recommended will depend upon the steepness and rockiness of your land. Seeding establishment using herbicides in lieu of physical tillage practices can be an alternative on steep, rocky fields so as to minimize risk of soil erosion.

3. Timing and method of seeding. This will vary by area of the country. Traditionally, new seedings have been band seeded with a drill in the spring using a companion grain crop such as oats. The companion crop does compete with the forage seedling for light and plant nutrients but may provide a harvestable crop which reduces the cost of establishment. Seeding in the spring without a companion crop requires use of herbicides for weed control. This method may require more cash outlay but usually results in better establishment of the forage seeding. Summer and fall seedings may be established without herbicides if proper tillage techniques are used to control weed growth.

4. Forage species selection. Much consideration should be given to planning your grazing management program as related to seasonal growth of the various forage species and seasonal demand for forage by your beef cow herd. Cool-season grasses produce relatively more during spring and early summer, whereas warm-season grasses start growth later in the season but produce relatively more growth during summer. Most grasses commonly used in pasture mixes tend to be cool-season grasses; e.g., bromegrass, timothy, orchardgrass, fescue, and reed canarygrass. Legumes commonly used are alfalfa, bird's-foot trefoil, red clover, white clover, ladino, alsike, etc. To enable better management of pasture mixes, it is suggested that the mixture in one seeding be composed of no more than two or three separate species.

5. Grazing management. In the seeding year, the new crop should not be put under stress. However, a harvest which is completed about 4-6 weeks before the first killing frost may be beneficial. In subsequent years, some system of rotational grazing should be developed.

Providing Forage for Winter Feeding

1. Hay feeding systems. As mentioned before, winter hay requirements range from 1.5 to 3.5 tons per cow unit. The required amount will vary with the length and severity of winter, forage quality, use of crop residues, winter grazing, size of cow, when the cow calves, and her level of milk production. Winter hay requirements for a 50-cow beef herd are presented below (Table 6-5). Hay alone is not adequate for all classes of beef cattle. Young stock, herd sires, and in some cases, lactating cows, require supplemental grain in addition to hay. If the hay is low in quality, it may be necessary to add a protein supplement.

Although performance of the beef cow is less sensitive than the dairy cow to feed quality, a high quality feed should neverthe-

Table 6-5
50-Cow Herd Winter Feed Budget Using Hay as Roughage[1]

Item	No.	Hay (T.)	Grain (T.)	Salt-Min. lb.
Preg. & lactating cows	40	102.0	--	1080
Preg. & lactating heifers	10	25.5	--	270
Open yrlg. heifers	13	14.0	5.85	351
Mature herd sire	1	2.7	0.90	27
Young herd sire	1	1.8	1.08	27
Total, T or lb.	--	146.0	7.83	1775

[1] 180-day winter feeding period.

Table 6-6
50-Cow Herd Winter Feed Budget Based on Corn Silage[1]

Item	No.	Corn Silage (T.)	Soy (T.)	Salt-Min. lb.
Pregnant & lactating cows	40	186.0	2.40	1080
Pregnant & lactating heifers	10	46.5	1.26	270
Open yrlg. heifers	13	46.8	1.29	351
Mature herd sire	1	6.3	0.13	27
Young herd sire	1	7.2	0.18	27
Total, T or lb.	--	292.8	5.26	1775

[1] 180-day winter feeding period.

less be your goal. Moreover, harvesting high quality hay dictates the need to cut the forage at late bud stage for legumes and early bloom for grasses. Harvesting hay early increases the total energy and protein yield provided by each acre.

2. Corn silage feeding systems. For producers who grow corn, Table 6-6 is a winter feed budget based on corn silage as the only forage.

3. Crop residues. Since winter feeding is the most expensive period for the beef cow, steps should be taken to minimize this cost without adversely affecting beef cow performance. If you calve in the spring, use crop residue and aftermath during the cow's dry period to decrease the need for harvested hay. Possible residue sources are corn stalk fields, bean stubble, and regrowth since the last cutting of hay. For example, grazing one acre of corn stalks can provide from 30 to 60 days maintenance ration for one dry beef cow. After the first 30 days of grazing, it may be necessary to supplement the stalks with some hay or other harvested feed. Stalks are deficient in calcium, phosphorus, vitamin A, and protein, so provisions should be made to supplement these nutrients, especially after the first 30 days. Grazing crop residues provide a relatively inexpensive energy source, require less labor than feeding from storage, and are an easy manure distribution technique. One invisible cost associated with this technique is the possible detrimental effect on next year's crop yield from those fields which are grazed and trampled by the cows.

6-6 BEEF COW HERD MANAGEMENT CALENDAR

This is an outline of suggested management practices to be followed at various stages of the beef herd's annual production cycle.

This calendar is based on a winter-spring calving schedule. Most of the management suggestions are also applicable to a fall calving program. The most significant difference between spring and fall calving is the winter feeding of the cow. Not all the practices outlined here will fit everyone's management program. Each producer will have to alter them to fit his needs.

January (July, if Fall Calving)

1. If forage quality is low, consider sending in a sample for analysis to determine possible supplement needs. Energy, protein, phosphorus, and vitamin A are nutrients most apt to be deficient.
2. Grain or corn silage is normally the cheapest energy source. NPN is usually the cheapest crude protein source, but natural protein supplements are better utilized than NPN if cattle are on low quality forage. Di-cal and bonemeal are common phosphorus sources. Vitamin A may be fed or injected (injections last about 100 days).
3. Be sure herd has an adequate water supply. Beef cattle need 5 to 11 gals/head/day in coldest weather, 13 to 20 gals in hottest weather.

4. Provide some protection, such as a windbreak, during extremely cold weather to reduce energy requirements. A woodlot is ideal.
5. Get ready for calving season.

February (August, if Fall Calving)

1. For some herds, calving season is here. Be sure you have the following items:
   a. Calf pulling equipment if you anticipate problems
   b. 7% iodine to disinfect newborn calf's navel
   c. Eartags for identification of calves
   d. Bo-Se shots if white muscle disease is apt to be a problem
   e. Vitamin A shots for cows and calves if forage quality is low
   f. Dehorning paste or equipment
   g. Castration equipment
   h. Frozen colostrum stored up in advance of calving season for calves that fail to nurse within first few hours after birth
   i. Drugs for scours and respiratory problems
   j. Leptospirosis and vibriosis vaccine for cows if these diseases are prevalent in your area.
2. Feed gestation ration up to calving if cows are in good condition. If not, gradually switch to lactation ration 30 days prior to calving.

March (September, if Fall Calving)

1. March and April are heavy calving months. Be prepared as follows:
   a. Calving area should be as clean and dry as possible.
   b. Check herd at frequent intervals. Be ready to assist any cow not making progress after 2 to 3 hours of labor.
   c. Give first-calf heifers extra attention.
   d. In extremely cold weather, have heat lamp or other supple-

mental heat ready for chilled, weak calves. But don't leave cows and calves locked up in tight quarters for more than 1 or 2 days.

    e. For maximum disease prevention, get colostrum into calf as soon after calving as possible (4 hours, or sooner).

2. Be prepared to give fluids to scouring calves that become dehydrated. Consult veterinarian for advice.
3. Plan your spring fertilizer needs. Some dealers give a discount on early orders. On straight grass, 50 lbs nitrogen/acre can give an extra ton of hay. For legumes and legume-grass mixes, use phosphate and potash but no nitrogen. Soil test should determine exact needs. Consult your local extension office for advice.

April (October, if Fall Calving)

1. Prepare for pasture season; plan fly control program early.
2. Beware of grass tetany. Provide supplemental magnesium if it is a problem in your area.
3. Get ready for breeding season if you haven't started already:

    a. If you use AI, order semen and check your equipment. Be sure breeding corral is in working order.

    b. If breeding naturally, make sure you have enough bulls: 10-15 cows per yearling bull; 20-25 cows per 2-year-old bull; 30-35 cows per mature bull.

    c. Have phosphorus source in form of free-choice mineral mix; phosphorus is important for maximum fertility.

    d. Yearling British heifers should weigh a minimum of 600 lbs and exotic heifers a minimum of 700 lbs before being bred.

    e. If lactating cows are thin and not cycling, feed more energy.

4. If IBR, leptospirosis, and vibriosis are problems in the area, vaccinate cows and heifers no later than 3 weeks prior to breeding season. *Never* vaccinate pregnant cows with live virus IBR vaccine because they will abort.
5. If pastures are weedy, consider spraying in the spring.

May (November, if Fall Calving)

1. Breed heifers one heat period before cows so they have extra time to recover from calving next year.
2. If you use AI, consider using heat detection aid like a gomer bull with chin-ball marker.
3. Beware of bloat on heavy alfalfa or ladino pasture stands; beware of grass tetany on lush grass stands fertilized with nitrogen.
4. Blocks with poloxalene can help prevent bloat. To prevent grass tetany, daily intake of magnesium is necessary. Consult veterinarian.
5. To take advantage of early summer grass, graze 1/3 to 1/2 of your acreage intensely for 6–8 weeks and make hay on remainder. After that, graze entire acreage.
6. If calves are 2–4 months old, vaccinate heifers for Bangs and all calves for blackleg-malignant edema. Consider implanting calves to boost growth; castrate and dehorn before peak fly season.
7. Get ready to make hay. Be sure equipment is working properly. Cut early before plants become too mature and fibrous. Harvesting forage early is the key to nutritional quality. Don't wait too long!

June (December, if Fall Calving)

1. After 1st cutting hay, consider fertilizing to maximize 2nd cutting.
2. For maximum carrying capacity, rotate pastures if possible. Don't overgraze pastures, because recovery is slow.
3. Be sure cows on pasture have free access to phosphorus source.
4. Control flies with backrubbers, dust bags, spray, or insecticidal salt-mineral mix.
5. If too many females return to heat, consider: (1) your bull; (2) your nutrition; (3) reproductive disorders such as IBR, vibrio, lepto, cystic ovaries, or uterine infection. Consult veterinarian.

140   Cow Herd Management

6. If you creep feed, whole oats and shelled corn are economical ingredients. Crimping or grinding is not necessary. Adding protein supplement is not necessary.

July (January, if Fall Calving)

1. Remove bulls after 60 days of breeding.
2. Control pinkeye: (1) reduce flies with backrubbers, dust bags, spray, or insecticidal salt-mineral mix; (2) clip grazed-over pasture so tall, coarse grasses do not irritate eyes; (3) treat pinkeye by injecting 1 cc antibiotic in each eyelid, and shut out irritating sunlight by gluing patch over eye with backtag cement or by putting animal in dark quarters.
3. If it looks like pastures will run out in late summer, get ready to provide emergency feed such as leasing a neighbor's idle pasture or planting a summer annual like Sudan grass. Start supplemental feeding *before* pastures are gone so as to extend grazing as long as possible.
4. Calves born after mid-May don't do well in summer heat and have lighter weaning weights. Stop breeding cow herd by August 1st to avoid late calves.

August (February, if Fall Calving)

1. New forage seedings may be established in late summer after tillage in early summer to control weeds. High-yielding quality forage is the key to beef production. Consult local extension office for information.
2. Vaccinate *heifer* calves (not bulls) for Bangs between 2 and 6 months of age. Vaccinate all calves for blackleg-malignant edema and leptospirosis if these diseases are a problem in your area.
3. If calves aren't doing as well as expected, have veterinarian check droppings for worm eggs.
4. Be prepared to supplement the herd if pastures dry up.

September (March, if Fall Calving)

1. Line up supplies and drugs for fall roundup and weaning. Consider the following:

a. Finish dehorning and castrating calves that were missed earlier.
   b. Buy ear tags to identify replacement heifers and cows.
   c. Worm calves and cows if needed.
   d. Apply pour-on chemical for grub and lice control.
   e. Vaccinate calves for IBR, PI_3, and pasteurella.
2. Get facility ready for working cattle. If you don't have one, contact local extension office for simple corral plan.
3. Plan your marketing program. Alternatives include: (1) special feeder calf sales; (2) weekly auction markets; (3) private treaty sales to cattle feeders; (4) private treaty sales to cattle dealers; (5) feeding out the calves yourself.
4. Seedstock producers should prepare to have their calf crop weighed and graded through their state or breed performance testing program.

October (April, if Fall Calving)

1. It's fall roundup time for many herds:
   a. Wean and market calf crop.
   b. Select replacement heifers.
   c. Vaccinate replacements for IBR, PI_3, and pasteurella.
   d. Worm herd if necessary.
   e. Apply pour-on for grub and lice control.
   f. Pregnancy test herd and cull open cows; you can't afford to winter empty cows.
2. Know what your calves are worth. Check newspaper and radio reports. Many states have market news services that can be telephoned for up-to-the-minute livestock prices.
3. Cull problem cows and marginal producers. Keep early calvers ahead of late calvers so as to sell more weight next fall.
4. Use diverted acres and crop residues as feed for the cow herd. After weaning, nutrient requirements of the cow herd are reduced.
5. Fall is a good time to take soil tests and topdress hay and pasture fields with potash and phosphate, if necessary.

November (May, if Fall Calving)

1. If you have access to cornstalk fields, consider the following:
   a. Grazing at rate of 1 to 2 acres per cow for 60 days. It may be necessary to supplement with protein and energy after first 30 to 45 days.
   b. Harvesting dry stalks as large stacks or bales. Beware; if too damp (over 40 percent moisture), they will spoil.
   c. If stalks are too damp, you can ensile them but add enough water to bring moisture up to at least 60 percent. Chop stalks fine for maximum compaction in the silo.
   d. NPN compounds such as urea are not as well-utilized on low-energy residues like corn stalks. When supplementing, try to stay with natural sources of crude protein.
2. Wean summer calves before hard winter weather sets in. They will do better on grain plus hay or on corn silage than if left on their mothers (unless you creep-feed).
3. If you keep calves to feed over winter, aim for 1.25 lbs gain/day on heifers and 1.5 lb/day on steers. Full-fed legume-grass hay plus 5 lbs grain or 30 lbs corn silage plus 1 lb soy plus free-choice hay will do the job.

December (June, if Fall Calving)

1. If cows are still grazing crop residues, be sure they have enough to eat. If not, supplement accordingly.
2. For best results, divide herd into groups for winter feeding: (1) weaned heifer calves; (2) first and second-calf heifers and old thin cows; (3) the rest of the dry herd; (4) lactating cows with fall calves; (5) herd sires.
3. Check winter rations. Use the following as a guide:
   a. Weaned heifer calves: Full-feed hay + 5 lbs grain or 4 lbs corn + 1-1/2 lbs supplement.
   b. Two-year-old pregnant heifers, thin 3-year-olds, and thin old cows: Full-feed hay or 4.5 lbs corn + 1-1/2 lbs supplement.
   c. Dry pregnant mature cows in good condition: 17-25 lbs hay or 15 lbs straw + 10 lbs hay or 40 lbs corn silage.

d. Lactating cows: Full-feed hay or 60-75 lbs corn silage + 1.5-2.5 lbs supplement.
   e. Herd sires: Full-feed hay or silage (add grain to diet according to condition).
   f. If feeding haylage instead of hay, figure 1 lb hay same as 1.5 to 3 lbs haylage, depending upon moisture.
   g. In those areas where fescue is adapted and winters are relatively open, make maximum use of this grass for winter grazing.
4. Put priorities on winter forage supply, approximately as follows:
   a. Feed lowest quality forage to mature dry cows during early winter.
   b. Feed highest quality forage to young stock and to lactating cows.
   c. Feed medium quality forage to dry cows during late pregnancy and to mature herd sires.

7

Selling, Buying, and Managing Feeder Cattle

7-1 DEVELOPING A MARKETING STRATEGY

Timing of cattle purchases has a large impact on cattle profits. Prices of feeder and finished cattle fluctuate greatly, and poor planning in purchasing feeder cattle can be disastrous. The major forces influencing the price of cattle are the national cattle inventory, disposable income of consumers, cost of feed, and supply of other red meats. These factors interact to cause what is known as "the cattle cycle." It is initiated when cattle production is sufficiently profitable to encourage producers to retain large numbers of the annual heifer crop for breeding purposes and to reduce culling of the cow herd. This reduces the supply of feeder cattle and slaughter cows, further expanding profits and accelerating the expansion phase of the cycle. Over several years this results in expansion of the cow herd and, later, in a surplus of feeder cattle. When this point is reached, prices fall. Eventually more than the normal number of cows are culled, further accelerating the downswing in prices. This liquidation phase will continue until a shortage of feeder cattle is created, leading to higher prices and initiating the expansion phase of the next cycle.

We have experienced eight cattle cycles since the first one bottomed out in 1895. Table 7-1 shows the change in cattle numbers during these cycles, and the duration of the expansion and liquidation phases of each cycle. The length of the cycle is determined to a large degree by the reproductive cycle of a beef cow. It is approximately two years before a heifer saved at weaning, rather than sold, weans her first calf. Then it is another 9 months to 2 years before her calf is finished and added to the beef supply. The expansion phase has lasted from 6 to 8 years, indicating that this is the time required to overexpand our beef herd.

Cattle numbers and prices are basically determined by the consumer demand for beef. For more than 30 years, consumers have

Table 7-1
Cattle Cycles in the United States

Years	Duration	Cyclical Upswings (expansion phase)	
		Changes in Cattle Numbers	
		Total Cattle and Calves	Cows
1896-1903	8 years	+ 35.0%	---
1912-1917	6 years	+ 31.2%	---
1928-1933	6 years	+ 29.7%	+ 27.1%
1938-1944	7 years	+ 31.1%	+ 27.8%
1949-1954	6 years	+ 25.7%	+ 23.5%
1958-1964	7 years	+ 19.5%	+ 12.7%
1967-1974	8 years	+ 21.2%	+ 19.3%
		Cyclical Downswings (liquidation phase)	
1890-1895	6 years	- 18.0%	---
1904-1911	8 years	- 16.2%	---
1918-1927	10 years	- 21.5%	- 8.3%
1934-1937	4 years	- 12.3%	- 12.7%
1945-1948	4 years	- 10.2%	- 10.1%
1955-1957	3 years	- 5.6%	- 7.5%
1965-1966	2 years	- 0.2%	- 2.7%
1975-1978*	4 years	- 13.3%	- 12.5%

*Projected through 1978

Source: DeGraff, "Beef Production and Distribution," University of Oklahoma Press, 1960, p. 39, plus subsequent data from U.S. Department of Agriculture.

Starting date of 1890 is arbitrary.

spent 2.4 to 2.6 percent of their disposable income for beef. Steady increases in consumer disposable incomes coupled with a steady increase in population have led to dramatic increases in per capita and total beef consumption (Figure 7-1). For a long period of years prior to this period, per capita consumption of beef averaged approximately 60 pounds. Per capita consumption grew as incomes and cattle numbers increased until 129 lbs/capita were produced and consumed in 1976. Most observers feel that this trend in per capita consumption will not continue—that beef consumption will reach a maximum, regardless of disposable income. However, further increases in population could increase the demand for beef as long as disposable income stays high.

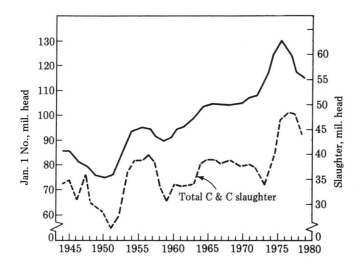

Figure 7-1
U.S. Cattle Numbers and Slaughter of Cattle and Calves

Within each cycle, the price a cattle feeder will pay for feeder cattle depends primarily on the price of feed grains and the market for finished cattle. When feed grain supplies are high and their prices are depressed, higher prices will be paid for feeder cattle without a change in the price of finished cattle, as the cost of gain is reduced. When finished cattle are in short supply and their price is high, the cattle feeder will bid higher for feeder cattle because he anticipates a higher price for them when finished.

All of these factors must be taken into account to develop a marketing strategy. A cattle feeder must develop a system for buying feeder cattle and selling finished cattle to avoid financial disaster. The most common practice used is to buy, and thus sell, many times a year to average out feeder cattle costs and finished cattle prices. The following are the primary strategies commonly used.

1. Buy cattle at least five different months/year, with the largest number being purchased in the fall months when the feeder cattle supply is the greatest. Then cattle are "topped out" of the pens as they reach slaughter condition. This will usually result in finished cattle being sold in all 12 months, as the initial weight, potential slaughter weight, and performance will vary within each group purchased.
2. Buy feeder cattle when finished cattle are sold. This results in feeder cattle being purchased on the same market as finished cattle. Feeder cattle prices usually follow finished cattle price. Thus, a high finished cattle price is averaged out by high feeder prices, but the cattle feeder is usually more optimistic and is in a better financial position to take the risk. Conversely, when finished cattle are low, feeders are purchased at a lower price, reducing the risk.
3. Buy feeder cattle often enough, and with enough variation in type and weight, to result in finished cattle to sell each week.

Feeder cattle producers who do not finish cattle can spread their marketing by using one or more of the following practices.

1. Have two calving seasons. Many large producers now calve in the fall as well as in the spring to spread marketing of their calves.
2. Background the calves on a growing ration for 120-180 days after weaning. The calves can be sold at any point during this period when the price is favorable.
3. Winter the calves on a high forage ration, then graze them the following summer and sell as yearlings. Many do this to sell more weight per cow. Fewer cows are kept, however, as part of the forage produced is needed for the yearlings.
4. Place the calves or yearlings in a commercial feedlot and retain ownership.

The economics of these alternatives are examined in Chapter 11.

7-2 SELECTING THE MOST DESIRABLE TYPE OF FEEDER CATTLE

Over a period of time, the average price a feeder calf producer receives for his calves will be based on his ability to produce those animals that are profitable to the cattle feeder.

Selection of feeder cattle is a big decision for the cattle feeder. Should he feed 500 lb steer calves? Heifer calves? Thin 600 lb yearling steers? Should he consider a combination of feeder types to reduce marketing risks by selling over a longer marketing period? Although more cattle feeders are buying and selling cattle several times during the year to spread price risks, feeder cattle movement is still at a peak in the fall.

What can a cattle feeder afford to pay for a feeder animal? How large does the purchase price differential between types have to be before he should consider switching from a program he has traditionally used?

What criteria should cattle feeders use in choosing among alternatives? For a continuously operating feedlot—a feedlot that tends to be near capacity at all times—he wants to maximize his net returns per year to the facility. Net return/day/head capacity should be as large as possible. On the other hand, the cattle feeder who feeds one lot per year wants to maximize the net returns per animals sold—turnover rate is less important to him.

Net returns/head are determined by gross margin (sale value less purchase value) less feed, veterinary, interest on the feeder, marketing, and yardage costs. Feed costs make up a lion's share of the cost, currently nearly 75 percent. Efficiency of feed conversion, therefore, is a key factor in assembling a bid chart among alternative feeder types. And, 8 out of 10 of the nonfeed dollars are related to time an animal is on feed. Knowledge of the relative performance of alternative feeder types obtained through research and experience is used to determine which are good buys and to avoid the bad ones. It is important for the feeder calf producer to be aware of the factors influencing prices paid for various types of feeder calves so he can produce those types that will have the highest sale value.

Heifers vs. Steers

Theoretically, if steers and heifers are fed until they are equal in fatness, heifers would be about 200 lbs lighter than comparably sized steers at slaughter. Their feed efficiencies would be similar.

In practice, heifers—particularly from smaller-framed cattle—are fed until they are fatter than steers in order to end up with acceptable carcass weights. And, cow-calf producers keep their best heifers for replacements, selling the balance as feeder calves. Perhaps the heifers culled are less efficient in feed utilization, due to management or genetic factors. An additional problem is that many pregnant heifers are sold as feeders.

Break-even charts are often based on selling heifers 100 lbs lighter than steers, and since the heifers often end up fatter, their feed/gain is higher than for steers. Also, they typically sell for a lower price/cwt at the same final grade due to lighter carcass weight and lower dressing percentage. As a result of these factors, heifers are usually worth 80-85 percent as much/lb as steers.

Frame size. As discussed previously, large-framed cattle reach equal fatness at heavier weights compared to small-framed cattle.

When fed to the same grade, carefully controlled research has shown that there is little difference among frame sizes in feed required/lb gain. Larger-framed cattle gain more rapidly, and require less feed/lb gain when fed to the same weight as smaller-framed cattle. However, they must be fed to heavier weights to have the same grade. This means a higher average weight to maintain, and more feed/lb gain for fat gain to reach equal grade. These factors offset the higher rate of gain of larger types.

Larger-framed feeder cattle do have some other advantages, however. Their heavier weight at the same grade gives packers a lower slaughter cost/lb carcass weight obtained, since many slaughter costs/head are fixed. This is especially true for larger-framed heifers; they can reach acceptable slaughter weights without being excessively fat. An additional advantage of larger-framed cattle is that if the demand for good grade beef is increased, they will still have acceptable weights (1000-1200 lbs) if slaughtered at that grade.

Breed effects. Recent carefully controlled experiments at Michigan State, Ohio State, the University of Wisconsin, the U.S. Meat Animal Research Center and other stations have shown that differences between beef breeds in feed efficiency are small *when compared at the fatness of low choice grade.* However, research (the University of California, Cornell, Michigan State, Oklahoma State, and other stations) has shown that Holsteins are 8-25 percent less efficient than beef breeds fed to the same degree of fatness. Further, other studies (Michigan State, Oklahoma State, and others) indicate that beef x Holstein cross calves are less efficient than straight beef breeds, the degree of lower efficiency depending on the percentage of Holstein breeding they contain. A larger proportion of the lower efficiency of the crossbreds may be due to the higher milk production of their Holstein cross dams. The heavier milking cows often are more efficient in producing weaning weights, even though the dam's feed requirements are higher. These data suggest a conflict between milk production of the dam and efficiency in the feedlot. Additionally, some data suggests that lifetime production of replacement heifers from high milking dams is lower.

Research indicates that there is variation among breeds in fat distribution at the same fatness. Angus have been shown to require less fat to reach a particular grade than some other breeds. Holsteins appear to need less total fat than some breeds to reach the same degree of marbling.

One of the big advantages of crossbreds for the cattle feeder is the combining of a larger frame size and growth rate of some breeds with the hardiness and ability to marble without excess fat of other breeds. Further, crossbred calves will have a small advantage (probably no more than 3-5 percent on the average) in feed efficiency due to heterosis (hybrid vigor).

Body condition. Cattle of the same frame size can be fleshy or thin at a given weight, depending on how they have been fed up to that time. Studies have shown that cattle that are fleshy will gain 5-10 percent more slowly and less efficiently than cattle in average condition. Those in thin condition will be more efficient than those in average condition by this same amount, as long as they are healthy and have not been stunted. Thus calves that have heavy milking dams and/or have been creep fed may be discounted if they are fleshy at weaning. Extra energy for nursing calves will be beneficial if they have the frame size to utilize that energy for muscle and bone growth.

Yearlings vs Calves

Yearlings, at the same degree of fleshiness as calves of the same body type would be expected to require more feed/cwt gain. This is because as animals increase in weight, the proportion of fat in the gain increases, which increases the energy requirements for weight gain. In practice, however, most yearlings placed in the feedlot have been on pasture previously, and thus may be thinner at a given weight than a calf placed in a feedlot at weaning. Thus, feed requirement calculations for yearlings must be adjusted for body condition. Yearlings, on the average, require 8 percent to 12 percent more feed/cwt gain than calves; but there is sometimes no difference when comparing a yearling with considerable compensatory potential and an average fleshed calf. They also are better able to withstand the stress of shipping and sorting.

Most cattle feeders feel yearlings must be fed to heavier weights than calves to grade choice. However, research results have not provided a clear-cut answer to the weight at which a yearling placed in the feedlot off grass will grade choice, as compared to a calf of similar body type placed in the feedlot at weaning. The result probably depends on their age or length of time they have previously been on a low level of nutrition. Perhaps long yearlings, or those over two years of age before being placed in the feedlot, are the ones that must be fed to heavier weights to be fat enough to grade choice.

Health

One of the greatest problems of the cattle industry is the transfer of feeder cattle from the farm or ranch to the feedlot. The stress associated with sorting, weaning, castration, and dehorning at the farm, mixing with other cattle and handling at the sale barn, and transporting and adapting to a new environment and feeding program at the feedlot causes heavy losses in shrink and poor performance due to sickness and death. Typically, a 15 to 30-day feeding period is required to recover pay-weights, with the death loss of ½ to 1 percent in yearlings and 1 to 2 percent in calves. It is not uncommon for losses in the feedlot to be much higher than this. The cost of shrink, sickness, death loss, and veterinary expenses typically runs $10–$15 head, costing the industry millions of dollars. Cattle buyers and cattle feeders soon learn where the calves with most problems originate and compensate by paying less for cattle from these sources or avoid them completely.

The desirability of different practices prior to shipment varies, depending on the buyer. However, practices that are usually beneficial to the cattle feeder include castration, dehorning, weaning at 3 to 4 weeks prior to sale, and teaching calves to eat from a feedbunk and drink from an automatic waterer. Worming is desirable where a problem. Treating for grubs is of particular value when the cattle will be shipped after the cutoff date for grub treatment. It is especially important to be sure heifers do not get bred. The cow-calf producer should certainly be appreciative of the problem associated with immature females calving. Uncertainty about pregnancy is one of the reasons why the price differential between heifers and steers is often wider than it should be. Handling cattle quietly and as little as possible and rapid movement to the feedlot reduces stress. Heavier calves (600 lbs and up) and yearlings are better able to withstand these stresses, and many feeders will not buy cattle under this weight due to health problems.

Economics

It is questionable if the herd owner can afford to produce thin feeder cattle that have the capacity for compensatory growth. However, he should be aware of the ability of his cattle to perform well and to grade choice at a desirable weight. Further, he should be able to consistently provide healthy animals that are castrated, dehorned, and are not pregnant. He should also make every effort to see that the cattle are handled carefully and efficiently until they reach the feedlot.

He must also be able to make some shifts in his operation as conditions change.

When cost of gain is less than the price of finished cattle, the cattle feeder can sell the weight gained in his feedlot for more than it costs him to put it on and will pay more for calves relative to yearlings. When cost of gain is higher than the price of finished cattle, he has to make a profit by selling the weight he buys for more than he pays for it by feeding the animal to a higher carcass grade. Under these conditions, he will pay relatively more/lb for yearlings than calves because he is buying more weight. Under these conditions, it may be desirable to keep fewer cows and keep the calves to yearling weight rather than use his feed supply to keep more cows and sell the calves at weaning.

Feeder Cattle Grades

New feeder cattle grades proposed in 1978 are expected to be implemented in the near future. The old feeder grades (prime, choice, good, and standard) were developed to designate potential quality grade when finished. Recent research, however, has shown little relationship between feeder calf and carcass grade. As discussed in the section on carcass grades (Chapter 12), marbling score determines the carcass grade. Under the new grading standards, the ability to marble is impossible to identify in a feeder calf, except for known breed differences that are discussed in that section. The new grades will reflect the expected weight at low choice, yield grade 3 for various frame sizes (Table 7-2). There will be three frame sizes; small, medium, and large (Figure 7-2). These will correspond to the frame codes outlined in Table 8-1 and Figure 8-1 (small, codes 1-3; medium, codes 4-6; large, codes 7-9), and performance for each type as in Tables 8-17 to 8-19. There will be three muscling scores for each frame size (Figure 7-2).

Table 7-2

Expected Weights at Low Choice Grade

	Steers	Heifers	Feeder grade frame type
Small frame British breeds	800-1000	640-800	Small
Average frame British breeds	1000-1100	800-880	Medium
Large frame British breeds	1100-1200	880-960	Medium
Average frame European breeds and Holsteins	1200-1300	960-1040	Large
Large frame European breeds	1300-1500	1040-1200	Large

Unthrifty cattle will be placed in a separate grade. They are types that appear to be stunted, sick, or parasitized.

The use of frame codes, muscling scores, and breed is the best system developed to date to describe feeder cattle.

7-3 MARKETING ALTERNATIVES AND SOURCES FOR FEEDER CALVES

There is not a best method for selling or buying feeder cattle. Because of the diverse nature of the beef cattle industry, the best system for sale or purchase of feeder cattle will vary with individual producers. This section will outline the methods for "trading" feeder cattle, and some of the advantages of each.

Most feeder cattle are sold either direct to cattle feeders or through auction sales. Within each of these categories, a number of methods are used to handle the sale.

Any order buyer, dealer, or company that buys or sells livestock must register with the packers and stockyards (P & S) division of the USDA. They can be prosecuted for unfair or illegal practices in conducting livestock transactions. Producers who experience such problems should contact the nearest P & S office.

FRAME SIZE

LARGE

MEDIUM

SMALL

Tall and long for age

Half inch of fat—
Steers, 1200 lbs or more
Heifers, 1000 lbs or more

Slightly tall and
slightly long for age

Half inch of fat—
Steers, 1000-1200 lbs
Heifers, 850-1000 lbs

Small frame and
shorter-bodied for age

Half inch of fat—
Steers, less than 1000 lbs
Heifers, less than 850 lbs

MUSCLING

No. 1

No. 2

No. 3

Very thickly muscled
throughout

Legs very wide apart

No evidence of non-beef
breeding

Slightly thick muscled
throughout

Legs slightly wide apart

Very high proportion
of beef breeding

Less thickly muscled
than No. 2

Legs closer together
than No. 2

Smaller proportion of
beef breeding than No. 2

Figure 7-2
Proposed Feedercalf Grades
(Courtesy of Texas A & M University, Gary C. Smith)

Direct Sales

Direct sales are handled either directly between cow-calf producers and cattle feeders, through order buyers and cattle dealers, or through terminal markets.

Producer to cattle feeder. This method is theoretically the best, because the stress of handling shrink and exposure to other cattle is reduced. If the cattle feeder is within a few miles, he will buy direct from small herds. If, however, the cow-calf producer is located a long distance away, he will want to buy at least a semiload (about 44,000 lbs) at one time, to minimize trucking costs. He can still buy from several small producers by pooling their cattle to get a load if they are all willing to sell to him.

The big advantage to the cattle feeder is reduced health problems, and knowing the source of cattle. Many cattle feeders develop a relationship with feeder calf producers so that they produce the kind of feeders desired and handle them prior to shipment according to the cattle feeder's specifications.

The cattle are usually purchased on a price based on weight at the nearest inspected scale less a pencil shrink, which is often about 3 percent. The biggest difficulty with this method is in establishing price. Most often the price is negotiated, based on both the feeder calf producers' and cattle feeders' knowledge of current feeder cattle prices. Most soon learn that the best price is the one fair to both; otherwise a mutual trust is not developed, making price negotiation difficult in subsequent years.

Order buyers and dealers. An *order buyer* acts as an agent for the cattle feeder; he locates and purchases cattle ordered of a specified type within a designated price range. The cattle feeder pays the producer for the cattle directly and pays the order buyer a commission. A *dealer* buys the feeder cattle, then resells them to the cattle feeder; the dealer is thus assuming the risk of a price change before he can resell them. Obviously, he takes this risk, hoping he can resell the cattle for more than he paid for them, in spite of shrink, trucking, and risk of death loss.

The proportion of feeder cattle handled through order buyers and dealers is increasing, due to the rising number of cattle needed by most cattle feeders and the increase in demands on the time of cattle feeders for management of their operation. Skilled order buyers and dealers can benefit both the feeder calf producer and cattle feeder.

7-3 Marketing Alternatives and Sources for Feeder Calves

Both the producer and feeder must be well informed on current prices, however. When auction markets are used, the price is established by competitive bidding.

Terminal markets. When calves are sold through a terminal market, a commission firm completes the transaction between the seller and buyer. He provides pens for the cattle, feed, and water, and negotiates a sale to a buyer for the cattle. He pays for the feeder cattle, and collects the purchase price from the buyer. For conducting these services, he collects a fee. The advantage of this system is to provide a means of bringing a larger number of feeder cattle to a common point where a buyer can collect the kind and number he needs. A producer and buyer can keep in touch with the commission firm who can keep both of them informed as to market trends and demands for cattle. The commission firm can keep feeder calf producers informed of the type of cattle that are selling best and how best to market their calves. The firm can call the cattle feeder when the type of cattle he wants are available.

The terminal markets are moving a lower proportion of feeder cattle than in past years. The reasons for this are not completely clear.

Forward contracting. This method of buying cattle ahead of the delivery date is utilized to spread buying risks and the pressure of having to obtain a large number of feeder cattle during the busy fall harvest season. Most forward contracting is done 30 to 60 days ahead of the delivery date. Plans are made for method of transportation, weighing procedures, and price. The price may be established at the time of the contract, or may be based on some specified market price at the time of delivery.

Auction Sales

Auction sales may be individually owned, or may be part of a corporation or cooperative in which several are owned throughout an area or state. Most of these sell all species, and have one sale a week. Many with large volumes have two or more weekly sales; some may have up to five sales per week. In areas containing large numbers of beef cattle, the producer is usually assured of a fair market price because several buyers are always present, resulting in competitive bidding. In areas where beef cattle are a minority and large numbers are not available, auction sales can be a disastrous means of selling feeder cattle. For example, in the northeast dairy cattle predominate, and the volume of cattle sold in auctions are cull dairy cows and veal calves. Feeder cattle buyers do not usually attend these sales, due to the small number of feeder cattle. Thus, often beef feeder cattle will sell for veal calf prices if sent to these sales.

Cattle are usually brought to the sale on the morning of sale day, and placed in pens. They are sold by individual owner groups, and the sale weight is usually taken just before they enter the sale ring. The auction collects the purchase price from the buyer, deducts a commission, and sends the proceeds to the seller. Calves or finished cattle removed from the pasture or feedlot and hauled directly to the sale the morning of sale day can be expected to have an average of a 3 percent shrink between weight at the farm and sale weight. Shrink will be discussed in more detail in the next section. In general, however, it is not wise to attempt to get the cattle to consume extra feed and water prior to shipment. An experienced buyer can detect overfilled cattle, and may overly discount the cattle to protect himself.

Some auction sales will sort and/or grade calves from several owners into uniform groups, especially during periods when volume is large. This is done for the convenience of the auction, as well as the buyer. Under these conditions, the cattle are weighed on arrival and identified with a number. This animal weight is used in combination with the sale price/cwt for the group to determine price for the cattle.

It is usually best to notify the market manager a week ahead of when you anticipate sending a load of cattle. He can contact prospective buyers, and may advise you which week or day would be best, depending on expected volume and current market conditions. This is especially important in areas where sale volume of beef cattle is small.

Special Feeder Calf Sales

Periodically auction sales will sponsor special feeder calf sales. Some of the cattle may be consigned by a producer and sold for him by the firm for a commission. Some may be sold without prior consignment. The firm may purchase feeder cattle for these sales. These are usually held separate from the weekly sales, often in the evening to accommodate farmer-feeders. These sales are usually restricted to feeder cattle within a specified weight and age range.

Many states have state cattlemen's association-sponsored feeder calf sales. These sales predominate in the eastern cornbelt, the northeast, and southeast, where small herds predominate.

They were developed to assemble a large enough volume of feeder cattle to attract volume buyers, and to improve selling methods to increase the price received for feeder cattle. They are held in the fall and spring, when feeder cattle volume is greatest and demand is highest. The sale is held at a local private or corporation-owned stockyard, or at a special facility built for these sales.

Cattle are consigned to the sale ahead of the date; most restrict the sellers to members of their organization. The cattle are sorted into uniform groups by sex, breed, weight, and grade. They are usually graded by trained state department of agriculture graders. The feeder calf producers usually are required to help with these sales, or are assessed an additional handling fee. These sales have been highly successful, especially where the cattle are grouped to the best advantage of the buyer and regulations have been carefully enforced. The regulations specify weight, breed, age, and health restrictions. The most successful sales have developed an excellent reputation, through a strong advertising program, following their cattle through the feedlot to see how they perform, and by improving the quality through careful sorting and grading and higher standards.

Handling Cattle Prior to Market

Discussions of management factors affecting the health and price of feeder and finished cattle, and determining when to sell cattle are in other parts of this section of the text. Several specific management practices, however, should be discussed or repeated here that apply to marketing all cattle to reduce stress and shrink and to improve the price received.

1. Do not overfill the cattle. This practice increases the stress on the cattle, and reduces the price.
2. Handle cattle quietly, carefully, and slowly during sorting, loading, transporting, and handling to minimize stress.
3. Wean, and if not done previously, castrate, and dehorn calves at least 30 days ahead of sale. Teach them to eat out of a feedbunk and to drink from a tank or automatic waterer.
4. Do not misrepresent the cattle. If you have off type or injured cattle, sell them privately or through an auction.

Market Information

The sources of market news information include local radio and newspapers, state department of agriculture market news service, federal market news, toll-free telephone tapes and publications, National Cattlemen's Association *Cattle Fax*, the *Drover's Journal*, and direct contact with order buyers and dealers, commission firms, local auction managers, and packers. Addresses for those available nationwide are listed in the section on marketing slaughter cattle. Local services can be located by contacting the nearest county agricultural agent.

Shrink

Weight loss occurs during shipment and handling of cattle. Shrink contains two components; excretory (loss of fill) and tissue (loss of nutrients in body tissue) shrink. Animals off feed and water for 12 hours or less usually have only excretory shrink. This shrink can be replaced in a short time when given access to feed and water. However, during handling and shipping, tissue shrink occurs also.

In an experiment at Iowa State University with 4685 feeder cattle, shrink averaged 7.2 percent when purchased directly from a rancher, but shrink averaged 9.1 percent when purchased from a sale yard. They found that there was 0.61 percent shrink for each 100 miles in transit. Table 7-3 shows the average expected shrink for feeder cattle, based on studies at Michigan State University and the University of Wyoming.

In the Iowa study, approximately half of the shrink was excretory and half was tissue loss. Calves required an average of 13 days and yearlings 16 days to recover weight lost due to shrink.

Other less defined conditions increase shrink. Excessive handling, crowding, extreme changes in temperature, wind, rain, snow, weaning, and temperament of cattle have been observed to affect shrink.

Managing feeder cattle to reduce shrink will be discussed later. However, weighing conditions should be taken into account. Divide the base "no shrink" price by *(100 - % expected shrink)/100* to adjust for shrink. For example, cattle priced at $50/cwt with no shrink are worth 50/0.97, or $51.55/cwt with a 3 percent shrink. A common practice is to use weights taken with cattle hauled to the nearest scale, less 3 percent "pencil" shrink. Thus, if cattle were price FOB the nearest scales at $50.00/cwt, they should be priced at $50 × 0.97, or $48.50 without a "pencil shrink."

There are a number of ways in which shrinkage can become excessive. In many areas it is customary to hold cattle in drylot overnight and weigh them early in the morning. In other places a pencil shrink, a percentage discount from the weights to determine sale weights is used. Depending on the area, pencil shrink varies from 2 percent to occasionally 5 percent.

Many ranchers and feedlots do not have weight scales handy and must move stock to where they can weigh them. These short trips from ranch or feedlot to the scale also involve shrinkage.

If "pencil shrink" is applied on weights obtained at a point some distance away, a double shrink will occur. It has been found that the larger shrink occurs during loading, unloading, and in the first hour of hauling. The shrink increases at a slower rate with additional miles. Thus, even though you may be a short distance from a scale, and if animals must be loaded and unloaded and hauled, the shrinkage will be significant.

Another situation which might increase shrink is when the animals are loaded and delivered as early as possible in the morning, before they are fed. An overnight stand could cause an additional 2 percent shrinkage. Consider the hypothetical situation in which a rancher or feeder loads his cattle early in the morning after an overnight stand and hauls them six miles to a scale. He could expect a 2 to 3 percent shrink loss because of hauling them those six miles and loading and unloading them. He could also expect a 2 percent shrinkage loss because of the overnight stand. If a 3 percent pencil shrink were added to this, it becomes obvious how shrinkage can become excessive. Consequently, the seller should plan his selling and shipping program carefully. The custom of specifying a pencil shrink was developed in lieu of the overnight stand without feed and water.

Table 7-3

Some Factors Affecting Shrink

Condition	Percent shrink
8 hr. off feed and water	3.3
16 hr. off water	2.0
16 hr. off feed and water	6.2
24 hr. off feed and water	6.6
8 hr. in moving truck	5.5
16 hr. in moving truck	7.9
24 hr. in moving truck	8.9

Source - Experimental data from Michigan State University and the University of Wyoming.

7-4 NEW FEEDER CATTLE: NUTRITION, HEALTH, AND TREATMENTS

A plan for handling newly arrived cattle needs to include minimizing stress, feeding, immunization, parasite control, castration and dehorning, and treatment of sick cattle.

Minimizing Stress

1. Avoid cattle that have been through several sales and are highly stressed. When possible know the source, how they are handled prior to shipment, and how long in transit.
2. Load cattle with a moderate fill of grass hay and water.
3. Cattle should be loaded quickly and quietly and moved to the feedlot as fast as possible. Insist that the trucker avoid any unnecessary stops.
4. Provide a good environment upon arrival.
    a. Small shallow lots with space for every animal to eat at one time minimize stress. This also allows detection of those animals that fail to come to the feedbunk when fed. Be sure that small calves can reach the feedbunk.
    b. Provide fresh and palatable feed in the feedbunk. It is desirable to feed several times a day to be sure that the feed is fresh. Frequent feeding and observation will also help detect sick cattle.
    c. The lot should be well drained and dry, but not dusty. It should provide cattle with a good place to rest without lung irritation. The starting lot should be located close to the working chutes and sick pens so that sick cattle can be handled with a minimum of disturbance. The sick pens should be small with a small shed to provide a dry, draft-free place for sick cattle. Small pastures provide a clean, dust-free environment, but sorting out sick cattle for treatment is difficult and creates additional stress for the rest of the cattle.
5. Cattle should be given routine immunizations upon arrival if they arrive strong and healthy. Groups that arrive containing many sick calves, however, should not receive routine immunizations until they have recovered. These calves have probably undergone stress and multiple point origin, and further stress is undesirable.

Nutrition of Newly Arrived Feeder Calves

A portion of the weight loss is loss of gut fill, but at least one-half of the shrink is from tissue loss of moisture, minerals, energy, and protein. These moisture and nutrient losses must be regained as rapidly as possible upon arrival in the feedlot to maintain the health of the animal. This involves proper feeding practices and nutritional management of new feeder cattle.

1. Energy level and sources of energy. Get cattle to eat as soon as possible after arrival. Good quality hay is likely to be the feed that will be most readily consumed, and it is the safest feed to use because of the variation in origin, size, and prior feeding history. However, hay is becoming expensive and may not be necessary in most cases. In three experiments at the Ohio Beef Cattle Research Center and in two experiments at Michigan State University, there has been no advantage to feeding chopped hay on arrival and then switching to corn silage, as compared to feeding silage and soybean meal based supplement. Calves fed chopped hay ate more and gained faster initially, but performance was lower during the change to silage and soy. Gains of 400-500 lb calves at the Ohio and Michigan stations on silage plus soybean meal supplement with no hay have been 1.5 to 2.6 lbs per head daily for the first 4-6 weeks. This is a desirable level of performance. These experiments suggest that in most cases calves and yearlings can be started on silage and soy supplement with no hay. Severely stressed or sick calves will probably do better if they are started on high quality hay.

The desirable starting ration for lightweight calves may be different. In three experiments at the California station with 300-350 lb calves, best results were obtained with a ration that was about 70 percent concentrates. One of the best rations for starting lightweight calves, or any feeder cattle that you expect may not eat well, is 1½ percent of body weight of shelled corn plus a full feed of alfalfa-brome hay. Experienced feedlot operators like to spread high quality hay on top of the regular ration for the first few days to attract young calves to the feedbunk.

2. Effect of energy level on health. In the California, Ohio, and Michigan studies, health was not related to energy level fed. In all experiments, the highest levels of energy gave the fastest gains, but had no influence on treatment required or days sick.

The cardinal rules of not increasing concentrate too rapidly and not making sudden feed changes would, of course, apply.

3. Protein requirements. Few studies have been conducted on the protein requirements of new feeder cattle. However, studies at the Ohio station suggest that the ration can be balanced according to the weight of the cattle and energy level of the ration with no special adjustments for the starting period. Thus, the protein requirements outlined in Chapter 8 should be adequate for new feeder cattle.

4. Sources of protein. Most studies have shown that performance is not as good the first 4-6 weeks when the protein supplement contains urea, as compared to soybean meal or other sources of natural protein. In the Ohio studies, urea supplements gave 20 percent lower gains and feed efficiency during the first 4 weeks. However, the urea-fed cattle gained about 50 percent faster and more efficiently than those fed only untreated silage and minerals.

In these same studies cattle fed urea-treated silage properly supplemented with vitamins and minerals gave performance nearly equal to those fed untreated silage plus soy supplement. In studies at Michigan State University with ammonia-treated silage, however, calves have benefited from feeding ½ to ¾ lb of soy supplement with the treated silage for the first four weeks.

5. Vitamins and medicated supplements. Vitamin A requirements are increased during stress. The starter supplement should provide 40,000 to 60,000 IU per head daily. One million IU per head could be injected as an alternative to providing vitamin A in the feed. Most studies have shown that feeding a low level of antibiotics such as Aureo S700 or neoterramycin will increase gains and feed efficiency during the first 4-6 weeks. However, these products will not prevent disease outbreaks nor are they adequate for treatment of sick cattle.

6. Water. Provide water for the cattle 3 to 4 hours after arrival. Although water is important in stress, the most immediate need is for energy and other nutrients, and cattle that have had a fill of water first may not consume enough hay to meet these needs. The water tank should be close to the feeding area and should be large enough so they can find it easily. The noise of running water also helps to attract the cattle to the water and is helpful in getting them to drink. Water tanks should be cleaned before new cattle arrive.

7. Changing from a low energy to a high energy ration. Most problems with cattle going off feed when the concentrate level is being increased occur when the ration reaches 67-70 percent concentrate. Most cattle feeders will be feeding corn silage, which is 40-50 percent grain, plus supplement prior to increasing the grain content. These cattle could be increased to concentrate equal to 1 percent of body weight by the addition of corn over a one or two-day period. The ration would then be the equivalent of about 70 percent concentrate. Increases of grain beyond this level should be at the rate of 1/3 to 1/2 lb per head daily to avoid animals that go off feed, bloat, or founder. The final high grain ration should contain 6-10 lbs of corn silage or 2-3 lbs of hay per head daily to prevent digestive disorders.

8. Guideline rations for newly arrived feeder cattle. A. Silage plus supplement: One pound of soybean meal supplement/20 lbs of corn silage. Two pounds of good quality hay per head daily can be fed until calves consume 15-20 lbs of silage if problems with consumption occur. The supplement given below or a similar supplement will balance this ration.

Ingredient	*Lb./1000 lb*
Soybean Meal	930
Dicalcium Phosphate	25
Limestone	15
Trace Mineral Salt	30
Vitamin A—40,000 IU/lb. or inject 1 million IU	
% Protein	41
% Calcium	1.5
% Phosphorus	1.2

After the first 4-6 weeks, follow the feeding schedule given in Tables 10-7 and 10-11.

B. Shelled corn plus hay: 1 percent of body weight of shelled corn and a full feed of early cut alfalfa-brome hay plus trace mineral salt free choice. If grass hay is fed, feed one pound per head daily of a 40 percent all natural supplement containing about 2 percent calcium and 0.5 percent phosphorus.

C. Ground ear corn plus hay: 1 lb of ground ear corn per 100 lbs of body weight and a full feed of alfalfa-brome hay plus trace mineral salt free choice. If grass hay is used, feed 1-1/4 lbs per head daily of a 40 percent all natural supplement containing about 2 percent calcium and 0.5 percent phosphorus.

Health Programs and Treatment Procedures

There is no single vaccination or control program which is best for all feedlots. The age, size, condition, health, and prior history will affect your decisions related to handling a particular group of cattle. It helps to know when a group of cattle were weaned, how they had been handled, and their vaccination history. L. H. Newman, Extension Veterinarian at Michigan State University, helped prepare these guidelines for a feedlot health program.

A number of excellent texts are available on feedlot diseases. The reader should consult one of them for detailed description, symptoms, and treatment procedures for specific diseases. This section will describe only the most common health problems encountered in the feedlot and how to deal with them.

Most cattle arriving in the feedlot should receive any injections you plan to give them as they come off the trucks, or as soon thereafter as possible. These cattle are most likely to become ill during their first three weeks in the feedlot. It simply does not make sense to excite and handle a group of cattle, some of which may be in the incubation stages of illness, that are just starting to settle down, eat, and drink. In approximately three weeks the cattle can be handled a second time or given injections which were not given upon arrival.

Most cattle should receive infectious bovine rhinotracheitis (IBR) parainfluenza 3 (PI_3), and leptospirosis vaccine upon arrival at the feedlot, regardless of prior vaccination history. There is an advantage in purchasing calves which have received IBR, PI_3, leptospirosis, and pasteurella vaccines at least three weeks prior to the time they were weaned. Calves which have not been dehorned and/or castrated may not have received any vaccine or bacterin. If you receive a group of cattle with a number of calves that need to be dehorned and castrated, you probably should include bacterins for blackleg and malignant edema (*Clostridium chauvei septicum*) in your vaccination program.

Most cattle should receive one to two million units of vitamin A by injection when they enter the lot. Vitamins D and E are also present in most of these intramuscular products. Bovine virus diarrhea (BVD) vaccination may not be necessary in all feedlots. However, the economic loss when this disease affects a group of cattle which are totally at risk may be severe. In a feedlot in which BVD is known to be a problem BVD vaccination should be accomplished upon arrival. Most feeders that have not had prior problems with BVD, if they elect to vaccinate for BVD, should vaccinate their cattle three weeks after arrival. Cattle should not be vaccinated for BVD in the face of a disease outbreak.

Some feedlots have a history of problems with enterotoxemia and/or the "blackleg-like" diseases. In these feedlots it may be advantageous to vaccinate the cattle for *Clostridium chauvei, septicum, novyi sordellii,* and *perfringens* B, C, and D. These organisms are all available in a single product (bacterin) which should be administered three weeks after arrival.

Lice, mange, and grubs must be controlled if optimum gains are to be obtained. Pour-on organophosphate products for control of cattle grubs and lice are satisfactory if they can be used prior to the cut-off date for the area where the cattle originated. This is best accomplished three weeks after the cattle arrive in the lot, but can be done upon arrival if waiting will result in going beyond the cut-off date for a given group of cattle. Spraying or dipping should be done after the cattle have been in the lot for three or more weeks. Spraying, dipping, or lice pour-ons are the only satisfactory methods for external parasite control during periods when the organophosphate grub control products should not be used.

Most calves will benefit from deworming. In some feedlots it is possible to collect fecal samples from ten calves upon arrival and estimate the parasite load from the parasite egg counts. Identification of the parasite eggs may also indicate which anthelmintic product would be most effective. Worming is most easily accomplished by feeding the anthelmintic after the ration has stabilized and all cattle are eating well. Boluses, drenches, or injections may be administered approximately three weeks after the cattle enter the lot. However, when boluses or drenches are administered care must be taken to avoid injury to the throat area.

Rarely should antibiotics or sulfa drugs be administered to arriving cattle. There are some situations where, based on the previous history of the feedlot or the health of the arriving cattle, it may be desirable to use feed containing aureomycin and sulfamethazine or drinking water containing sulfathiazide. Routine injection of antibiotics to all cattle is undesirable in most situations.

Branding may be accomplished upon arrival in the feedlot or when the cattle are handled approximately three weeks later. Calves should be castrated three weeks after entering the lot if they are doing well. Calves may also be dehorned at this time; however, dehorning often sets the calves back and may not be economically advantageous.

Handling Sick Cattle

One of the most serious problems in the feedlot is early detection and treatment of sick animals. Early detection and treatment is essential; as little as twelve hours may mean the difference between a rapid recovery and a chronic poor-doing calf that results in an economic loss.

A schedule of surveillance and observation for signs of illness should be established and initiated on the day of arrival. The entire group of cattle should be closely observed for signs of illness at least three times a day for a minimum of two weeks. The best time to observe cattle is following feeding.

It is essential that one person be delegated the responsibility to observe the cattle and to identify those animals that are to be handled individually. He must be familiar with the common signs of illness. Some of the signs of illness which we look for (many of these signs are often present in the same calf) include:

1. An elevated body temperature (103.5°F or higher)
2. Drooping of one or both ears
3. A drooping head or carrying the head in an abnormal position
4. Reluctance to rise
5. Reluctance to move about
6. Failure to come to the feed bunk
7. A gaunt appearance
8. A stiff gait; dragging the hind feet when walking
9. Abnormal discharges from eyes or nose
10. Dull or sunken eyes
11. Respiratory signs including: a dry crusted muzzle, pus discharging from nose, a harsh dry cough, and/or rapid or labored breathing
12. Diarrhea; occasionally containing mucus or blood
13. Dehydration and a rough hair coat

Sick cattle must be treated individually as soon as clinical signs appear. These cattle will not consume adequate antibiotics or sulfa drugs to treat the condition by placing drugs in either the feed or water. They must be treated for at least three consecutive days regardless of the amount of improvement shown at the end of 24 hours. Animals that are missed when they show the first signs of illness and are not detected until twelve hours later may require treatment over a period of five days or more. When large numbers of animals are becoming ill, or if over 25 percent of the cattle are sick at any one time, treatment of those cattle that are not yet sick by placing sulfathiazole in the drinking water or a product such as Aureo S-700 in the feed may be indicated.

Treatment of disease is one area in which the use of a professional can be an exceptionally good investment. Although we can provide a number of guidelines, you should rely upon your veterinarian to help you select the proper drugs, dosages, and treatment schedule. Veterinary assistance is extremely important in arriving at a diagnosis, utilizing diagnostic tests, and performing postmortem examinations. And these, in turn, are essential to establish an accurate diagnosis and arrive at effective treatment regimens.

Bovine respiratory disease (BRD) is the most common disease complex encountered early in the feedlot period. The therapeutic agents most commonly employed in the treatment of BRD include penicillin, penicillin and dihydrostreptomycin in combination, penicillin-dihydrostreptomycin and a corticosteroid in combination, and sulfamethazine. Treatments which include a corticosteroid often result in a more dramatic response than treatments which do not include the corticosteroid. However, because the corticosteroid has a depressing effect upon the immune system of the calf, it is important that the corticosteroid be used only the first one or two treatments and that treatment without the corticosteroid be continued for at least two or three additional days.

We most often elect to treat with either antibiotics or sulfa drugs. There are some situations in which a veterinarian may, based on his professional judgment, elect to use antibiotics and sulfas together. The dosage of sulfamethazine, either orally or intravenously, necessary for effective treatment, is one and one-half grains per pound of body weight the first day followed by one grain per pound of weight for an additional two to four days. This would be 1-1/2 fifteen gram boluses per 225 pounds of body weight the first day followed by one bolus per 225 pounds each additional day. The 12-1/2 percent solution contains 2 grains per cc; the 25 percent solution contains 4 grains per cc; hence, a 500 pound calf would receive intravenously 375 cc of the 12-1/2 percent solution the first day, followed by 250 cc each additional day. Under no circumstances should sulfa drugs be continued beyond five days, or those dosages be exceeded, without making a change in the sulfa utilized or combining more than one sulfa. Damage to the kidneys can be caused by excessive dosages of a sulfa or prolonged use of the sulfa. Adequate water must be provided to cattle on sulfa drugs.

The most common antibiotic injected by feedlot operators is a combination of penicillin and dihydrostreptomycin. These products are generally least expensive, most easily administered, and most effective. If they are not doing the job, you need professional help. Most products of this nature contain 200,000 units of penicillin per cc of product. In order to be effective this product must be given at a dosage rate of 2 cc per hundred pounds of body weight *twice a day*. Lower dosages or dosages given less frequently result in too low a level of the antibiotics being maintained in the blood stream. As with the sulfa drugs, early detection of the sick animal, early treatment, and maintaining an adequate blood level of the antibiotics for at least three days are essential to satisfactory treatment. The likelihood of relapses, chronically poor-doing cattle, chronic pneumonia, and organisms which gain resistance to the antibiotic being used are increased when inadequate levels of the therapeutic agent or inadequate duration of treatment are employed.

Drugs are not the solution to all problems. Switching drugs back and forth may not allow a product time to be effective (especially if you missed the calf when it first became ill). Postmortem examination of all dead animals, histopathology, culturing tissues, and drug sensitivity tests may save you thousands of dollars. Do these things first, not when all else has failed. Your veterinarian can help you set up the treatment program that will be most effective in your lot based on this kind of information. He can tell you which drugs to use, what dosage is best, and how often and how long you must continue treatment.

The utilization of sick pens, recovery pens (to get the animals back on feed), and individual marking systems, all under the supervision of one individual who works closely with the veterinarian, markedly enhance the likelihood of a successful treatment program.

Summary of Health Management for Feedlot Cattle

1. New feeder cattle
    a. Vaccinate within the first 48 hours for IBR, leptospirosis, PI_3, and pasteurella. Inject Vitamin A, implant with a growth stimulant, ear tag or brand.
    b. Use pour-on or dip to control lice, mange, and grubs. After cut-off date, use pour-on for lice only. Compounds used for lice, tick, and grub control include ronnel, coumaphos, ruelene, trichlorfon, famphur, prolate, fenthion, methoxychlor, toxaphene, dioxathion, Malathion®, ciodrin, delnav, pyrethrins.

c. At 3-4 weeks after arrival, give pasteurella booster and vaccinate for blackleg, worm, castrate, and dehorn if needed. Compounds used for worming include phenothiazine, levamisole hydrochloride, thiabendazole, and organic phosphate.

d. Observe at least three times daily for the first three to four weeks for signs of sickness. Treat sick cattle as described previously.

2. Cattle on feed

   a. Acute indigestion—caused by overeating. Animal may kick at stomach, stagger, and/or go down. Give ½ lb baking soda/500 lbs body weight by stomach tube. Place on hay for a few days, then slowly bring back to a full feed.

   b. Founder—hooves get long and feet get tender, unsteady gait, due to overconsumption of grain. Treat with mineral oil and magnesium oxide, antihistamine, and cortical steroids. May be necessary to sell animals that have reduced performance. Increase roughage level, then start back on feed slowly. If a continual problem, don't increase grain more than ½ lb/head daily to get on full feed, always include 10-20 percent roughage in the ration, and never let the cattle run out of feed.

   c. Bloat—usually due to overconsumption of grain, legumes, or potatoes. Bloat can be caused by adhesions in stomach or constrictions in esophagus. Prolonged high levels of antibiotic or sulfa drugs can cause bloat. For treatment of bloat caused other than from legumes, relieve gas by inserting a hose down the throat into the rumen. (When inserting hose, do not force if resistance is felt or if hose can get into lungs. A gurgling noise will indicate the hose is in the rumen. The release of gas sounds like a rush of air from an air compressor.) Mineral oil can be given. If caused by legumes, give a surface tension reducer. If problem persists, a permanent trochar can be inserted, or the animal should be sold. Prevention is the same as for founder.

   d. Lameness—can be caused by mechanical injury (wire, nails, sprain, etc), founder, or foot rot. Check foot for wire or a nail. If a sprain, lameness usually disappears within a few days. Foot rot is caused by an infection of the tissue around and within the hoof. Swollen feet, necrotic areas between toes or at the heels, and lameness are signs of foot rot. To treat, the affected area is cleaned, and with more valuable animals, the foot is treated with antibiotics and bandaged. Antibiotics are injected for at least three days. Keeping cattle out of mud is the best prevention. Low level feeding of antibiotics and/or ethylene diamine dihydroiodide (EDDI) will aid in prevention of foot rot. Compounds used for treatment include oxytetracycline, penicillin, streptomycin, chlortetracycline, Tylosin, sulfapyridine, sulfamethazine, or sulfamerazine.

e. Bursitis or abscesses—swelling or wounds caused by bruising or injury; usually in brisket, shoulder, and hips. Often caused by bottom board of feedbunk being too high. In severe cases, the wound is lanced (at the bottom for drainage); the necrotic tissue is removed; and the wound is cauterized with iodine.

f. Buller steers—can be caused by indigestion, calculi, implants, or for no observable reason. Isolate individuals or put in with pen of heifers. Can remove the implant, smear the tail and rump with F-R-S, then return to original lot during morning feeding.

g. Lump jaw—the presence of hard tissues in the head. Iodine compounds given in the feed are used for prevention and treatment.

h. Polioencephalomalacia (Polio)—Blindness, off feed and water, down and/or convulsions, and sore joints are symptoms. Massive doses of tetracycline and thiamine are given IV.

i. Pinkeye—redness around the eye, ulceration on the eyeball, and cornea rupture are signs. A topical aerosol is applied in mild cases. In severe cases (where more than half of the cornea is cloudy) ½ cc of Methagan and ½ cc of penicillin is injected subconjunctivally.

j. Allergic reactions—sometimes occur after a vaccine or antibiotic injection. Difficult breathing, staggering, or animal is down. Immediately inject 5 cc of epinephrine solution intravenously.

k. Uterine infections—an odorous vaginal discharge appears, often caused by an abortion. Place 8 Terramycin® calf scour pills (250 mg/pill) in the uterus and inject antibiotics.

l. Urinary calculi—caused by kidney stones blocking the urinary tract. Signs are swollen underline, kicking at the belly, straining. If the bladder is not tense, administer penicillin daily and have the veterinarian examine. If the bladder is tense (determined by rectal examination) but the steer is not showing signs of toxicity or does not have swelling around the sheath, it should be slaughtered that day. If the animal is showing signs of toxicity, surgery must be performed as quickly as possible.

m. Diarrhea—can be caused by coccidiosis, salmonellosis, BVD, or overeating. If overeating, increase roughage level, then work back up on feed. If elevated temperature, severe diarrhea, putrid smell, blood in feces, abdominal pain, the following products are used for treatment: streptomycin, neomycin, tetracycline, nitrofurazone, ampicillin, and electrolytes.

Summary of Commonly Available Commercial Products Containing Compounds Used for Treating Cattle

	Compound	*Commercial Product*
Antibiotics	Chlortetracycline	Aureomycin® Chloratel 10 Chlorachel
	Erythromycin	Erythromycin Injectable Gallimycin Injectable
	Oxytetracycline	Terramycin® Biocycline Oxy Oxytet 50 Super Booster Crumbles Super Sweet Oxyject
	Penicillin	Crysticillin® 300 A.S. Penicillin
	Penicillin-Dihydrostreptomycin	Combiotic Penstrep Distrycillin Pro-Penstrep Durbiotic

	Tylosin	Tylosin
	Neomycin	Neomix
		Neomix plus
		Biosol
		Biosol-M
	Neomycin + Oxytetracycline	Super TN
		Neo-Terramycin
	Nitrofurasine	NF 2 Puffer
		Furacin® Soluble Powder
	Ampicillin	Ampicillin
Sulfas	Sulfamethazine	Metzol 25%
		Sulmet
		Spanbolet
		TS 543
	Sulfathiazole	Sulyte
		T Sulfa Boluses 699
		T Sulfa Boluses 888
	Sodium Sulfathiazole	Meralite
		Hog and Cattle Sulfa
		Sultrol-E
		Sul-Thi-Zol
		Flue-Medic
	Sulfadimethoxine	Alban Boluses P.O.
	Sulfabromomethazine Sodium	Sulfabrom
	Sulfaquenoxaline	S.Q. Solution 20%
	Chlortetracycline-Sulfamethazine	Aureo S 700
Organic Iodine		Multidine
(foot rot and lumpy jaw)		EDDI Poladide-20
		EDDI Poladide-40
		EDDI
		Biodine
		Ancho-Dyne
Sodium Iodide		Sodium Iodide Injection
		Sodium Iodide Solution
		Sodium Iodide 20% Solution
		Iodosol

Ethylenediamine Dihydroiodide   "Hi Boot" Salt

Pinkeye Treatment
    Furazolidone   Topazone
                                   Furox Aerosol

    Methol Violet   Pink Eye Spray
                                   Spra-Optic
                                   Box Optic
                                   Trisulfamol

Internal Parasites
    Phenothiazine   Pharmaceutical Grade
                                          Phenothiazine
                                   Phenothiazine Wettable
                                        Powder
                                   Bar Fly Phenothiazine
                                   Phenothiazine Boluses
                                   Phenothiazine Drench
                                   Phenothiazine Drench with
                                        Lead Arsenate

    Coumaphos   Baymix Crumbles
                                   Baymix 11.2 Feed Premix

    Levamisole   Tramisol

    Thibendazole   Thibenzole, TBZ
                                   Omnizole

    Haloxon   Loxon

    Levamisole HCL   Cattle Wormer with
                                   Tramisol

External Parasites (lice, ticks, grubs)
    Ronnel   Korlan 2 Pour-on
                                   Korlan 24E
                                   Tiolene 18
                                   Korlan Smear

    Famphur   Warbex

    Coumaphos   Co-Ral 25% W.P.
                                   Co-Ral E.L.I.
                                   Co-Ral Pour-on
                                   Zipcide Cattle Dust Bag

    Fenthion   Pfizer Dust Bag
                                   Tiyuvan Pour-on

Toxaphene	Livestock Insecticide Rub
	Dry Cide
	Lintox-X
	Copper Tox
Trichlorfon	Neguvon Pour-on
	Ox-130
Prolate	GC-118
Malathion®	Mange and Lice Control
	Malathion® Compound
Gardona	Rabon Livestock Dust
Ruelene	Ruelene 25E
	Ruelene 35D
	Ruelene 12R
Delnav	Delnav Extra
	DEL Tox
Ciodrin	Simax
	Pest-O-Kill
	Lindane 20%
	Livestock Insecticide Powder
	Zipcide Dust Bags
Dioxathion	Cutter-Tox-D
Co-Ral	Pfizer Dust Bag
	Supersweet Cattle Duster

8

Nutrient Requirements of Beef Cattle

The nutrient requirements of cattle depend on their weight, stage of growth or reproduction, and rate of growth. All of the nutrients are usually required in proportion to each other. Thus, it is uneconomical to feed quantities of any nutrient in excess of the most limiting nutrient.

Energy and protein are the nutrients required in the greatest quantity, and therefore are the most costly. The amounts needed of each of these should be established first, and then the level of minerals and vitamins should be added in the correct amounts to ensure maximum use of the energy and protein.

To feed cattle for optimum performance, their nutrient requirements, the availability of these nutrients from various sources, and how to build least cost feeding programs must be understood. This section was developed to provide this information in a form that can be easily understood and utilized, based on experience working with various sizes of cattle operations. General nutrient requirements for all cattle will be discussed first, followed by those specifically for growing and finishing cattle. Then specific nutrient requirements and guidelines for feeding the breeding herd will be discussed.

8-1 IMPACT OF FRAME SIZE ON NUTRIENT REQUIREMENTS

The nutrient requirements of growing cattle at any weight depend on their frame size or mature weight. Large-frame cattle mature at heavier weights and thus are less fat at the same actual weight as small-frame cattle. Table 8-1 gives weights at which various frame sizes, breeds and types of cattle are equivalent in percent of fat and protein, as estimated by D. G. Fox and co-workers. This weight

Table 8-1
Weights at Which Various Frame Sizes of Growing Cattle Have Similar Nutrient Requirements

		Empty body composition, %						
Fat	14.9	17.2	19.5	21.8	24.2	26.5	28.8	
Protein	19.5	19.1	18.6	18.1	17.6	17.1	16.5	
			— Shrunk weight, lb —					
Frame code			Steers					Breed and type
1	400	480	560	640	720	800	880	Small-frame British
2	425	510	595	680	765	850	935	
3	450	540	630	720	810	900	990	
4	475	570	665	760	855	950	1045	Average-frame British
5	500	600	700	800	900	1000	1100	
6	525	630	735	840	945	1050	1155	Large-frame British
7	550	660	770	880	990	1100	1210	Average-frame European
8	575	690	805	920	1035	1150	1265	British x European
9	600	720	840	960	1080	1200	1320	Large-frame European, Holstein
			Heifers					
1	320	385	450	510	575	640	705	Small-frame British
2	340	410	480	540	610	680	750	
3	360	435	510	575	645	720	795	
4	380	455	535	610	685	760	840	Average-frame British
5	400	480	560	640	720	800	880	
6	420	500	585	670	755	840	920	Large-frame British
7	440	525	610	705	790	880	965	Average-frame European
8	460	550	640	735	830	920	1010	British x European
9	480	575	670	770	865	960	1055	Large-frame European, Holstein
			Bulls					
1	480	575	670	770	865	960	1055	Small-frame British
2	510	610	715	815	920	1020	1120	
3	540	650	755	865	970	1080	1190	
4	570	685	800	910	1025	1140	1255	Average-frame British
5	600	720	840	960	1080	1200	1320	
6	630	755	880	1010	1135	1260	1385	Large-frame British
7	660	790	925	1055	1190	1320	1450	Average-frame European
8	690	830	965	1105	1240	1380	1520	British x European
9	720	860	1010	1150	1300	1440	1585	Large-frame European, Holstein

8-1 Impact of Frame Size on Nutrient Requirements 179

will be referred to as their "equivalent weight"; the weight at which various types of cattle are similar to an average-frame steer in body composition and thus nutrient requirements. These frame sizes and codes will be used to indicate adjustments in nutrient requirements needed. Figure 8-1 shows examples of cattle of eight different frame sizes, to use as a guide in determining the frame size of the cattle in question.

The relationship of frame size to nutrient requirements and carcass composition will be discussed in detail in sections on energy and protein requirements and buying and selling cattle. New grade standards for feeder cattle using small, medium, and

Four cattle types, with frame codes in parentheses (left to right): small Angus (1), Chianina crossbred (9), average Angus (4) and Holstein (9).

Four additional cattle types at 28-30% body fat: Charolais crossbred (9), Holstein crossbred (9), selected Hereford (7) and unselected Hereford (3).

Figure 8-1
Relationship of Breed and Type Within Breed to Frame Code

large frame size as criteria are under consideration. These will allow relating requirements and weight at the fatness of low choice to a feeder calf grade.

Feed composition values, energy, and mineral requirements are based in part upon modifications of those published in the National Research Council publication "Nutrient Requirements of Beef Cattle" (fifth revised edition). Energy requirements for different cattle types, protein requirements, and adjustments for different conditions were developed by D. G. Fox and co-workers based on recent research results at various universities.

This section of the text is more detailed than other sections. Feed costs represent 70-80 percent of the cost of producing beef, and are the area where individual producers can have the greatest impact in reducing costs of production. Proper ration formulation and selecting the most profitable feeding system are major keys to a successful beef operation.

8-2 WATER REQUIREMENTS AND WATER QUALITY FOR BEEF CATTLE

The water requirements of cattle are influenced by a number of physiological and environmental conditions. These include such things as the rate and composition of gain, pregnancy, lactation, physical activity, type of ration, salt and dry matter intake, and environmental temperature.

The minimum requirement of cattle for water is a reflection of that needed for body growth, for fetal growth or lactation, and of that lost by excretion in the urine, feces, or sweat, or by evaporation from the lungs or skin. Anything influencing these needs or losses will influence the minimum requirement.

The amount of urine produced daily varies with such things as activity of the animal, air temperature, and water consumption, as well as with certain other factors. The antidiuretic hormone, vasopressin, controls reabsorption of water from the kidney tubules and ducts, and thus it affects urine excretion. Under conditions of restricted water intake, an animal may concentrate its urine to some extent by reabsorbing a greater amount of water than usual. While this capacity for concentration of the urine solutes is limited, it can reduce water requirement some. When an animal consumes a diet high in protein or in salt or containing substances having a diuretic effect, the excretion of urine is increased and so is the water requirement.

The water lost in the feces depends largely on the diet and the species. For instance, substances in the diet which have a diuretic effect will increase water loss by this route, and cattle excrete feces of a high moisture content while sheep excrete relatively dry feces.

The amount of water lost through evaporation from the skin or lungs is not obvious to us, but it is important and in some cases it may even exceed that lost in the urine. If the environmental temperature and/or physical activity increase, water loss through evaporation and sweating increases.

From a practical point of view, all of these factors and their interplay make the minimum water requirement difficult to assess. And still another matter adds to this difficulty. Since feeds themselves contain some water and since the oxidation of certain nutrients in feeds produces water, not all must be provided by drinking. Feeds such as silages, green chop, or pasture are usually very high in their moisture content while grains and hays are low, and high energy feeds produce much metabolic water while low energy feeds produce little. These are obvious complications in the matter of water requirements. Fasting animals or those on a low protein diet may form water from the destruction of body protein or fat, but this is of minor significance.

Water requirements have been measured in a practical way by many investigators by determining voluntary water intake under a variety of conditions. In brief, the results of these studies imply that thirst is a result of need and that animals drink to fill this need. The need results because of an increase in the electrolyte concentration in the body fluids which activates the thirst mechanism. There is experimental evidence which supports this reasoning.

As this discussion suggests, water requirements are affected by many factors and it is impossible to list specific requirements with accuracy. However, the major influences on water intake in beef cattle on typical rations are dry matter intake, environmental temperature, and lactation, and Table 8-2 has been designed with this in mind. It is a guide only, and it must be used with considerable judgment.

Livestock Water Quality

A successful livestock enterprise requires a good water supply, in terms both of quantity and quality. While shortage is obvious to the stockowner, he sometimes needs the help of a laboratory in evaluating the quality of a supply.

Most ground or surface waters are satisfactory for livestock.

Table 8-2
Estimated Daily Water Intake of Cattle

Month	Mean Temp.	Cows Nursing Calves[1]	Cows Bred Dry Cows & Heifers	Bulls	Growing Cattle			Finishing Cattle			
					400Lb.	600Lb.	800Lb.	600Lb.	800Lb.	1000Lb.	1200Lb.
	°F	GAL	GAL	GAL	GAL	GAL	GAL	GAL	GAL	GAL	GAL
Jan.	36	11.0	6.0	7.0	3.5	5.0	6.0	5.5	7.0	8.5	9.5
Feb.	40	11.5	6.0	8.0	4.0	5.5	6.5	6.0	7.5	9.0	10.0
Mar.	50	12.5	6.5	8.6	4.5	6.0	7.0	6.5	8.0	9.5	10.5
April	64	15.5	8.0	10.5	5.5	7.0	8.5	8.0	9.5	11.0	12.5
May	73	17.0	9.0	12.0	6.0	8.0	9.5	9.0	11.0	13.0	14.5
June	78	17.5	10.0	13.0	6.5	8.5	10.0	9.5	12.0	14.0	16.0
July	90	16.5	14.5	19.0	9.5	13.0	15.0	14.5	17.5	20.5	23.0
Aug.	88	16.5	14.0	18.0	9.0	12.0	14.0	14.0	17.0	20.0	22.5
Sept.	78	17.5	10.0	13.0	6.5	8.5	10.0	9.5	12.0	14.0	16.0
Oct.	68	16.5	8.5	11.5	5.5	7.5	9.0	8.5	10.0	12.0	14.0
Nov.	52	13.0	6.5	9.0	4.5	6.0	7.0	6.5	8.0	10.0	10.5
Dec.	38	11.0	6.0	7.5	4.0	5.0	6.0	6.0	7.0	8.5	9.5

[1] Cows nursing calves during first 3 to 4 months after parturition – peak milk production period.

Table prepared by Paul Q. Guyer, University of Nebraska

Some are not, however, resulting in poor performance or even death in animals confined to them.

What makes waters unsatisfactory for livestock? Very often it is excessive salinity—too high a concentration of dissolved salts of various kinds. Of lesser importance is nitrate content, and on rare occasions alkalinity or other factors may become involved.

Salinity. Water is a very good solvent, and all natural waters contain dissolved substances. Most of these are inorganic salts, the calcium, magnesium and sodium chlorides, sulfates, and bicarbonates predominating. Occasionally these are present in such high concentrations that they cause harmful osmotic effects resulting in poor performance, illness, or even death in animals confined to them. The various salts have slightly different effects, but these differences are of no practical significance. Thus, while sulfates are laxative and may cause some diarrhea, their damage to the animal seems no greater than that of chlorides, and magnesium salts seem no more of a problem than calcium or sodium salts. Further, the effects of the various salts seem additive, which means that a mixture of them seems to cause the same degree of harm that a single salt at the same total concentration does.

A number of observations have been made relative to saline livestock waters, some of them verified by experimental findings. At high salt concentrations that are somewhat less than toxic, increasing salinity may actually cause an increased water consumption, even when at first the animals refuse to drink for a short period of time. On the other hand, at very high salinities animals may refuse to drink for many days, followed by a period where they drink a large amount at one time and become suddenly sick or die. Older animals are more resistant to harm from salinity than are the young. Anything causing an increase in water consumption such as lactation, high air temperatures, or exertion also increases the danger of harm from saline waters. Animals do seem to have the ability to adapt to saline water quite well, but abrupt changes from waters of low to waters of high salts concentrations may cause harm while gradual changes do not. Whenever an alternate source is available to them, even every two or three days, livestock will avoid excessively saline waters. And finally, when animals suffering from the effects of saline water are allowed water from a source of low salts content they make a rapid and complete recovery.

Salt is sometimes used in feeds to regulate their intake. Special care to supply a drinking water of low salt content should be taken in these instances.

Nitrates. The poisoning of cattle by nitrates was first observed prior to 1900, and there have been many cases since. As a rule, it results from their eating forages of high nitrate content. The nitrates are not themselves very toxic, but in the rumen the bacteria reduce them to nitrites which then get into the bloodstream. There the nitrites convert the red pigment, hemoglobin, which is responsible for carrying oxygen from the lungs to the tissues, to a dark brown pigment, methemoglobin, which will not carry oxygen. When this conversion is about 50 percent complete, the animal shows signs of distress suggesting a shortage of breath, and at 80 percent or more conversion the animal usually dies from a type of suffocation.

Nonruminants may convert small amounts of ingested nitrate to nitrite in their intestines, but the amount so converted is not harmful. It has been found that nitrates in the diet may interfere in the conversion of carotene to vitamin A under some circumstances, but an impressive amount of experimental data shows this to be of no practical significance. Further, the experimental evidence suggests that nitrate poisoning does not occur in livestock and that the young are no more susceptible to the acute type than are older animals. Nitrates are occasionally found at toxic levels in water. Nitrites are also found in water on many occasions, but not at levels dangerous to livestock. As a rule, reports of water analyses include nitrites with the nitrates.

Alkalinity. Many and perhaps most waters are alkaline. This is fortunate since, if they were acid, they would corrode pipes and plumbing. Only in a very few instances have they been found too alkaline for livestock. Alkalinity is expressed either as pH or as titratable alkalinity in the form of bicarbonates and carbonates. A pH of 7.0 is neutral, below that is acid and above that is alkaline. Most of our waters have pH values between 7.0 and 8.0, which means that they are very mildly alkaline, and this further means that they contain only bicarbonates and no carbonates. As the pH goes up, the waters become more alkaline, and at values of around 10, waters are very highly alkaline and they contain carbonates. Most waters have alkalinities of less than 500 parts per million, and these are not harmful.

Excessive alkalinity in their water can cause physiological and digestive upset in livestock. The level of alkalinity at which it begins to become troublesome and its precise effects have not been thoroughly studied. Therefore, the establishment of guidelines to the suitability of alkaline waters for livestock is difficult.

Other factors. On rare occasions, natural waters may contain or become contaminated with certain toxic elements such as arsenic, mercury, selenium, cadmium, etc, or radioactive substances. While these may harm animals that drink these waters, our major concern is that they do not accumulate in the meat, milk, or eggs, making them unsafe for human consumption. These are analyzed for only when there is good reason to suspect their presence at excessive levels.

Persistent organic pesticides have been found as contaminants in most surface waters. However, their concentration is so small in these waters (because of their low solubility in water) that they have been found to be no problem to livestock.

Occasionally, heavy algal growths occur in stagnant or slow-flowing bodies of water. A few species of these can, under some circumstances, be toxic. We have no tests for these toxins, and at present we can only recommend avoiding using any stagnant source of water for livestock.

While we have no meaningful laboratory methods to measure it, filth in livestock waters must obviously be avoided. A reasonable effort should always be made to provide animals a clean and sanitary supply.

Interpreting a Water Analysis

Salinity. A guide to the use of saline waters for livestock is presented in Table 8-3. Considerable judgment should be exercised

8-2 Water Requirements and Water Quality for Beef Cattle   185

in using this guide. It has built into it reasonable margins of safety, and adherence to it should prevent deaths or economic losses with rare exceptions.

Nitrates. Comments relating to the use of waters containing nitrates are shown in Table 8-4. In using this table, it is important to take into account the way in which the nitrate content is expressed on the report of analysis. Some express it in parts per million (ppm) of nitrate nitrogen ($NO_3\,N$). Others express it as

Table 8-3

Guide to the Use of Saline Water for Livestock

Total dissolved solids (milligrams/liter or parts/million)*	Comments
Less than 1000	From the standpoint of its dissolved solids, this water should be excellent for all classes of livestock.
1000 to 2999	This water should be satisfactory for all classes of livestock. Those waters approaching the upper limit may cause some watery droppings in poultry, but they should not adversely affect the health or production of the birds.
3000 to 4999	This water should be satisfactory for livestock. If not accustomed to it they may refuse to drink it for a few days, but they will in time adapt to it. If sulfate salts predominate, they may show temporary diarrhea, but this should not harm them. It is, however, a poor to unsatisfactory water for poultry. It may cause watery feces, and particularly near the upper limit it may cause increased mortality and decreased growth, especially in turkey poults.
5000 to 6999	This water can be used for livestock except those that are pregnant or lactating, without seriously affecting their health or productivity. It may have some laxative effects and be refused by the animals until they become accustomed to it. It is unsatisfactory for poultry.
7000 to 10,000	This is a poor livestock water that should not be used for poultry or swine. It can be used for older, low-producing ruminants or horses that are not pregnant or lactating with reasonable safety.
Over 10,000	This water is considered unsatisfactory for all classes of livestock.

*Electrical conductivity expressed in micromhos per centimeter at 25° can be substituted directly for total dissolved solids without introducing a great error in interpretation.

parts per million of nitrate (NO_3) or of sodium nitrate ($NaNO_3$). The relationship between these various methods of expressing it are as follows: 1 ppm of nitrate nitrogen = 4.43 ppm of nitrate or 6.07 ppm of sodium nitrate. With livestock waters having a total dissolved solids content of less than 1000 ppm or a conductivity of less than 1400 micromhos/cm at 25°C, there is no need to make a nitrate determination.

Table 8-4

Guide to the Use of Waters Containing Nitrate for Livestock and Poultry

Nitrate content* (ppm nitrate nitrogen)	Comments
Less than 100**	Experimental evidence to date indicates that this water should not harm livestock or poultry.
100 to 300	This water should not by itself harm livestock or poultry. When feeds contain nitrates, this water could add greatly to the nitrate intake to make it dangerous. This could be of some concern in the case of cattle or sheep during drought years and especially with waters containing levels of nitrates that approach the upper limits.
Over 300***	This water could cause typical nitrate poisoning in cattle and sheep, and its use for these animals is not recommended. Because this level of nitrate contributes significantly to salinity and also because experimental work with levels of nitrate nitrogen in excess of this are meager, the use of this water for swine, horses or poultry should also be avoided.

* includes nitrite nitrogen.
** Less than 443 ppm of nitrate or less than 607 ppm of sodium nitrate.
*** Over 1329 ppm of nitrate or over 1821 ppm of sodium nitrate.

Alkalinity. Waters with alkalinities of less than 1000 ppm are considered satisfactory for all classes of livestock and poultry. Above that concentration, they are probably unsatisfactory, although for adults they may do little harm at concentrations less than about 2500 ppm unless carbonates are present in excess over bicarbonates.

Miscellaneous. Waters may in some instances supply a portion or even all of an animal's requirement for certain minerals. As a general rule, however, their contribution with respect to minerals is of no practical significance.

Hard waters have often been suggested as a cause of urinary calculi (kidney stones or water belly). Experimental evidence shows that this is not true, however, and hardness might, in fact, actually contribute to the prevention of certain types of calculi formation.

The results of water analyses have been expressed in a number of ways. Some of these ways and their interrelation are shown below:

- One *part* per million (ppm) means one pound per million pounds of water. For all practical purposes, *milligrams per liter* (mg/l), *milligrams per kilogram* (mg/kg), and *parts per million* (ppm) mean the same thing.
- One *grain per gallon* is equivalent to about 17 *parts per million*.

Highly saline waters are often mistakenly referred to as "alkali" waters. They may or may not be highly alkaline, and usually they are not. Sometimes they are referred to as hard waters. If most of their salinity is in the form of sodium salts, however, they may actually be soft waters, as hardness is due largely to calcium and magnesium.

8-3 MINERAL REQUIREMENTS OF BEEF CATTLE

Minerals are needed by cattle for both maintenance and growth, due to their involvement in various enzyme systems and chemical reactions that occur in the body tissue. They also are deposited in bones and teeth, and are therefore needed for bone and teeth formation. The amounts that are needed are primarily dependent on stage of growth or reproduction. The amount and kind of minerals that need to be fed in addition to the major feeds in the ration, however, depend on the kind and source of feedstuffs and the soil mineral deficiencies in various areas. The only minerals in addition to salt that are usually deficient in rations for beef cattle are calcium, phosphorus, potassium, and sulfur. In certain areas, iodine, cobalt, and selenium may also be deficient. Feeding trace mineral salt will usually safeguard against deficiencies of minerals other than calcium, phosphorus, potassium, and sulfur unless the soils in an area are known to be specifically deficient in one or more trace minerals. For example, it is known that many midwestern and eastern soils are deficient in selenium or iodine, and special supplementation of these minerals is needed in many specific areas.

Most protein supplements will contain enough trace minerals to satisfy trace mineral requirements, and if a protein supplement is fed to balance the ration for protein, it is not likely that feeding another source of trace minerals would be needed.

As a general guide in determining the need for supplemental calcium, phosphorus, or potassium, grains tend to be high in phosphorus and low in calcium. In addition, corn grain, oats, wheat, and brewers' grains are low in potassium for growing and finishing cattle. Most forages except corn or sorghum silage or mature, weathered grasses tend to be high in calcium, and nearly all forages are high in potassium. Most forages are low in phosphorus, however, especially mature, weathered grasses, and they need to be supplemented with additional phosphorus under most conditions.

The only situation where supplemental sulfur is needed is when part or all of the supplemental protein comes from nonprotein nitrogen and the protein has to be synthesized by bacteria in the rumen of cattle from urea or other nonprotein nitrogen compounds. Under these conditions the protein supplement should contain additional sulfur, and in most supplements of this type the addition of 4 lbs of elemental sulfur per ton should be adequate.

The following discussion summarizes the mineral requirements of cattle, their major functions and deficiency symptoms, and supplementation normally required under most conditions. This discussion will be followed by a table that summarizes the mineral requirements of beef cattle and some general free choice mineral supplement formulations.

Major Minerals

Calcium. Calcium is required in the greatest quantity for bone growth. It is also needed in constant amounts for various metabolic functions in the animal, including formation and maintenance of the skeletal structure, blood clotting, and muscle tone and contraction. Therefore, requirements are highest for young, lightweight cattle and decrease to a constant amount as bone growth ceases and maturity is reached, with increases needed for reproduction and lactation. A severe deficiency in very young animals results in rickets. In older animals the usual deficiency symptoms are reduced performance and lameness and stiffness of joints.

Grains, grain by-products, and supplements high in natural protein tend to be low in calcium, and forages tend to be high in calcium. Therefore, when the ration is high- or all-concentrate

8-3 Mineral Requirements of Beef Cattle

or contains mostly corn silage and/or grains, supplemental calcium will be needed. Ground limestone (calcium carbonate) is usually the most economical source in supplying only calcium. It contains about 38 percent calcium and its content of calcium has a high availability. If both calcium and phosphorus are need, then dicalcium phosphate may be the most convenient and economical source.

Formulating the ration to contain a total of 0.40 percent calcium for growing cattle over 600 lbs and 0.55 percent for cattle under 600 lbs (100 percent dry matter basis) should meet the requirements of growing cattle under most conditions. The rations of beef cows being fed for maintenance or gestation should contain a minimum of 0.18 percent calcium on a dry basis, and the ration of lactating beef cows should contain 0.25 to 0.44 percent calcium, on a dry basis.

Phosphorus. Many of the functions of phosphorus in the body are similar to those of calcium, and the greatest quantity is also required for bone growth. Also, constant amounts are required for various metabolic functions such as metabolism of nutrients and energy transfer and storage in the body and maintaining the proper acid-base balance in the blood. As in the case of calcium, the greatest amounts are therefore needed during early growth and decrease to a constant amount as bone growth ceases and maturity is reached, with increases needed for reproduction and lactation. Symptoms of phosphorus deficiency are decreased appetite and reduced performance, followed by a depraved appetite (chewing objects and eating soil) and eventually lameness and stiffness of joints. However, the chewing of objects and eating of soil by cattle on all-concentrate rations or animals held in close confinement should not be confused with a phosphorus deficiency. Lower fertility can also result from a lack of phosphorus.

Most grains are high in phosphorus, and most forages are low in phosphorus. Thus phosphorus supplementation is usually needed in high roughage growing or wintering rations or for beef cows fed low quality forages.

Dicalcium phosphate, defluorinated rock phosphate, steamed bonemeal, or mineral supplements containing both calcium and phosphorus are usually the most convenient and economical sources of supplemental phosphorus, particularly where calcium is also needed. (They contain 23-32 percent calcium and about 18 percent phosphorus.) In situations where the calcium level is already excessive, however, it may be better to use monosodium or disodium phosphate (about 22 percent phosphorus) or similar supplements which contain only phosphorus.

Formulating the ration to contain 0.30 to 0.35 percent phosphorus for growing cattle, 0.18 percent during gestation and 0.25 to 0.39 percent during lactation should be adequate to meet the phosphorus requirements under most conditions.

Care should be taken to avoid phosphorus levels greatly in excess of those given for growing cattle, as an excess phosphorus intake has been shown to cause urinary calculi, particularly if there is an inbalance in the calcium:phosphorus ratio. The desired ratio has been suggested to be between 1:1 to 2:1 due to three factors. First, the calcium:phosphorus ratio is about 2:1 in the body. Secondly, if the ratio is too wide, insoluble tricalcium phosphate compounds may be formed in the digestive tract, resulting in a reduced absorption of both calcium and phosphorus. Thirdly, a ratio outside of these limits has been suggested to be a cause of urinary calculi. This ratio will usually be maintained, however, in formulating most rations where calcium or phosphorus, or both, are below the minimum requirement. The ratio is set from 1.3:1 to 1:1 in the requirements given, as the phosphorus content in feeds and its availability appears to be more variable than calcium.

Some rations such as a high- or all-alfalfa hay ration will have a high ratio of calcium to phosphorus, with the calcium level greatly exceeding the requirements and the phosphorus level being barely adequate. These rations have not been shown to be detrimental to cattle, however, especially in mature cattle or those fed for slow growth.

Potassium. Potassium is required for muscular activity, maintenance of the acid-base balance in the blood, and for osmotic pressure of body fluid. Forages are high in potassium, whereas shelled corn, oats, wheat, and brewer's grains contain borderline or inadequate levels of potassium for growing cattle. Therefore, most rations contain adequate levels of potassium except for high- or all-concentrate rations based on the above grains. Growing and finishing rations should contain 0.6 to 0.8 percent potassium (dry matter basis). Potassium chloride (contains about 50 percent potassium) is usually the most economical means of providing supplemental potassium. Potassium sulfate may be used if supplemental sulfur is also needed.

Deficiency symptoms are nonspecific, but are associated with reduced feed consumption, growth rate and feed efficiency, and stiffness and emaciation.

Magnesium. Magnesium is closely related to calcium and phosphorus in metabolism and maintenance. It is needed in several enzyme systems and for muscle relaxation. (Requirement is 0.04 to 0.20 percent of ration DM.) Most feeds appear to contain adequate amounts of magnesium. However, small grain pastures or rapidly growing grasses may be deficient in available magnesium in the spring and sometimes in the fall. A deficiency results in a disease commonly called grass tetany or grass staggers. Symptoms are a staggered gait, inability to stand or relax muscles, convulsions, and eventually death.

A free choice mineral mixture containing 25 to 35 percent magnesium oxide has been at least partially successful in preventing grass tetany in cattle grazed on tetany-prone pasture. Injections of magnesium chloride have been used successfully to treat cattle showing symptoms of grass tetany, resulting in a relatively rapid rate of recovery if treated promptly at the first sign of symptoms.

Salt (Sodium chloride). The minerals in salt are among the most deficient in livestock rations. Sodium is needed for maintaining the acid-base balance in the blood, osmotic pressure of body fluid, normal muscle tone, and in gastric juice. Chlorine is essential in water balance, osmotic pressure regulation, in acid-base balance in the blood, and in hydrochloric acid in gastric juice. Salt also stimulates appetite, and when deprived of salt, feed intake may rapidly decrease. A salt deficiency results in an abnormal appetite for salt as evidenced by chewing and licking various objects. A prolonged deficiency of salt results in lack of appetite and decreased production.

Most rations should contain 0.25 to 0.5 percent salt (100 percent dry matter basis) or feed free choice as loose or block salt. Iodized salt is usually recommended where an iodine deficiency is known to exist. Also, it may be advisable to use trace mineralized salt to furnish additional trace minerals when no other source of supplemental trace mineral is being fed.

In some areas where soil salinity is a problem due to heavy manure applications, salt levels may have to be reduced below the level recommended here.

Sulfur. Sulfur is a component of certain amino acids in protein. The nitrogen to sulfur ratio is 15:1 in beef protein. Natural protein in feedstuffs provides adequate amounts of sulfur. Deficiencies can occur when high-urea supplements are fed rather than natural protein supplements, and about 3 to 4 lbs of inorganic sulfur should be fed per 100 lbs of urea.

Deficiency symptoms are nonspecific and are the same as for a protein deficiency (slow growth, poor feed efficiency). The requirement is 0.1 to 0.15 of ration DM.

Trace Minerals

Cobalt. Cobalt is needed to synthesize vitamin B_{12} which is used for hemoglobin formation (carries oxygen in the blood), and in the utilization of propionic acid (important energy source produced from fermentation in the rumen). Cobalt-deficient animals appear to be starved, and loss of appetite is an early sign of a deficiency.

Cobalt is not deficient under most conditions, except when poor quality roughages are fed.

The requirement for cobalt is 0.05–0.1 parts per million (ppm) in the ration dry matter.

Copper. Copper is also needed for hemoglobin formation as well as in several enzyme systems. Deficiency symptoms are depraved appetite, stunted growth, rough hair coat, diarrhea, and anemia. Forages generally contain 3 to 4 times the amount of copper needed and grains are somewhat lower in copper than most forages. In rations where molybdenum and sulfur are high, the copper requirement may be increased two to three times, however. This situation may occur where soils have excess levels of molybdenum or sulfur. When this situation occurs, supplementing with copper will often overcome the problem.

A copper deficiency may occur in calves fed only milk for long periods of time or older cattle consuming only forages grown on copper deficient soils found in areas of the southeastern United States.

The requirement for copper is about 4 ppm.

Iodine. Iodine is needed for synthesis of thyroxine, which is a hormone involved in the control of the rate of metabolism. A deficiency of iodine results in a condition called goiter, which is an enlargement of the thyroid gland, and reduced metabolic rate. A deficiency most often shows up in calves, as evidenced by the birth of weak, goitrous, or dead calves. Visible deficiency symptoms seldom occur in feedlot cattle, however. Iodine-deficient areas occur primarily in the northwest and in the Great Lakes region.

Deficiencies may be prevented by feeding salt containing 0.007 percent of stable iodine or by having the ration dry matter contain 0.1 ppm of iodine.

Iron. Iron is needed for hemoglobin, and a deficiency results in anemia. It is also needed in several enzyme systems.

Adequate amounts are present in most feedstuffs. Anemia is most likely to occur when milk is the major source of nutrients, however, as it is low in iron. Deficiency symptoms in young cattle start out as a loss of appetite and progress to scouring, prolonged rapid heart rate after exercise, low blood hemoglobin, and anemia.

The requirement for iron is about 10 ppm in the ration dry matter.

Manganese. Manganese is involved in several enzyme systems, primarily as an activator. A deficiency results in reproductive disorders in the adult cow such as delayed estrus, reduced fertility, abortions, and calves born with deformed legs and weak and shortened bones. A deficiency also results in poor growth of calves. There appears to be no need for supplemental manganese except possibly when all-concentrate rations based on corn and nonprotein nitrogen are fed. The manganese requirement is 1 to 10 ppm.

Molybdenum. Molybdenum is needed in some enzyme systems, but the requirement is quite small. It is not advisable to supplement beef cattle rations with molybdenum until more information is available on specific symptoms of deficiency, as cattle are extremely sensitive to excessive molybdenum. Toxic levels (20 to 50 mg per lb of ration) interfere with copper metabolism and thus increase copper requirements. Symptoms of toxicity are the same as those for copper deficiency plus severe scours and loss of condition. Molybdenum toxicity in cattle can be overcome by increasing the copper level in the ration to 1 gram per head daily.

Molybdenum toxicity occurs only occasionally in cattle and appears to be an area problem.

Selenium. Selenium-containing compounds are thought to act as a carrier of vitamin E and/or function in the absorption and retention of vitamin E in the body. Thus, it apparently increases the biological activity of vitamin E in the blood and body tissues.

The symptoms of a deficiency are similar to those of a vitamin E deficiency; the most common are white muscle disease, heart failure and paralysis in calves. A hollow or swayed back is typical.

Most feeds grown in the midsection of the United States contain 0.5 to 2.0 ppm, which is 5 to 20 times greater than the amount needed, and selenium toxicity has been reported in many areas of the Great Plains (above 5 ppm). Feeds grown in many areas in the western or eastern sections of the United States, however, apparently contain much lower levels of selenium (0.1 to 0.2 ppm) and may often be below the requirement (0.1 ppm).

Four procedures have been effective in preventing white muscle disease in calves:

1. Giving selenium supplements as a drench.
2. Subcutaneous or intramuscular injection with sodium selenite and tocopherol (Vitamin E).
3. Using selenium as a feed additive to dam and/or her calf.
4. Placing selenium in fertilizers applied to pastures.

Fishmeal and linseed meal are high in selenium and can be fed to prevent white muscle disease.

Zinc. Zinc is required for functioning of several enzyme systems. A severe deficiency in young calves results in parakeratosis, which is characterized by skin lesions, inflamed nose and mouth, roughened hair coat, and stiffness of joints. A mild deficiency in feedlot cattle results in reduced gains without other noticeable outward symptoms, and may occur in some high grain rations.

Requirements are 20 to 30 ppm in cattle rations.

Determining the Amount of Supplemental Minerals

The following outline will show how to balance a ration for minerals. Then several free choice mineral mixes will be given that can be used where force feeding a separate mineral supplement is not feasible (Table 8-8).

1. To determine the daily allowance of major minerals, multiply the expected dry matter intake from Table 8-6 times the percent requirement from Table 8-5. For example, 800-lb cattle can be expected to consume about 18 lbs dry matter per day. The daily requirement for calcium for growing cattle would be $18 \times 0.0045 = 0.08$ lb.
2. To determine the daily allowance of minor minerals:

    a. $\text{Lbs required} = \dfrac{\text{expected intake} \times \text{ppm required}}{1{,}000{,}000}$

For example, the 800-lb animal consuming 18 lbs would require (18 × 30)/1,000,000 = 0.00054 lbs zinc per day.

b. Grams required per day = lb required per day × 454. (From above example, 0.00054 × 454 = 0.25 grams.)

c. Milligrams required per day = lb required × 454,000. (From above example, 0.00054 × 454,000 = 250 milligrams.)

3. To determine the amount of mineral supplement needed, first subtract the requirement from the amount furnished by the ration to get the amount of supplemental mineral needed. Then divide this deficiency by the percent of the mineral in the supplement, and the answer multiplied by 100 gives the amount of the mineral supplement needed.

Example: The above 800 lb cattle are being fed 2 lbs alfalfa hay, one lb of a 40 percent protein supplement (2 percent calcium), and a full feed of shelled corn. The amount of supplemental minerals needed can be determined as shown in Table 8-7.

Vitamin Requirements

Vitamins are organic substances that are required in very small quantities for various metabolic functions in the animal. Twenty-five to thirty have been identified, but cattle can synthesize in the rumen the amounts required of all of these vitamins except for two to three that must be included in the ration. Even though the supplemental amounts needed of these vitamins (A and possibly D and/or E) are quite small, a deficiency of them can have a drastic effect on the animal.

The requirements for the various vitamins are discussed below, along with a description of their function and deficiency symptoms.

Vitamin A. Vitamin A is required for maintaining the skin, lining of the mouth, eye, gut, genital tract, in bone formation, and in functioning of the eye in the dark. The vitamin A requirement can be met from provitamin A or carotene in feedstuffs or by oral or injected vitamin A supplements. One milligram of carotene is equal to 400 international units (IU) of vitamin A. Minimum requirements per pound of ration dry matter are as follows: growing and finishing steers, 1000 IU; pregnant heifers and cows, 1275 IU; lactating beef cows and breeding bulls, 1775 IU. Some studies show that vitamin A reduces heat stress, and it may be advisable to

Table 8-5
Summary of Mineral Requirements
(All on 100% Dry Matter Basis)

Major Minerals		Trace Minerals[1]	
Calcium		Cobalt	0.1 PPM
300 - 600 lb.	0.45 — 0.55%		
600 lb. and up	.28 — .45%	Copper	4 PPM
Gestation	0.18%		
Lactation	0.25 — 9.44%	Iodine	0.1 PPM
Phosphorus			
300 - 600 lb.	0.30 — 0.35%	Iron	10 PPM
600 lb. and up	.23 — 27%		
Gestation	0.18%	Manganese	10 PPM
Lactation	0.25 — 0.39		
Potassium	0.6 — 0.8%		
Magnesium	0.04 — 0.2%	Selenium	0.1 PPM
Salt	0.25 — 0.5%		
Sulfur	0.15%	Zinc	30 PPM

[1] 1 part per million (**PPM**) or $\dfrac{1}{1,000,000}$ or 1 lb. in 1 million lb or 0.0001%

Table 8-6
Expected Daily 100% Dry Matter Intake of Beef Cattle

Body weight, lb.	300	400	500	600	700	800	900	1000	1100	1200
Expected daily dry matter intake	9.0	11	12.5	14.4	16.2	18.0	20	21.5	23	24

Table 8-7
Calculating Mineral Needs

Ingredient	Lb. Consumed	Lb. Dry Matter	% Calcium	Amount of Calcium	% Phosphorus	Amount of Phosphorus
Alfalfa hay	2	1.8	1.25	.022	0.23	.004
Protein supplement	1	.9	2.00	.018	1.00	.009
Shelled corn	18	15.3	0.02	.003	0.40	.061
Totals		18.0		.043		.074
		Amount needed:	18 × .0045 =	.081	18 × .003 =	.054
		Deficiency or excess:		−.038		+ .02

Table 8-8
Free Choice Mineral Mixes

Free-choice mineral mixes can be formulated from commonly available ingredients, as indicated in each mixture below. However, commercial mineral mixes that contain a similar amount of calcium and phosphorus can be used for the type of ration indicated.

Mixture 1 — For feeding with rations containing mostly hay or pasture.

	Amount	Calcium	Phosphorus
Dicalcium phosphate	100 lb.	23%	18.5%
Trace mineralized salt	100 lb.	—	—
Total	200 lb.	11.5%	9.25%

Mixture 2 — For feeding with corn silage or other rations low in calcium and phosphorus.

Feeding limestone	100 lb.	38%	—
Dicalcium phosphate	100 lb.	23%	18.5%
Trace mineralized salt	100 lb.	—	—
Total	300 lb.	20.3%	6.2%

Mixture 3 — For feeding with high grain rations.

Feeding limestone	200 lb.	38%	—
Trace mineralized salt	100 lb.	—	—
Total	300 lb.	25.3%	0%

Mixture 4 — For feeding on pasture in areas where grass tetany is a problem.

Magnesium oxide	25
Dicalcium phosphate	25
Trace mineralized salt	25
Ground corn	25
	100

increase the vitamin A level to 1500 IU per pound of ration dry matter for growing and finishing cattle during the summer. Doubling or tripling the vitamin A level during periods of stress such as in newly arrived feeder cattle may be advisable.

Cattle store vitamin A in the liver and body fat during times of abundant intake from pastures and other feeds, and these reserves can reduce the requirement for supplemental vitamin A or meet the needs of older cattle for as long as six months. Vitamin A is destroyed in feeds during storage. In all hays and other forages the vitamin value decreases after the bloom stage and much of the carotene is destroyed by oxidation during field curing. Thus deficiencies are likely to occur in early winter in cattle previously grazed on weathered forages or in late winter in cattle previously grazed on green pastures in the summer and then fed stored feeds. The injection of one million IU of vitamin A palmitate intramuscularly or intraruminally will provide sufficient vitamin A for 2 to 4 months in growing and breeding beef cattle. Feeding 20,000 to 40,000 IU of supplemental vitamin A per head daily to cattle not on pasture will meet the needs for supplemental vitamin A under most conditions.

A mild deficiency results in reduced feed intake and poor gains but no outward symptoms. More severe deficiencies result in night blindness, muscular incoordination, staggering gait, and convulsive seizures. Other symptoms are diarrhea, lameness in the hock and knee joints, and swelling of the brisket area. An animal deficient in vitamin A may become more susceptible to pinkeye, and many of the eye problems attributed to a vitamin A deficiency may actually be due to pinkeye. In breeding bulls, sexual activity and semen quality is reduced. In beef cows, ability to become pregnant is impaired. In pregnant cows, abortion or birth of dead, weak, or blind calves and retained placentas can occur.

Vitamin D. Beef cattle usually receive adequate quantities of vitamin D by synthesizing it in their own bodies during exposure to direct sunlight or from sun-cured hay. Cattle being fed in confinement, however, may not receive adequate vitamin D, and their ration should include 125 IU of vitamin D per pound of dry ration or 3000 IU per head daily, or about 1/10 of the level of vitamin A. Vitamin D deficiency in calves results in rickets. The symptoms of a deficiency of vitamin D are those of a calcium and phosphorus deficiency as the principal action of vitamin D is to increase the absorption of calcium and phosphorus from the intestine. Vitamin D also has a direct effect on the calcification process in bone. Clinical symptoms, usually preceded by a decrease in blood calcium and inorganic phosphorus, are poor appetite, decreased growth rate, digestive disturbances, stiffness, labored breathing, irritability, weakness and, occasionally, tetany and convulsions.

Later, enlargement of the joints and bowing of the legs occur. Symptoms develop more slowly in older cattle. Vitamin D deficiency can result in the birth of deformed or dead calves, as well as lowered fertility.

Vitamin E. Except possibly with certain types of grain processing, under most conditions natural feedstuffs appear to supply adequate quantities of vitamin E for adult cattle. However, calves deficient in vitamin E will show symptoms of white muscle disease, usually between the ages of 2 and 12 weeks. The most common symptoms are heart failure and paralysis; a hollow or swayed back is also typical. Vitamin E facilitates the absorption and storage of vitamin A, and a deficiency of vitamin E may result in a vitamin A deficiency even though the ration contains adequate levels of vitamin A. A specific role for vitamin E in metabolism has not yet been discovered, but in general vitamin E serves as a physiological antioxidant, and its biochemical roles in the body appear to be related to its antioxidant capability. Abnormally high levels of nitrates may produce a vitamin E deficiency. One way to prevent vitamin E deficiency in calves is to supplement the ration of the cow during the last 60 days of gestation and during the first part of lactation with at least 25 to 30 IU of vitamin E per pound of ration or 500 to 600 IU per head daily. Another way is to inject the calf with selenium-vitamin E at birth. Incidence of vitamin E deficiency is usually lower where the cows have been receiving 2-3 lb of grain mixture the last 60 days of pregnancy.

Where grains are heat processed, research shows that it may be advisable to provide supplemental vitamin E. The National Research Council indicates that the requirement for vitamin E is about 7 to 25 IU per pound of ration dry matter or 250 IU per head daily for growing and finishing cattle and feeding this level may be advisable under these conditions.

Vitamin K. Vitamin K is synthesized in the rumen of cattle in adequate amounts under most feeding conditions. Vitamin K is involved in the blood clotting mechanism, and the symptom of a deficiency is excessive bleeding. Cattle fed moldy sweet clover hay may show symptoms of a vitamin K deficiency because it contains dicoumarol, which interferes with the normal activity of vitamin K in blood clotting. This problem is often called sweet clover poisoning or bleeding disease, and mild cases of this disease can be effectively treated by administering vitamin K_3 and not feeding the moldy sweet clover hay.

B Vitamins. B vitamins are synthesized by rumen microorganisms in adequate quantities and usually a source of B vitamins is not needed in the ration of ruminants. A dietary source of B vitamins (thiamine, biotin, niacin, pyridoxine, pantothenic acid, riboflavin, and vitamin B_{12}) may be needed by calves during the first 8 weeks of life prior to the development of a functioning rumen, but these needs are usually met by milk supplied by the cow during early lactation. In most cases the various B vitamins function as constituents of cellular enzyme systems and are necessary for the metabolism of nutrients. Unusual feeding conditions such as a severe protein deficiency could impair rumen fermentation to such an extent that sufficient quantities of B vitamins would not be synthesized, but such deficiencies have not been clearly established for beef cattle.

8-4 GROWTH STIMULANTS AND FEED ADDITIVES

A number of growth and digestive stimulants are available that will improve performance. Development of several of these products has had a greater impact on improving efficiency than any other single factor. The effect of the various products available on daily gain and feed efficiency are summarized in Table 8-9.

Growth Stimulants

Growth stimulants act by increasing protein deposition without changing protein and energy intake. Animals given these compounds will be less fat at any given weight than if they had not received the compound.

Several of these are implants. They are placed under the skin with a needle in the middle of the back side of the ear. A strong chute with a headgate attached is necessary to restrain the steer or heifer. A noselead can be used to hold the head, or a strong person doing the implanting can hold the head from moving with his hip. The needle should be slid full length under the skin, then pulled back slightly before inserting the implant pellets to avoid crushing them. If crushed, the life of the implant will be reduced by half or less.

Table 8-9
Recommended Growth Stimulants and Feed Additives

Growth stimulants	Feed intake %	Change in Daily gain %	Change in Feed efficiency %	How given	Dosage/day or period	Withdrawal period	Active ingredient
For Steers							
Diethylstilbestrol (DES)	1-3	+15.0	+8.5	Implant	30 mg/100 days	120 days	DES
Synovex S	1-3	+15.0	+8.5	Implant	1 dose/100 days	60 days	20 mg estradiol + 200 mg progesterone/implant
Ralgro	1-3	+10.4	+6.8	Implant	36 mg/100 days	65 days	Zeralanone
For Heifers							
DES	NC*	+11.0	+7.0	Implant	30 mg/100 days	120 days	DES
Synovex H	NC*	+11.0	+7.0	Implant	1 dose/100 days	60 days	20 mg estradiol 200 mg testosterone/implant
Ralgro	NC*	+6.6	+5.9	Implant	36 mg/100 days	65 days	Zeralanone
MGA	NC*	+11.2	+7.6	Feed	.35 mg/day	48 hours	Melengestrol acetate
Antibiotics							
Aureomycin	NC*	+3-4%	+3-4%	Feed	1. To improve performance 70-80 mg	None	Chlortetracycline
Terramycin	NC*	+3-4%	+3-4%	Feed	2. Prevent liver abcess 70-80 mg	None	Oxytetracycline
Bacitracin	NC*	+3-4%	+3-4%	Feed	3. Reduce diarrhea and foot rot 70-100 mg	None	Bacitracin
Gallimycin	NC*	+3-4%	+3-4%	Feed	4. Reduce shipping fever 35 mg	48 hours	Gallimycin
					5. Prevent anaplasmosis, 50 mg/100 lb body wt.	48 hours	
Rumensin	-9	NC*	+10	Feed	20-30 gm/ton DM	None	Monensin

* NC = No change

Suckling calves implanted at 90 days will have 20-30 lbs heavier weaning weights. After weaning, steers and heifers to be fed for slaughter should be implanted at weaning, then reimplanted about every 100 days. Cattle not reimplanted on time may have a lower performance after the implant expires than a nonimplanted steer or heifer. Thus, the benefits of one implant given initially to a steer fed 200 days will be lost between 100 and 200 days on feed.

All implants have a withdrawal period. This means that the implant cannot be given within that period of time prior to slaughter. The best procedure is to use one with a short withdrawal period for the last implant, and reimplant based on the time left in the feeding period. For example, a steer expected to be on feed for 160-180 days is implanted on arrival with diethylstilbestrol (DES). He should be reimplanted at about 90 days with Synovex or Ralgro.

At present, the Food and Drug Administration (FDA) does not approve the use of more than one product at a time. Any sequence can be used, however.

Antibiotics

Antibiotics are organic substances produced by a fungus, bacteria, or algae. They inhibit or destroy microorganisms in the animal that are detrimental. They appear to give the greatest response in a poor environment or when the animal is under stress. Low level feeding (80-100 mg/hd/day) helps control foot rot and diarrhea, improves rate of gain and feed efficiency 3-4 percent on the average, and will prevent liver abscesses in cattle fed high grain rations. This level, however, will not prevent diseases such as shipping fever. Feeding 350 mgs antibiotic for the first 28 days after arrival in the feedlot has been shown to be beneficial in improving performance while getting acclimated and started on feed. However, this level will not prevent a disease outbreak either. High levels of antibiotic (1 gram/hd/day) for short periods can be helpful in getting off feed cattle back to normal feed consumption. However, the level should not be fed for more than 2-3 days.

Rumensin

Rumensin is a feed additive that reduces feed requirements approximately 10 percent, by giving the same rate of gain while reducing feed intake 10 percent. It is a fermentation product like antibiotics, but has little antibacterial effect. It functions in the rumen by its action on rumen microbes. The proportion of propionic acid increases relative to acetic and butyric acid. Propionic acid synthesis in the rumen is 70 percent more efficient and it is utilized more efficiently after being absorbed than are acetic or butyric acids. It also reduces feedstuff protein degradation in the rumen, allowing more to bypass to the small intestine. This improves efficiency of feed protein utilization with most feeds fed to cattle.

8-5   ENERGY UTILIZATION BY CATTLE AND THE USE OF ENERGY VALUES IN RATION FORMULATION

Digestion of feedstuffs in ruminants is primarily a fermentation process that occurs in the rumen. This allows ruminant animals to use both roughages and grains as sources of carbohydrates for energy. Part of the carbohydrates pass through the rumen and are digested in the abomasum and small intestine. Most carbohydrates in feeds are converted to either acetic, propionic, or butyric acid by rumen bacteria and protozoa. These short-chain fatty acids are then absorbed through the rumen wall into the bloodstream and eventually are used for energy in body tissue. This series of processes that occur in energy digestion and metabolism in ruminant animals is summarized in Figure 8-2.

The Relationship of Rations to Digestive Disturbances

The type of short-chain fatty acids produced in the rumen is a function of the type and amount of various bacteria and protozoa that make up the rumen microbial population which is primarily determined by the types of feedstuffs fed. This is because some types of bacteria and protozoa can digest cellulose and hemicellulose but not starch. Others can digest starch but not cellulose and hemicellulose, and some can utilize both, but can utilize one better than the other. Therefore, when the ration is changed to a grain ration from a forage ration, the population shifts toward more starch digesters. Starch digesting bacteria produce more propionic acid relative to acetic acid than do cellulose digesters. In addition, grain is mostly starch and is rapidly digested. Thus, a switch to a high grain ration results in a rapid production and increase in the rumen concentration of short-chain fatty acids and a rapid increase in rumen acidity. During sudden introduction of large amounts of grain in the ration, the starch digesting bacteria,

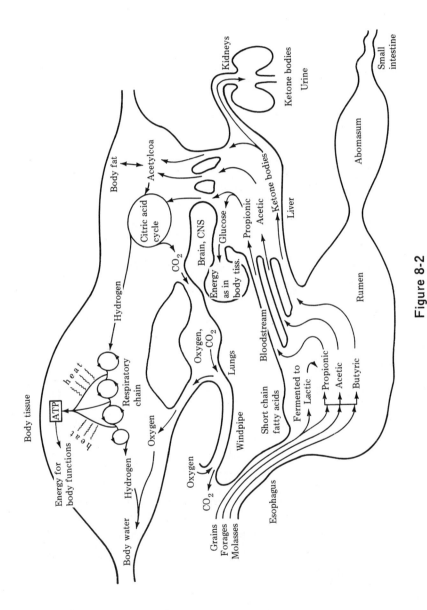

Figure 8-2 Energy Digestion and Metabolism in Ruminants

Streptococcus bovis, rapidly increase and 80-85 percent of the acid produced by this species is lactic acid. Lactic acid is normally converted to propionic acid in the rumen, and small quantities of lactic acid can be efficiently utilized. However, when an excessive quantity of lactic acid is produced, it accumulates in the rumen and excessive quantities are absorbed across the rumen wall into the bloodstream. The result can be an increase in rumen and blood acidity leading to an acute acid indigestion and symptoms of lactic acidosis such as posterior incoordination and dullness and sometimes death. High levels of lactic acid have also been associated with damage to the rumen wall, microbial penetration of the rumen wall and subsequent invasion of the liver, resulting in rumen wall ulcerations and liver abscesses. When cattle become adapted to high grain rations, a new microbial balance develops in which *Streptococcus bovis* is not exceptionally high and in which other species predominate. Cellulose and hemicellulose are digested more slowly than starch and soluble carbohydrates. This results in a lower total concentration of acids at any one time when rations high in cellulose are fed. Thus, one can switch rather rapidly from a grain to a forage ration, but not from a forage to a grain ration. A rapid switch to a grain ration without allowing the proper microbial balance to develop can result in digestive disorders such as acidosis, bloat, founder, and enterotoxemia, which are related to grain engorgement, and the rapid accumulation of acids, particularly lactic acid. The same things occur during infrequent feeding of grains or when animals have been "off feed." They gorge themselves in a short time when fed, resulting in a rapid shift in microbial population and accumulation of acids, particularly lactic acid. The gradual introduction of grains over a period of 2 to 3 weeks or more and then continuous feeding of high levels of grain will allow the proper balance of rumen microbial population to develop that is adapted to high grain feeding. Roughage level may then be kept at a low level in the ration if the feeder takes great care to keep the cattle on feed by avoiding abrupt changes in the ration and by using good feedbunk management.

Crude Fiber as a Source of Energy

The ability of the rumen bacteria and protozoa to digest a forage is greatly affected by its crude fiber content. Crude fiber refers to the feed content of the complex carbohydrates cellulose, hemicellulose, and lignin which are found in plant walls. Feeds high in crude fiber are usually less digestible than those high in starches and sugars because of their content of lignin, which is nearly indigestible. As a plant becomes mature, the lignin content increases, resulting in a lower proportion of digestible nutrients. Often the term roughage level is used to mean the same thing as crude fiber level because roughage type feeds such as forages are high in crude fiber and become more indigestible as they mature and become more lignified. It has become common practice to use the term roughage level in describing the crude fiber level of a finishing ration. Grinding forages tends to increase digestibility because the particle size is reduced and more of the plant cell contents are exposed to the rumen bacteria. Sugars and fat depress fiber digestibility, whereas protein increases fiber digestibility.

Fat as an Energy Source for Ruminants

Beef cattle cannot be fed as high levels of fat as can swine because they are unable to digest large quantities of fat at one time. Research studies indicate that up to 5-8 percent fat can be added to beef rations, resulting in an 8-10 percent increase in feed efficiency because of the high energy value of fat. Levels higher than this or feeding fat with higher fiber rations, however, can depress performance. The use of fat in a ration depends on the relative cost of grains and fat as energy sources. Fat has a 10 percent higher digestibility and has 2.25 times the energy value of grains per pound and therefore is worth 2.5 times the cost of grain as an energy source if not fed in excess of the above limitations.

Expressing Energy Values of Feeds

Calories are used to express the energy value of feedstuffs. One calorie is the amount of heat required to raise the temperature of 1 gram of water 1° Centigrade. 1 kcal = 1000 calories. 1 Mcal or therm = 1 million calories.

The combustion of a foodstuff in the presence of oxygen results in the production of heat. The heat produced by this combustion is measured as calories and is called the total or gross energy content of the feed. The body obtains energy for its physiological processes (beating of the heart and functioning of other organs, muscular work, or for synthesizing new body tissue) by combusting glucose and fatty acids obtained from feed, as previously discussed. The amount of energy that is stored by the animal can be determined by measuring the amount of fat and protein stored and then calculating the calories (potential heat) that the muscle and fat tissue contain. By knowing the energy content of the fat and muscle tissue deposited by cattle during growth, the net energy required to obtain a given amount of weight gain can be calculated. Thus, the energy requirements of cattle are expressed in terms of calories. The energy value of a feed depends on the amount and proportions of carbohydrates, fat, and protein it contains.

Through various digestive and metabolic processes, about 60 percent of the total combustible energy in grains and about 80 percent of the total combustible energy in roughages is lost as feces, urine, gases, and heat. The size of these losses and the main factors that affect them are summarized in the following illustration.

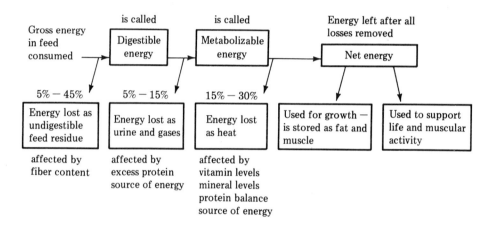

The major objective of formulating rations is to put feedstuffs together in the combination that will give least cost performance. The useful energy in feeds must be precisely known before a least cost ration can be accurately calculated. Therefore, it becomes of considerable economic significance to select the system that best describes the useful energy in feedstuffs and the one that most consistently relates these feed values to the requirements of beef cattle.

Gross energy (GE). GE represents the total combustible energy in a feedstuff, and is expressed as kcal or Mcal per lb. GE is determined directly by burning a small amount of the feedstuff and measuring the heat liberated. GE does not vary greatly over a wide range of feeds, except for those high in fat. For example, 1 lb of corn cobs contains about the same amount of GE as 1 lb of shelled corn. Therefore, GE does little to describe the useful energy in feeds for beef cattle.

Digestible energy (DE). DE is that portion of the GE in a feed that is not excreted in the feces, and is expressed as kcal or Mcal per lb. DE is determined by measuring the energy that is contained in the feces produced from a given amount of the feed. Then DE is calculated by subtracting the energy lost in the feces from the GE and then dividing this result by the pound of feed fed. It is assumed that DE is energy absorbed into the animal. This is not true as secretions from the body and tissue from the intestinal tract are eliminated in the feces, resulting in energy in the feces from body tissues and secretions.

In addition, gas (mostly methane) produced during fermentation in the rumen is an energy loss and is eliminated from the digestive tract and is not absorbed. Grain processing generally increases digestibility. The major factor affecting variability in DE values among feedstuffs is fiber content for the reasons previously discussed. DE more accurately describes the useful energy in a feedstuff than does GE. DE does not take into account the energy lost as urine, gas, or heat.

Metabolizable energy (ME). ME represents that portion of the GE that is not lost in the feces, urine, and gas and is expressed as kcal or Mcal per lb. The ME value of a feed is determined similarly to DE with the energy lost as urine and gas being measured and deducted from the GE value, in addition to the fecal energy loss. Then the energy remaining is divided by the pound of feed fed to calculate the ME value of the feed per pound. ME adjusts for urinary and gas energy losses in addition to fecal energy loss and therefore more accurately describes the useful energy in a feed than does GE or DE. However, ME also does not take into account the energy lost as heat. The major factor affecting variability in ME values of feedstuffs is also fiber content. In addition, feeding excess protein or nonprotein nitrogen will decrease the ME value of a feed.

Total digestible nutrients (TDN). TDN is commonly used to represent the same energy portion in a feed as DE and is expressed as a percent. TDN is the proportion of the energy sources (fiber, protein, sugars, and starches and fat) in a feed that are not excreted in the feces. TDN is determined in a digestion trial similar to DE, except the feed fed and feces produced are chemically analyzed to determine their content of fiber, protein, sugars, and starches (NFE) and fat. TDN is then calculated by adding up the pounds of these energy sources not excreted in the feces and then dividing this result by the pounds of feed fed. In this calculation the fat weight is multiplied by 2.25, as fat contains 2.25 times as much energy per unit as do the other components. TDN is related to both DE and ME. It adjusts for fecal energy losses as do both DE and ME. Unlike DE, it also adjusts for some urinary energy losses by including protein at the same weight as fiber and nitrogen-free extract. It does not adjust for energy lost as gas, but ME does. TDN is used in practice as an equivalent to DE since both adjust for fiber digestibility but not for gas or heat loss. The major factor affecting variation in TDN values of feeds is fiber content of the feed.

Net energy (NE). NE represents the energy fraction in a feed that is left after the fecal, urinary, gas, and heat losses are deducted from the GE and is expressed as kcal or Mcal per lb. In addition, a feed has two net energy values. The net energy value of a feed for maintenance (NE_m) represents the energy in the feed per pound that is available for supporting the animal's maintenance functions (beating of the heart and functioning of the other organs and muscular activity). The energy value of a feed for growth (NE_g) represents the energy in a feed per pound that is available for supporting growth of body tissues and is actually deposited as protein and fat tissue gain in beef cattle. The NE_m value of a feed is calculated by first determining the amount of the feed required to keep an animal of a given size at a constant weight. This amount of feed has an energy value equal to the animal's maintenance requirement. For example, a 600 lb steer requires 5.21 Mcal of net energy daily for maintenance. If 10 lbs of alfalfa hay were required to maintain this weight, then this hay would have a NE_m value of 5.21/10 = 0.52 Mcal per lb.

The NE_g value of a feed is calculated by determining the energy deposited as fat and protein in body tissue from the amount of the feed consumed above that utilized for maintenance. To make this determination, representative cattle are first slaughtered to determine the energy contained in their body tissues at the start of the experiment. Then all remaining cattle are slaughtered at the end of the experiment, and the Mcal gained from the feed fed is equal to the Mcal in the body tissues at the end of the trial minus that calculated to be in the body tissues at the start. For example, in the same 600 lb steer if an average of 18 lbs of alfalfa hay were consumed daily and 10 lbs were required daily for maintenance, then 8 lbs per day were used for gain. If the steer gained 2.0 Mcal as body tissue per day, then the alfalfa hay has a NE_g value of 2/8 = 0.25 Mcal per lb.

There is considerable variation among feeds in NE_m and NE_g values. All feeds have higher net energy values for maintenance than growth. Roughages are lower in both NE_m and NE_g than concentrates for two reasons. The first is, because of their content of fiber, less of the energy is digestible and more is lost as feces. The second is that during the process of transforming the energy in the feed into a form usable by the animal (fermentation in the rumen and breakdown of the fatty acids produced) a higher proportion of the ME is lost as heat from roughages than from grains. For example, corn contains 1.34 Mcal of ME per pound and has a heat loss of approximately 0.55 Mcal per pound. Timothy hay contains 0.75 Mcal of ME per pound and has a heat loss of 0.42 Mcal per pound, when both the corn and timothy hay are fed in a typical growing ration. Even though there was more total heat produced per pound of corn fed, the proportion of the ME lost as heat from the corn is 0.55/1.34, or 41 percent, and the proportion of the ME lost as heat from the timothy hay is 0.42/0.75, or 56 percent. In this case there was about 1.3 Mcal of heat produced for every Mcal of net energy consumed from the timothy hay, but there was only 0.7 Mcal of heat produced for every Mcal of net energy consumed from corn. The proportion of the heat produced from a ration that is usable during cold weather primarily depends on the total net energy intake, wind velocity, and air temperature. A higher proportion of heat production per unit of net energy intake is detrimental during hot weather. Thus, this system tells us that high levels of roughages are better utilized during the winter than during the summer, that feeds vary in the amount of NE they contain per pound of TDN or per Mcal of DE or ME and that feeds have a different value for maintenance than for gain.

NE adjusts for all of the variables that DE, ME, and TDN do. In addition, it adjusts for the energy lost as heat. Furthermore, it separates the energy value into usefulness for maintenance or gain.

The major factors that cause variation in net energy values in feedstuffs are the same as for the other systems, and in addition the source of energy, as discussed previously, affects the proportion lost as heat.

Applying Energy Values in Formulating Rations

DE, TDN, and ME are related, and one value is used to estimate the others as the major factors that affect one also affect the others. One lb TDN = 2 Mcal of DE, and ME is approximately equal to 82 percent of DE. Therefore, 1 lb of TDN = 2 Mcal of DE = 1.64 Mcal of ME. It is more difficult to directly relate these to

NE, however, because of the extreme variation in heat loss from one feed to another. This is illustrated by the fact that 1 lb of TDN from corn contains more NE than does 1 lb of TDN from hay, primarily because TDN does not take into account the difference in heat loss per unit of TDN or ME, as previously discussed.

NE adjusts for the variation in heat loss as well as the other energy losses, making NE values more consistent from one feed to another. In addition, it separates the feed energy value into NE_m and NE_g, further reducing the variation in results from one feed to another. For example, shelled corn has a NE_m value of 0.92 Mcal and a NE_g value of 0.60 Mcal per pound and fair quality alfalfa hay has a NE_m of 0.46 Mcal and a NE_g of 0.16 Mcal per pound. This alfalfa is worth 0.46/0.92, or 50 percent as much as the corn for maintenance, but only 0.16/0.60, or 27 percent as much as the corn for growth. This separation allows a more accurate prediction of gains from a given combination of feedstuffs. In addition, this indicates that to get the most out of these two types of feedstuffs, the hay should be used in the maintenance or wintering type ration and the corn in the growing and finishing ration where both types of feeding programs are being used. Thus, it appears that the net energy system allows us to more correctly estimate the useful energy in a feedstuff than does DE, ME, or TDN, and results in the best evaluation of the true economical value of a feedstuff for growth.

Using DE, TDN, or ME to formulate rations for maintaining cattle such as beef cows whose energy requirement is primarily for maintenance most of the time will work reasonably well, however, especially in colder climates. There is less total heat produced when feeding at a maintenance level and most of it can likely be used to maintain body temperature in cold weather. Also, the amount of feed fed to beef cows is usually estimated, and using DE or TDN is usually adequate to estimate the amount needed under most conditions.

Net Energy Requirements of Growing and Finishing Cattle

The net energy requirements of growing cattle are related to body weight, stage of growth the animal is at for a given weight, rate of gain, sex, environmental conditions, age, and previous nutritional history.

The net energy requirements for growth (NE_g) is the actual amount of energy that will be deposited as fat or muscle at a given weight and rate of gain. As cattle increase in weight the proportion of fat in the gain increases. As fat contains 2.25 times as much energy as a similar amount of protein in muscle tissue, more than twice as much energy is required to synthesize a pound of fat. In addition, more water is deposited with protein than with fat during growth. Therefore, daily net energy requirements for growth are related to the amount gained per day and the proportion of fat in the gain made at a given weight, which in turn is related to the eventual mature weight of the cattle. Heifers mature earlier than steers and therefore deposit more fat in the gain at a similar weight. Steers mature earlier than bulls, and thus deposit more fat at the same weight as bulls as discussed previously. Cattle with a relatively large mature size such as the Charolais, Chianina, Holstein, and Simmental breeds will deposit less fat in the gain at the same weight as moderate-sized cattle such as the Angus, Hereford, or Shorthorn breeds, but will have the same energy requirements for gain at the same equivalent weight.

The net energy requirements for maintenance (NE_m) are directly related to the actual weight of the cattle, and represent the amount of actual energy needed to support life (beating of the heart and functioning of the organs, maintaining body temperature, and for muscular activity). Since they are related to the actual weight of the cattle they are likely similar at a given weight for various breeds of cattle, regardless of eventual mature size.

When appropriate feed net energy values and these net energy requirements are used along with a growth stimulant, experience has shown that projected gains will usually be within 5 to 10 percent of actual gains on average commercial cattle in feedlots under average environmental conditions. Thus, this system can be used to estimate rate and cost of gain on a given combination of feedstuffs with a reasonably high degree of accuracy, enabling cattle feeders to project break-even costs and profits on cattle when they are purchased, as well as the time and feed required to reach market weights.

Using Net Energy Requirement Tables

Tables 8-10, 8-11, and 8-12 were developed to be used with feed composition values discussed later to formulate rations and predict gains. The following brief description may be useful as a simple guide in using these tables. The examples will be given for an average size 600 lb steer consuming 14.5 lbs of ration dry matter that contains 0.70 Mcal NE_m and 0.45 Mcal NE_g per pound.

1. Divide the net energy required daily for maintenance (NE_m) for the weight of the cattle by the NE_m value of the ration per lb to get lbs of the ration required daily for maintenance

(5.21/0.70 = 7.5 lbs of ration for maintenance for the example steer).

2. Subtract lbs needed for maintenance from total lbs consumed to get lbs of ration left for gain (14.5 - 7.5 = 7 lbs left for gain for the example 600 lb steer).
3. Multiply lb left for gain times NE_g value of the ration per lb to get energy available for gain per day (7 × 0.45 = 3.15 Mcal left for gain).
4. Look down the NE_g column under the weight and sex of the cattle for the value nearest the energy left for gain, and find the expected rate of gain in the lb daily gain column across from this value (3.15 is near 3.19, which gives a 1.8 lb per day gain for the example 600-lb steer).

The expected rate of gain can be determined on various combinations of the feeds available, and then the cost of the ration plus overhead costs can be divided by the expected rate of gain to find the least cost ration. Using these values for ration formulation and finding the most profitable feeding system will be discussed in later sections.

Adjusting Net Energy Requirements for Various Conditions

- *Frame size.* Use the actual weight of the cattle to find the NE_m requirement. Then use the equivalent weight of an average-frame steer or heifer to find the NE_g requirement. For example, Table 8-10 shows a 720 lb frame code 9 steer is equivalent to a 600 lb average frame steer in requirements. The NE_m requirement for a 720 lb weight is 5.85 + 0.2 (6.47 - 5.85) or 5.97 Mcal. The NE_g requirement for a 2.5 lb/day gain for this large-frame 720 lb steer is in the same column as a 600 lb average-frame steer. The requirement is 4.59 Mcal/day.

- *Breed.* Research results indicate that feed requirements for various beef breeds and crosses are similar at the same equivalent weight. Holsteins, however, have been 12–25 percent less efficient at the same equivalent weight. To adjust for this, multiply the NE_m and NE_g requirement by 1.12. For example, a 720 lb Holstein would require 5.97 × 1.12 = 6.69 Mcal NE_m and 4.59 × 1.12 = 5.14 Mcal NE_g for a 2.5 lb/day gain.

- *Environment.* Factors such as extreme cold (wind and/or temperature), rain, and mud and extreme abrupt changes in weather affect maintenance requirements.

Mud and cold rain can increase maintenance energy costs by

Table 8-10
Net Energy Requirements for Steers

				Energy Requirements for Maintenance				
Weight, lb	300	400	500	600	700	800	900	1000
NE_m, Mcal/day	3.10	3.85	4.55	5.21	5.85	6.47	7.06	7.65

Frame size				Net energy requirements for gain Weight, lb				
1 (small)	240	320	400	480	560	640	720	800
5 (average)	300	400	500	600	700	800	900	1000
9 (large)	360	480	600	720	840	960	1080	1200

Daily gain, lb				NE_g, Mcal/day				
.5	.59	.72	.47	.83	.93	1.02	1.12	1.21
.6	.73	.87	.59	1.00	1.12	1.24	1.35	1.46
.7	.84	1.02	.68	1.17	1.31	1.45	1.58	1.71
.8	.98	1.17	.79	1.34	1.51	1.67	1.82	1.97
.9	1.10	1.33	.88	1.52	1.71	1.89	2.06	2.23
1.0	1.24	1.48	1.00	1.70	1.91	2.11	2.30	2.49
1.1	1.39	1.64	1.12	1.88	2.11	2.33	2.55	2.76
1.2	1.51	1.80	1.21	2.06	2.31	2.56	2.79	3.02
1.3	1.66	1.96	1.33	2.24	2.52	2.78	3.04	3.29
1.4	1.78	2.12	1.43	2.43	2.73	3.01	3.29	3.56
1.5	1.93	2.28	1.55	2.62	2.94	3.25	3.55	3.84
1.6	2.06	2.45	1.65	2.81	3.15	3.48	3.80	4.12
1.7	2.21	2.61	1.78	3.00	3.37	3.72	4.06	4.40
1.8	2.34	2.78	1.88	3.19	3.58	3.96	4.33	4.68
1.9	2.50	2.95	2.01	3.39	3.80	4.20	4.59	4.97
2.0	2.63	3.12	2.14	3.58	4.02	4.45	4.86	5.26
2.1	2.79	3.30	2.24	3.78	4.25	4.69	5.13	5.55
2.2	2.95	3.47	2.37	3.98	4.47	4.94	5.40	5.84
2.3	3.08	3.65	2.48	4.18	4.70	5.19	5.67	6.14
2.4	3.25	3.83	2.61	4.39	4.93	5.45	5.59	6.44
2.5	3.38	4.01	2.72	4.59	5.16	5.70	6.23	6.74
2.6	3.55	4.19	2.86	4.80	5.39	5.96	6.51	7.05
2.7	3.69	4.37	2.97	5.01	5.63	6.22	6.80	7.35
2.8	3.86	4.56	3.11	5.22	5.87	6.48	7.08	7.67
2.9	4.00	4.74	3.22	5.44	6.11	6.75	7.37	7.98
3.0	4.18	4.93	3.36	5.65	6.35	7.02	7.67	8.30
3.1	4.32	5.12	3.47	5.87	6.59	7.29	7.96	8.61
3.2	4.50	5.31	3.62	6.09	6.84	7.56	8.26	8.94
3.3	4.68	5.51	3.76	6.31	7.09	7.83	8.56	9.26
3.4	4.83	5.70	3.88	6.54	7.34	8.11	8.86	9.59

Table 8-11
Net Energy Requirements for Heifers

Weight, lb NE_m, Mcal/day	300 3.10	400 3.85	500 4.55	600 5.21	700 5.85	800 6.47	900 7.06
Frame size			Net energy requirements for gain Weight, lb				
1 (small)	240	320	400	480	560	640	720
5 (average)	300	400	500	600	700	800	1000
9 (large)	360	480	600	720	840	960	1200
Daily gain, lb			NE_g, Mcal/day				
.5	.52	.64	.78	.90	1.01	1.11	1.22
.6	.64	.80	.95	1.09	1.22	1.35	1.47
.7	.74	.92	1.12	1.28	1.44	1.59	1.73
.8	.87	1.08	1.29	1.48	1.66	1.83	2.00
.9	.97	1.21	1.46	1.68	1.88	2.08	2.27
1.0	1.11	1.38	1.64	1.88	2.11	2.33	2.55
1.1	1.24	1.54	1.82	2.09	2.34	2.59	2.83
1.2	1.35	1.68	2.00	2.30	2.58	2.85	3.11
1.3	1.49	1.86	2.19	2.51	2.82	3.12	3.40
1.4	1.61	2.00	2.38	2.73	3.06	3.39	3.70
1.5	1.75	2.18	2.57	2.95	3.31	3.66	4.00
1.6	1.87	2.32	2.77	3.17	3.56	3.94	4.30
1.7	2.02	2.51	2.97	3.40	3.82	4.22	4.61
1.8	2.14	2.66	3.17	3.63	4.08	4.51	4.92
1.9	2.29	2.85	3.37	3.87	4.34	4.80	5.24
2.0	2.42	3.01	3.58	4.11	4.61	5.09	5.57
2.1	2.57	3.20	3.79	4.35	4.88	5.39	5.89
2.2	2.74	3.40	4.01	4.59	5.16	5.70	6.23
2.3	2.87	3.57	4.22	4.84	5.44	6.01	6.56
2.4	3.03	3.77	4.44	5.09	5.72	6.32	6.90
2.5	3.17	3.93	4.66	5.35	6.01	6.64	7.25
2.6	3.33	4.15	4.89	5.61	6.30	6.96	7.60
2.7	3.47	4.32	5.12	5.87	6.59	7.28	7.96
2.8	3.65	4.54	5.35	6.14	6.89	7.61	8.32
2.9	3.79	4.71	5.59	6.40	7.19	7.95	8.68
3.0	3.97	4.94	5.82	6.68	7.50	8.29	9.05
3.1	4.11	5.12	6.06	6.95	7.81	8.63	9.43
3.2	4.30	5.34	6.31	7.23	8.12	8.98	9.81
3.3	4.48	5.51	6.56	7.52	8.44	9.33	10.19
3.4	4.63	5.76	6.81	7.80	8.76	9.68	10.58

Table 8-12
Net Energy Requirements for Bulls

Weight, lb	\multicolumn{7}{c}{Energy requirements for maintenance}

Weight, lb	480	600	720	960	1080	1200	1320
NE_m Mcal/day	5.21	5.93	6.64	7.35	8.02	8.69	9.33

Frame size	\multicolumn{7}{c}{Net energy requirements for gain — Weight, lb}
1 (small)	385	480	757	670	770	865	960
5 (average)	480	600	720	840	960	1080	1200
9 (large)	575	720	860	1010	1150	1300	1440

Daily gain, lb	\multicolumn{7}{c}{NE_g, Mcal/day}
.5	.59	.72	.83	.93	1.02	1.12	1.21
.6	.73	.87	1.00	1.12	1.24	1.38	1.46
.7	.84	1.02	1.17	1.31	1.45	1.58	1.71
.8	.98	1.17	1.34	1.51	1.67	1.82	1.97
.9	1.10	1.33	1.52	1.71	1.89	2.06	2.23
1.0	1.24	1.48	1.70	1.91	2.11	2.30	2.49
1.1	1.39	1.64	1.88	2.11	2.33	2.55	2.76
1.2	1.51	1.80	2.06	2.31	2.56	2.79	3.02
1.3	1.66	1.96	2.24	2.52	2.78	3.04	3.29
1.4	1.78	2.12	2.43	2.73	3.01	3.29	3.56
1.5	1.93	2.28	2.62	2.94	3.25	3.55	3.84
1.6	2.06	2.45	2.81	3.15	3.48	3.80	4.12
1.7	2.21	2.61	3.00	3.37	3.72	4.06	4.40
1.8	2.34	2.78	3.19	3.58	3.96	4.33	4.68
1.9	2.50	2.95	3.39	3.80	4.20	4.59	4.97
2.0	2.63	3.12	3.58	4.02	4.45	4.86	5.26
2.1	2.79	3.30	3.78	4.25	4.69	5.13	5.55
2.2	2.95	3.47	3.98	4.47	4.94	5.40	5.84
2.3	3.08	3.65	4.18	4.70	5.19	5.67	6.14
2.4	3.25	3.83	4.39	4,93	5.45	5.59	6.44
2.5	3.38	4.01	4.59	5.16	5.70	6.23	6.74
2.6	3.55	4.19	4.80	5.39	5.96	6.51	7.05
2.7	3.69	4.37	5.01	5.63	6.22	6.80	7.35
2.8	3.86	4.56	5.22	5.87	6.48	7.08	7.67
2.9	4.00	4.74	5.44	6.11	6.75	7.37	7.98
3.0	4.18	4.93	5.65	6.35	7.02	7.67	8.30
3.1	4.32	5.12	5.87	6.59	7.29	7.96	8.61
3.2	4.50	5.31	6.09	6.84	7.56	8.26	8.94
3.3	4.68	5.51	6.31	7.09	7.83	8.56	9.26
3.4	4.83	5.70	6.54	7.34	8.11	8.86	9.59

25 to 40 percent. Mud increases the energy requirements for activity and decreases feed intake as cattle often will not travel through mud to reach feed until they are excessively hungry. Also exposure to cold rain in the winter reduces the insulation barrier of the hair coat, resulting in excessive body heat losses. Often cattle will compensate for short periods of poor environmental conditions when followed by more suitable weather, but prolonged exposure to poor environmental conditions such as mud and cold rain can greatly increase total energy requirements.

Cold temperatures alone apparently do not necessarily greatly increase the proportion of feed consumed that is needed for maintenance, however, as cattle are able to utilize the large amount of heat increment normally available from the types of rations they are fed to maintain body temperature during cold weather. Calculations from Canadian studies show that the critical temperature (i.e., the temperature at which body tissue or energy normally available for gain is used to maintain body temperature) is 0°F or below for cattle on an average intake of a ration that contains about 0.45 Mcal NE_g per lb DM (dry matter) and as low as -20°F on rations containing .60 Mcal NE_g per lb DM, without any wind.

The Canadian studies show that wind greatly increases energy requirements, however, and protection from wind is much more important than protection from cold. The critical temperature changes about 1°F for each 1 mph of wind. Also, it appears that abrupt, severe weather changes affect cattle performance due to readaptation required and adjustments in feed intake that occur. Cattle tend to consume additional feed just prior to a weather change and then reduce intake after the change has occurred. Adjustments in management that reduce the effect of mud, rain, wind, and severe weather changes will contribute an overall reduction in energy requirements for maintenance.

Table 8-13 gives adjustment factors for various environmental conditions.

Table 8-13
Adjustment for Environmental Conditions

Environmental code	Lot condition	Multiplier for NE_m
1	Outside lot, no wind protection, poor drainage and frequent deep mud	1.30
2	Above lot with less mud.	1.25
3	Outside lot with some mud and wind protection	1.20
4	Outside lot, wind protection, minimum mud	1.15
5	Outside lot, well mounded, bedded during adverse weather	1.10
6	Outside lot in dry climate or cemented barn with outside lot	1.05
7	No mud, shade, good ventilation, no chill stress	1.00

For example, the Holstein steer fed in a code 4 environment would require 6.69 × 1.15 = 7.69 Mcal NE_m/day.

- *Growth stimulants.* If growth stimulants (DES, Ralgro, Synovex, MGA) are not used, the energy deposited/lb of gain is increased. If a growth stimulant is not used, still use the actual weight to determine NE_m requirement. However, multiply the weight by the factor in Table 8-14 to determine the weight to use to determine NE_g requirements.

Table 8-14
Adjusting NE_g Requirement for Not Using a Growth Stimulant

Steers	Heifers
Multiplier for weight	
1.18	1.13

For example a 600-lb steer requires 5.21 Mcal NE_m daily and 4.59 Mcal NE_g for a 2.5 lbs/day gain. If DES or Synovex is not used, however, the gain requirement is the same as for a 600 × 1.18 = 708 lb steer. NE_g requirement is thus increased to 5.16 Mcal/day.

8-6 PROTEIN REQUIREMENTS AND FEED PROTEIN UTILIZATION

A critical step in ration formulation is to meet the protein requirement, due to the usual deficiency of protein in most feedstuffs used primarily for growing and finishing cattle. Supplemental protein is expensive, but adequate quantities must be fed to make up for the deficiency and variability in protein content of ration ingredients. To avoid underfeeding protein and thus poor performance or overfeeding and thus increased ration costs, rations must be formulated to contain an adequate level of protein. Proper balancing for protein involves proper use of two components.

1. Protein requirements are the sum of the protein needed for maintenance, body tissue, growth, and rumen fermentation. These depend on the weight, stage of growth, rate of gain, and type of feeds in the ration.

2. The amount and form of protein in ration ingredients, and the variation in protein content in the feeds used determine whether the animal can meet its requirements from the ration being fed.

Protein Requirements

Requirement for maintenance. Protein in body tissue is being degraded and replaced continuously. Cattle require protein for maintenance to replace or repair protein in skeletal and organ muscles. The weight of vital tissue is called the metabolic weight of the animal. As an animal increases in weight the growth of this tissue increases at a proportion of the total increase in weight, and is assumed to be the 3/4 power of the actual weight. Protein requirements for maintenance of body tissue are estimated to be 0.88 grams/kilogram of metabolic weight.

Protein is also needed for maintenance in cattle to meet the requirements of rumen microbes. Adequate protein must be available for them to grow as they ferment the feed consumed to short-chain fatty acids, as discussed in the energy utilization section. To meet this need the ration protein level should not drop below 5 percent in the dry matter.

Requirements for gain. Protein requirements for growth are directly related to the amount of protein being deposited daily in body tissue. Thus, even if ration energy is adequate for a 3 lbs/day gain, it may be limited to 2 lbs/day if protein intake is only adequate for that rate of tissue growth. At early stages of growth, a high proportion of the weight gained is protein. As an animal matures, the proportion of protein in the gain decreases. For example, an average-frame steer (frame code 5) gaining 2 lbs/day will deposit 0.39 lb protein at 390 lbs, but at 1000 lbs will only deposit 0.24 lb protein. Thus, this steer needs 0.39 lb net protein for gain (NP_g)/day at 300 lbs but only 0.24 lb NP_g/day at 1,000 lbs. A larger-frame steer will be heavier when he has the same requirement. A large frame (code 9) steer at 360 lbs will have the same NP_g requirement as the frame code 5 steer at 300 lbs.

Protein requirement tables. Table 8-15 summarizes the net protein requirements for maintenance and gain (NP_m and NP_g). The requirements for maintenance and gain must be added to get the total NP needed daily. For example, a 600 lb average-frame steer gaining 2 lbs/day needs 0.12 lb NP_m + 0.31 NP_g = 0.42 lb NP/day. Thus, once the expected rate of gain is determined based

Table 8-15
Net Protein Requirements for Growing and Finishing Cattle

Weight, lb	Net protein requirements for maintenance						
	400	500	600	700	800	900	1000
NP_m, lb/day	.09	.11	.12	.13	.15	.16	.17

	Net protein requirements for gain, lb/day						
Steers			Weight, lb				
Frame 1	320	400	480	560	640	720	800
5	400	500	600	700	800	900	1000
9	480	600	720	840	960	1080	1200
Heifers							
Frame 1	255	320	385	450	510	575	640
5	320	400	480	560	640	720	800
9	575	480	575	670	770	865	960

Daily gain, lb	NP, lb/day						
1.0	.18	.17	.16	.15	.14	.13	.12
1.1	.20	.19	.17	.16	.15	.14	.13
1.2	.22	.20	.19	.18	.17	.16	.14
1.3	.24	.22	.20	.19	.18	.17	.15
1.4	.26	.24	.22	.21	.20	.18	.17
1.5	.27	.26	.24	.23	.22	.20	.18
1.6	.29	.27	.25	.24	.23	.21	.19
1.7	.31	.29	.27	.26	.24	.22	.20
1.8	.33	.31	.28	.27	.26	.24	.21
1.9	.35	.32	.30	.28	.27	.25	.22
2.0	.37	.34	.31	.30	.29	.26	.24
2.1	.38	.26	.33	.32	.30	.28	.25
2.2	.40	.37	.35	.34	.32	.29	.26
2.3	.42	.39	.36	.35	.33	.30	.27
2.4	.44	.41	.38	.37	.35	.31	.28
2.5	.46	.42	.39	.38	.36	.33	.29
2.6	.48	.44	.41	.39	.37	.34	.31
2.7	.49	.46	.42	.41	.39	.35	.32
2.8	.51	.48	.44	.42	.40	.37	.33
2.9	.53	.49	.46	.44	.42	.38	.34
3.0	.55	.51	.47	.45	.43	.39	.35
3.1	.57	.53	.49	.47	.45	.41	.37
3.2	.59	.54	.50	.48	.46	.42	.38
3.3	.60	.56	.52	.50	.48	.43	.39
3.4	.62	.58	.53	.51	.49	.45	.40
3.5	.64	.59	.55	.53	.50	.46	.41

on the energy feeds fed, the net protein consumed from the basic ration can be compared with the requirement to determine if supplemental protein is needed. Adjustment should be made for frame size. For example, a 720 lb large-frame steer needs 0.14 lb NP_m. His NP_g requirements are the same as the 600 lb steer. If his expected gain is 2 lbs/day, his daily NP needs are 0.14 lb NP_m + 0.31 lb NP_g = 0.45 lb NP_g/day.

8-6 Protein Requirements and Feed Protein Utilization

Table 8-16 contains average requirements for crude protein to meet the NP needs. These values were calculated from NP values by assuming that 70 percent of the crude protein consumed is lost during metabolism, based on a summary of recent experiments. These are only useful as guides, however. This loss varies considerably, depending primarily on the degree of fermentation, heat damage, or solubility of the protein in the feeds being fed.

Table 8-16
Crude Protein Requirements for Growing and Finishing Cattle

Weight, lb.	\multicolumn{7}{c}{Crude protein requirements for maintenance}

Weight, lb.	400	500	600	700	800	900	1000
CP_m, lb/day	.30	.37	.40	.43	.50	.53	.57

Crude protein requirements for gain, lb/day

Steers	\multicolumn{7}{c}{Weight, lb}
Frame 1	320	400	480	560	640	720	800
5	400	500	600	700	800	900	1000
9	480	600	720	840	960	1080	1200
Heifers							
Frame 1	255	320	385	450	510	575	640
5	320	400	480	560	640	720	800
9	575	480	575	670	770	865	960

Daily gain, lb	\multicolumn{7}{c}{CP_g, lb/day}
1.0	.60	.57	.53	.50	.47	.43	.40
1.1	.67	.63	.57	.53	.50	.47	.43
1.2	.73	.66	.63	.60	.57	.53	.47
1.3	.80	.73	.67	.63	.60	.57	.50
1.4	.87	.80	.73	.70	.67	.60	.57
1.5	.90	.87	.80	.77	.73	.67	.60
1.6	.97	.90	.83	.80	.77	.70	.63
1.7	1.03	.96	.90	.87	.80	.73	.67
1.8	1.10	1.03	.93	.90	.87	.80	.70
1.9	1.17	1.07	1.00	.93	.90	.83	.73
2.0	1.23	1.13	1.03	1.00	.97	.86	.80
2.1	1.27	1.18	1.10	1.07	1.00	.93	.83
2.2	1.33	1.23	1.17	1.13	1.07	.97	.87
2.3	1.40	1.30	1.20	1.17	1.09	1.00	.90
2.4	1.47	1.37	1.27	1.23	1.16	1.03	.93
2.5	1.53	1.40	1.30	1.26	1.20	1.10	.96
2.6	1.60	1.47	1.37	1.30	1.23	1.13	1.03
2.7	1.63	1.53	1.40	1.37	1.30	1.17	1.07
2.8	1.70	1.60	1.47	1.40	1.33	1.23	1.10
2.9	1.76	1.63	1.53	1.47	1.40	1.27	1.13
3.0	1.83	1.70	1.57	1.50	1.43	1.30	1.16
3.1	1.90	1.76	1.63	1.57	1.50	1.37	1.23
3.2	1.96	1.80	1.67	1.60	1.53	1.40	1.27
3.3	2.00	1.86	1.73	1.67	1.60	1.43	1.30
3.4	2.06	1.93	1.76	1.70	1.63	1.50	1.33
3.5	2.13	1.96	1.83	1.76	1.67	1.53	1.37

Using these tables will give protein levels similar to those recommended by the NRC. The use of recent research data to adjust for frame size and feed quality, however, should allow ration formulation to meet protein needs to be accomplished more accurately.

Feedstuff protein utilization. Part of the protein in feeds is degraded to ammonia in the rumen; most of the undegraded fraction is digested to amino acids in the small intestine. Part is undigested and passes out in the feces. The amount degraded in the rumen depends primarily on the solubility of the protein. For example, only about 56 percent of corn protein is degraded in the rumen, but nearly 80 percent of the protein in fresh alfalfa and 96 percent of dried milk protein is degraded in the rumen. The rumen bacteria utilize part or all of the ammonia to synthesize bacterial protein, which in turn is degraded to amino acids in the small intestine. The amino acids from bypassed feed protein and bacterial protein are then absorbed, and are used to synthesize protein in tissue. This process can be summarized as in Figure 8-3.

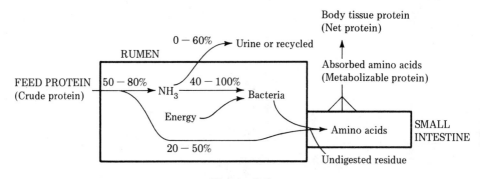

Figure 8-3
Protein Utilization by Cattle

The protein in bacteria has a high biological value because of its content of amino acids. The quality of many feed proteins is improved when the bacteria resynthesize them.

The overall objective is thus to optimize the proportion of degraded ammonia to stimulate bacterial growth but not to exceed their ability to utilize the ammonia produced. Most of the excess ammonia will be absorbed from the rumen and excreted in the urine. Having 30-40 percent of the protein bypass the rumen is usually beneficial because of the ruminal limits to synthesize protein. The ability to resynthesize protein depends on the energy available in the rumen. On the average, approximately 22 grams bacterial protein is produced/Mcal DE available in the rumen. Thus, the protein in feeds with a high soluble protein content can only be well utilized if the available energy level fed is high and/or the supplemental protein source is less soluble. It becomes obvious that these factors also determine the amount of urea or other ammonia-containing products that can be utilized. Determination of the protein supplementation needed for various rations is a complicated procedure. Further, highly accurate values for all of the factors involved such as rate of degradation and resynthesis under various conditions are not available. Thus, guidelines will be given for protein supplementation based on limited research data available and field experience.

The major factors affecting protein quality for cattle are heat damage and fermentation. Protein in haylage that is brown has been damaged by heating during fermentation. Ten to thirty percent of the protein will be unavailable under these conditions. Fermentation reduces protein quality due to degradation of part of the protein to unidentified fractions that are unavailable or only partly available.

Net protein values will be given for various feeds in the feed composition tables to adjust for these effects. These are based on limited research data. Although data is limited, using net protein requirements and feed values should be more accurate, especially when fermented feeds are a major part of the ration. Adjustment factors will be given that can be used to adjust crude protein values from feed analysis to net protein values. Some feed additives, such as Rumensin appear to improve the utilization of feed protein under most conditions. These products reduce feed intake but increase energy utilization. Thus the crude protein value for the feed should be increased about 10 percent when balancing a ration containing one of these products.

Guidelines for Protein Supplements

1. *Using crude protein feed values.* Crude protein requirements are given for those who wish to use them for balancing rations (Table 8-16). They will be reasonably accurate, except with fermented or heat-damaged feeds. Silage crude protein can be multiplied by 0.75 to adjust for its lower quality. Protein values for heat-damaged hay or haylage can be multiplied by 0.7 to 0.9, depending on the extent of damage. The color is a reasonable guide, if the forage was green at harvest. The greater the damage, the more brown the color; some may even be black.

Feeds should be analyzed for protein when in doubt. It is especially important to analyze haylage and corn silage when they are a major part of the ration.

2. *Safety factor.* If the feed is not analyzed and average protein values in feed composition tables are used, the feed crude protein value should be reduced to protect against lower than expected feed protein value. If silage and hay crude protein values in tables are multiplied by 0.85 and grains by 0.96, then the result used to balance the ration, an adequate safety factor will be provided 83 percent of the time. For example, if "book" value for corn silage is 8 percent protein, then a value of only 6.8 percent is used to balance the ration.

If an all-silage ration were being fed and it was assumed to contain 8 percent crude protein, only 1 lb of a 40 percent protein supplement would be fed to meet the requirements in Table 8-15. If adjusted for protein quality first (8×0.75), 1.6 lbs of the 40 percent protein supplement would be fed. If adjusted for both safety factor and quality ($8 \times 0.85 \times 0.75$), 2 lbs/day would be fed. This also illustrates the importance of feed analysis on samples that are representative of the feed to be used. One-third more supplemental protein would have to be fed to provide an adequate safety factor.

Use of Urea and Other Ammonia-Containing Compounds

The ability of rumen bacteria to synthesize protein from ammonia permits the use of ammonia-containing compounds (nonprotein nitrogen or NPN) to provide part or all of the supplemental protein needed. The amount that can be used depends primarily on the energy available in the rumen and the amount of ammonia that results from the degradation of protein from other ingredients fed. Also, new feeder cattle must be healthy and started on feed before they make efficient use of NPN. Complex methods are being developed to calculate the amount of NPN that can be fed with a particular ration. However, using the following guidelines will result in making effective use of NPN.

1. Do not feed urea to new feeder cattle until they are healthy and started on feed. This usually requires 3-4 weeks.
2. Make sure urea-containing supplements are thoroughly mixed. It is best to add enough ground grain so that the urea is not more than 20 percent of the supplement.

3. Mix urea-based supplements completely with the other ration ingredients being fed to appetite (full fed). These are usually silage and grain. This ensures a uniform consumption of small amounts of urea throughout the day for most efficient utilization. This practice will practically eliminate any possibility of urea toxicity.
4. Avoid feeding or minimize the amount of urea fed to calves under the equivalent weight of 600 pounds. Light cattle can utilize some NPN, but the level fed to them should not exceed 1/4 to 1/3 of the supplemental protein, or 0.5 percent urea in the total ration dry matter.
5. Avoid feeding urea to cattle of all weights being fed low-energy forages.

In university trials, calves (450-550 lbs, initial weight) fed corn-corn silage rations supplemented entirely with NPN (either urea or urea- or ammonia-treated corn silage) after they were started on feed have averaged 5-7 percent lower gains and feed efficiency. Under most conditions, feeding NPN rather than soybean meal was more profitable, even with the reduced performance. Analysis of these experiments showed that most of the reduced performance occurred before they reached an equivalent weight of 600 pounds. After that time, performance was nearly the same for those fed NPN or soybean meal. Thus, starting with an all natural supplement, then switching to a higher urea supplement is the best approach most of the time. Many feedlots will never feed more than 1/2 to 3/4 percent urea in the ration dry matter. This is often done to avoid overfeeding urea due to mixing errors, or unexplained, lower than expected performance with high urea supplements. Factors such as level of soluble nitrogen in the ration, health problems with the cattle, or reduced or excessive feed intake often occur after the cattle are on feed. Under these conditions, having some natural supplemental protein in the ration is observed to be beneficial.

Treating Silage with NPN

Adding NPN to corn silage at ensiling is preferred over feeding NPN at feeding time. The added NPN extends the fermentation and appears to reduce the breakdown of plant protein. Two to three times as much lactic acid is formed, which improves the overall efficiency of utilization of the energy in the silage. Spoilage is also likely reduced, due to the preservative effects of lactic acid.

226 Nutrient Requirements of Beef Cattle

This may be offset, however, by a lower performance in light calves unless soybean meal is fed until they reach an equivalent weight of 600 lbs or more.

Levels to use and the economics of NPN treatment are discussed in detail in Chapter 10. Under most conditions, it has been profitable to treat corn silage with NPN.

8-7 SUMMARY OF NUTRIENT REQUIREMENTS FOR GROWING AND FINISHING FEEDLOT CATTLE

Nutrient requirements for growing and finishing feedlot cattle are usually presented on a percentage basis, for convenience in ration formulation. For a percentage value to be accurate, however, dry matter intake must be accurately estimated. Factors affecting dry matter intake and estimating feed consumption are discussed in detail in Chapter 9. Values given there were adjusted for the effect of Rumensin, then were divided into the daily nutrients needed to calculate the percentage of protein, calcium and phosphorus for various weights and levels of grain feeding. Tables 8-17 and 8-18 contain these values. Average expected rate of gain is given, based on average expected feed intake and ration energy and protein levels.

Weight Groupings

Nutrient requirements are given for groups of cattle that have the same equivalent weight. Within a weight category, rate of gain is higher for larger-type cattle and is lower for smaller-type cattle. However, dry matter intake varies with body size, and because of the differences in body weight to maintain, feed efficiency is similar for cattle within each weight category.

All animals within a particular weight category have similar nutrient requirements, when expressed as a percentage of the ration dry matter. Variation in dry matter intake will adjust for differences in actual lb/day of nutrients needed due to differences in body size.

Five weight groupings are used in summarizing nutrient requirements. Rations 1-3 are for *equivalent weights* of 500 lbs, rations 4-6 are for 600 lbs, rations 7-10 are for 700 lbs, rations 11-14 are for 800 lbs, and rations 15-18 are for 900 lbs to market weight. Table 8-17 is for cattle started on feed as calves and Table 8-18 is for cattle started on feed as yearlings. The major difference

Table 8-17
Summary of Nutrient Requirements for Growing and Finishing Cattle Started on Feed as Calves

Telplan 44 Ration No. to use	Heifers						Steers						NEg Mcal/lb	NP %	Ca %	P %	Crude protein needed in corn & hay crop silage based ration
	Small		Average		Large		Small		Average		Large						
	ADG	DMI	ADG	DMI	ADG	DMI	ADG	DMI	ADG	DMI	ADG	DMI					
	320		400		480		400		500		600						
1	1.5	8.4	1.7	9.8	2.0	11.3	1.7	9.8	1.9	11.7	2.3	13.4	.46	3.67	.50	.37	13.6
2	1.7	8.4	2.0	9.8	2.2	11.3	2.0	9.8	2.3	11.7	2.6	13.4	.51	4.27	.50	.37	14.7
3	2.0	8.4	2.3	9.8	2.6	11.3	2.3	9.8	2.7	11.7	3.0	13.4	.57	4.87	.50	.37	15.5
3a	2.0	7.6	2.3	8.8	2.6	10.1	2.3	8.8	2.7	10.6	3.0	12.1	.63	5.38	.55	.41	16.2
3b	1.8	6.8	2.1	7.9	2.4	9.1	2.1	7.9	2.5	9.5	2.8	10.9	.66	5.58	.55	.41	15.4
	385		480		575		480		600		720						
	ADG	DMI	ADG	DMI	ADG	DMI	ADG	DMI	ADG	DMI	ADG	DMI					
4	1.5	9.6	1.7	11.3	1.9	12.9	1.7	11.3	1.9	13.4	2.2	15.4	.46	3.13	.44	.31	12.2
5	1.7	9.6	2.0	11.3	2.2	12.9	2.0	11.3	2.3	13.4	2.6	15.4	.51	3.58	.44	.31	12.9
6	2.0	9.6	2.3	11.3	2.6	12.9	2.3	11.3	2.7	13.4	3.0	15.4	.57	4.03	.44	.31	13.5
6a	2.0	8.7	2.3	10.1	2.5	11.6	2.3	10.1	2.7	13.9	3.0	15.4	.63	4.46	.49	.34	14.0
6b	1.9	7.8	2.2	9.1	2.4	10.4	2.2	9.1	2.5	10.9	2.8	12.5	.66	4.68	.49	.34	13.2
	450		560		670		560		700		840						
	ADG	DMI	ADG	DMI	ADG	DMI	ADG	DMI	ADG	DMI	ADG	DMI					
7	1.5	10.8	1.7	12.8	1.9	14.6	2.0	12.8	1.9	15.1	2.2	17.2	.46	2.72	.37	.29	11.6
8	1.7	10.8	2.0	12.8	2.2	14.6	2.0	12.8	2.3	15.1	2.6	17.2	.51	3.18	.37	.29	11.9
9	2.0	10.8	2.3	12.8	2.6	14.6	2.3	12.8	2.7	15.1	3.0	17.2	.57	3.58	.37	.29	11.9
10	2.0	9.8	2.3	11.6	2.6	13.2	2.3	11.6	2.7	13.7	3.0	15.6	.63	3.94	.41	.32	12.1
10a	1.9	8.8	2.2	10.9	2.4	11.9	2.2	10.4	2.5	12.3	2.8	14.0	.66	4.15	.41	.32	11.5
	510		640		770		640		800		960						
	ADG	DMI	ADG	DMI	ADG	DMI	ADG	DMI	ADG	DMI	ADG	DMI					
11	1.3	11.7	1.6	14.0	1.8	15.9	1.6	14.0	1.8	16.5	2.0	18.8	.46	2.48	.31	.26	10.3
12	1.6	11.7	1.8	14.0	2.1	15.9	1.8	14.0	2.1	16.5	2.4	18.8	.51	2.73	.31	.26	10.6
13	1.9	11.7	2.2	14.0	2.5	15.9	2.2	14.0	2.6	16.5	2.9	18.8	.57	3.15	.31	.26	10.7
14	1.9	10.5	2.2	12.6	2.5	14.3	2.2	12.6	2.6	14.8	2.9	16.9	.63	3.51	.34	.28	10.7
14a	1.8	9.5	2.1	11.3	2.3	12.9	2.1	11.4	2.4	13.3	2.7	15.2	.66	3.83	.24	.28	10.2
	575		720		865		720		900		1080						
	ADG	DMI	ADG	DMI	ADG	DMI	ADG	DMI	ADG	DMI	ADG	DMI					
15	1.3	12.6	1.6	14.9	1.8	16.9	1.6	14.9	1.8	17.6	2.0	20.0	.46	2.27	.31	.26	9.6
16	1.6	12.6	1.8	14.9	2.1	16.9	1.8	14.9	2.1	17.6	2.4	20.0	.51	2.50	.31	.26	9.7
17	1.9	12.6	2.2	14.9	2.5	16.9	2.2	14.9	2.6	17.6	2.9	20.0	.57	2.90	.31	.26	9.7
18	1.9	11.4	2.2	13.5	2.5	15.3	2.2	13.5	2.6	15.6	2.9	18.1	.63	3.23	.34	.28	9.3
18a	1.8	10.2	2.1	12.1	2.3	13.8	2.1	12.1	2.4	14.3	2.7	16.3	.66	3.29	.34	.28	9.5

Equivalent weight

Table 8-18
Summary of Nutrient Requirements for Growing and Finishing Cattle Started on Feed as Yearlings

Telplan 44 Ration No. to use	Heifers						Steers						NE Mcal /lb	NP %	Ca %	P %	Crude protein needed in corn & hay crop silage based ration %
	Small		Average		Large		Small		Average		Large						
	ADG	DMI	ADG	DMI	ADG	DMI	ADG	DMI	ADG	DMI	ADG	DMI					
	\-\-\- Equivalent weight \-\-\-						\-\-\- \-\-\- \-\-\- \-\-\- \-\-\- \-\-\-										
	385		480		575		480		600		720						
4	1.7	10.7	2.0	12.6	2.3	14.3	2.0	11.2	2.3	14.9	2.6	17.1	.46	3.22	.44	.31	12.2
5	2.0	10.7	2.3	12.6	2.3	14.3	2.3	11.2	2.6	14.9	3.0	17.1	.51	3.56	.44	.31	12.9
6	2.3	10.7	2.7	12.6	3.0	14.3	2.7	11.2	3.1	14.9	3.5	17.1	.57	4.09	.44	.31	13.5
6a	2.3	9.6	2.7	11.3	3.0	12.9	2.7	10.1	3.1	13.4	3.5	15.4	.63	4.55	.49	.34	14.0
6b	2.2	8.7	2.5	10.1	2.9	11.6	2.5	9.1	2.9	12.1	3.2	13.9	.66	4.79	.49	.34	13.2
	\-\-\- Equivalent weight \-\-\-																
	450		560		670		560		700		840						
7	1.7	12.0	2.0	14.2	2.3	16.2	2.0	14.2	2.3	16.8	2.6	19.1	.46	2.86	.37	.29	11.6
8	2.0	12.0	2.3	14.2	2.6	16.2	2.3	14.2	2.6	16.8	3.0	19.1	.51	3.10	.37	.29	11.9
9	2.3	12.0	2.7	14.2	3.0	16.2	2.7	14.2	3.1	16.8	3.5	19.1	.57	3.57	.37	.29	11.9
10	2.3	10.8	2.7	12.8	3.0	14.6	2.7	12.8	3.1	15.1	3.5	17.2	.63	3.97	.41	.32	12.1
10a	2.2	9.8	2.5	11.6	2.9	13.2	2.5	11.6	2.9	13.7	3.2	15.6	.66	4.16	.41	.32	11.5
	\-\-\- Equivalent weight \-\-\-																
	510		640		770		640		800		960						
11	1.6	13.0	1.9	15.4	2.1	17.7	1.9	15.6	2.2	18.3	2.4	20.9	.46	2.57	.31	.26	10.3
12	1.9	13.0	2.2	15.4	2.4	17.7	2.2	15.6	2.5	18.3	2.8	20.9	.51	2.79	.31	.26	10.6
13	2.2	13.0	2.6	15.4	2.9	17.7	2.6	15.6	2.9	18.3	3.3	20.9	.57	3.11	.31	.26	10.7
14	2.2	11.7	2.6	14.0	2.8	15.9	2.6	14.0	2.9	16.5	3.3	18.8	.63	3.45	.34	.28	10.7
14a	2.2	10.5	2.4	12.6	2.8	14.3	2.4	12.6	2.8	14.8	3.1	16.9	.66	3.72	.34	.28	10.2
	\-\-\- Equivalent weight \-\-\-																
	575		720		865		720		900		1080						
15	1.6	14.0	1.9	16.6	2.1	18.8	1.9	16.6	2.2	19.6	2.4	22.2	.46	2.32	.31	.26	9.6
16	1.9	14.0	2.2	16.6	2.4	18.8	2.2	16.6	2.5	19.6	2.8	22.2	.51	2.50	.31	.26	9.7
17	2.2	14.0	2.6	16.6	2.9	18.8	2.6	16.6	2.9	19.6	3.3	22.2	.57	2.76	.31	.26	9.7
18	2.2	12.6	2.6	14.9	2.8	16.9	2.6	14.9	2.9	17.6	3.3	20.0	.63	3.07	.34	.28	9.3
18a	2.2	11.4	2.4	13.5	2.8	15.3	2.4	13.5	2.8	15.6	3.1	18.1	.66	3.40	.34	.28	9.5

between these two tables is a 10 percent higher intake and thus a higher rate of gain for yearlings.

Ration Numbers

The ration number given in the right-hand column is used in Telplan 44, "Computerized Ration Formulation for Growing and Finishing Cattle." The number increases with the weight of cattle and concentration of energy in the ration. Those rations containing an a or b suffix are not normally fed to that particular weight of cattle, and are not programmed as such in the computer program. They can be formulated, however, by using the ration number given and entering from Tables 8-17 or 8-18 the appropriate NE_g, dry matter intake, and protein requirements on the input form.

Level of Grain Feeding and Protein Requirements

Nutrient requirements are given for five concentrations of net energy in the ration. Expected rate of gain increases as net energy concentration or level of grain feeding increases, up to a concentration of 0.63 Mcal NE_g/lb dry matter. At that point, intake declines as less roughage is fed, resulting in no change in daily gain. When all roughage is removed, intake further declines and daily gain decreases. Concentration of net protein needed is determined by dividing the lbs needed/day for maintenance and gain for the expected rate of gain by the expected intake. Thus as rate of gain increases, a higher concentration of protein is needed. Also when intake declines while the rate of gain stays the same, concentration of protein needed increases.

The level of crude protein needed to meet net protein requirements will vary with the quality of the feed protein. The most important effect is when the ration is based on silage. The levels of crude protein that result when the ration net protein needs are met in silage-based rations are given in Tables 8-17 or 8-18. Crude protein needs may be lower if dry roughage is used. However, they may be higher if ensiled high moisture corn is used, rather than dry corn. Using a supplemental protein source with a lower protein solubility, such as brewer's grain or treated soybean meal, may reduce supplemental protein needs. Research is underway at several experiment stations to identify and quantify these effects.

Other Factors Affecting Dry Matter Intake and Gain Projections

Feeding Rumensin. Expected intakes are based on the ration containing 33 grams Rumensin/ton dry matter. If Rumensin is not fed, or if the ration contains less than 20 grams/ton Rumensin, intakes will be 5-10 percent higher than those given in Tables 8-17 and 8-18. The concentration of protein and minerals needed will also decline 10 percent if Rumensin is not fed. The same protein intake will be obtained at a 10 percent lower concentration in the ration, because dry matter intake will increase 10 percent when Rumensin is not fed. Thus *the same lb/head/day of protein mineral supplement* should be fed, with or without Rumensin. Some research suggests that Rumensin increases feed protein utilization, in addition to improving feed efficiency 10 percent. However, until the conditions under which this improvement in protein utilization occurs are clearly defined, it is suggested that the guidelines given here be followed.

Thus, dry matter intakes, protein and mineral levels in Tables 8-17 and 8-18 are adjusted for the influence of Rumensin on dry matter intake. However, NE_g concentrations given were not, for the following reason. Telplan 44 and other computer programs are set up for these unadjusted values, and make the adjustment for Rumensin internally, by multiplying feed NE_m and NE_g levels by 1.10. However, the correct concentration of protein must be given as the program at present does not make adjustments on protein and minerals for the effect of Rumensin. For hand calculations, however, in predicting the rate of gain for NE_g, values given here should be increased 10 percent if Rumensin is fed.

Cattle condition, growth stimulants and environment. All gain projections are based on feeders purchased in average flesh condition, and receiving a growth stimulant. Fleshy calves will gain 5-10 percent slower on the same intake; thin calves will gain 5-10 percent faster. Decrease gains 12 percent if a growth stimulant is not used; implants must be given every 100 days to obtain the performance given. Decrease daily gains 5-10 percent for well drained, mounded unsheltered lots in the corn belt and other northern climates. In poorly drained unsheltered lots, gains could be 20-30 percent lower. The gains given are for cattle after they are on feed. Overall pay-to-pay gains and feed efficiency will be 5-10 percent lower when shrink and death loss are taken into account, as in Table 10-14.

Summary of other nutrient requirements. Other nutrients normally of concern in formulating beef cattle rations are potassium, salt, sulfur and vitamins. The ration dry matter should be formulated to contain 0.6-0.8 percent potassium and 0.25-0.30 trace mineralized salt.

When nonprotein nitrogen products, such as urea, are mixed in a protein mineral supplement or added to corn silage prior to ensiling, it may be desirable to add inorganic sulfur to the ration to ensure that the nitrogen-sulfur ration is normal. The normal range is 10:1 to 15:1. One must be careful to not add more supplemented sulfur than needed, however, because high levels of sulfur in the ration will likely reduce intake.

A good rule-of-thumb is to add 3.5 lbs of inorganic sulfur for every 100 lbs of urea or 0.3 lb of inorganic sulfur for every 5 lbs of anhydrous ammonia. The most commonly available, and usually least costly, sources of sulfur are calcium or potassium sulfate. To meet the above sulfur levels, add 20 lbs of calcium or potassium sulfate/100 lbs urea or 37 lbs/100 lbs of anhydrous ammonia.

The use of calcium sulfate to meet the sulfur requirement reduces the supplemental calcium requirements since calcium sulfate contains 20 percent calcium. As a consequence, for every 100 lbs of calcium sulfate used, limestone can be reduced by 53 pounds.

Calcium:phosphorus ratio should be at least 1:1. The ration should contain the following levels of vitamins/lb of dry matter: Newly arrived feeder cattle, 2,000-3,000 IU; cattle under heat stress, 1,500 IU; normal conditions, 1,000 IU. The ration should contain 120 IU Vitamin D/lb dry matter when fed to cattle in confinement.

Typical Ration Formulations for Various Weights and Energy Levels

Table 8-19 gives ration dry matter formulas that result when the requirements given in Tables 8-17 and 8-18 and average net energy and protein values from Table 9-1 are used. These can be used as guidelines, as well as for gain simulation (Telplan 56) to identify the most profitable feeding system and for feed budgeting purposes. For example, a feeder may compare using ration sequences 1, 4, 7, 11 and 18 vs. 1, 4, 10 and 18 or 1, 6, 10, and 18 to determine the optimum point to change to a high energy ration. He may also look at haylage vs. corn silage based rations, or different sequences of haylage and corn silage feeding.

A 40 percent protein supplement is used due to its wide usage. Appropriate levels of other supplements can be substituted for purposes of gain simulation. However, the complete ration should be balanced and the appropriate protein and mineral supplement selected for each ration after the rations (1–18) to be fed are identified. This can be done with Telplan 44 or by hand. Guidelines for protein and mineral supplement compositions are given in Tables 10-7 and 10-13.

Table 8-19

Typical Ration Dry Matter Formulas for Various Weights and Energy Levels When Feed Protein Values are Average

Ration No.	Shelled corn-corn silage			Ear corn-corn silage				Corn-legume hay		
	Shelled corn %	Corn silage %	40% Protein supplement %	Shelled corn %	Ear corn %	Corn silage %	40% Protein supplement %	Shelled corn %	Legume haylage %	40% Protein supplement %
1	0	85.0	15.0	0	0	85.0	15.0	45.0	55.0	0
2	13.2	69.4	17.4	0	38.8	46.0	15.2	58.0	39.1	2.9
3	41.7	40.2	18.1	0	85.1	0	14.9	67.6	22.7	9.7
3a	70.4	11.3	18.3	64.8	16.7	0	15.5	77.8	6.3	15.9
3b	84.5	0	15.5	84.5	0	0	15.5	84.5	0	15.5
4	0	88.7	11.3	0	0	88.7	11.3	45.0	55.0	0
5	15.3	71.2	12.5	0	43.8	46.0	10.2	60.0	40.0	0
6	45.3	42.4	12.3	0	86.6	0	9.0	73.9	23.9	3.5
6a	75.5	13.7	11.8	63.6	23.6	0	7.3	83.3	7.7	9.0
6b	90.9	0	9.1	90.9	0	0	9.1	90.9	0	9.1
7	0	91.6	8.4	0	0	91.6	8.4	45.0	55.0	0
8	18.6	72.6	8.8	0	47.7	46.0	6.3	60.0	40.0	0
9	48.0	44.0	8.0	0	95.6	0	4.4	75.2	24.8	0
10	77.5	15.5	7.0	69.1	28.0	0	2.9	87.6	8.7	3.7
10a	95.8	0	4.2	95.8	0	0	4.2	95.8	0	4.2
11	0	93.8	6.2	0	0	93.8	6.2	45.0	55.0	0
12	20.4	73.7	3.9	0	50.6	46.0	3.4	60.0	40.0	0
13	50.1	45.3	4.6	0	99.2	0	0.8	75.0	25.0	0
14	80.0	16.9	3.1	68.2	31.8	0	0	89.5	9.5	0
14a	99.5	0	0.5	99.5	0	0	0.5	99.5	0	0.5
15	0	95.7	4.3	0	0	95.7	4.3	45.0	55.0	0
16	22.0	74.6	3.4	0	53.2	46.0	0.8	60.0	40.0	0
17	51.9	46.4	1.7	0	100.0	0	0	75.0	25.0	0
18	80.9	18.1	0	66.3	33.7	0	0	90.0	10.0	0
18a	100.0	0	0	100.0	0	0	0.4	100.0	0	0

In practice, it is not usually practical for cattle feeders to have as many weight groupings as given here. Three practical groups for calves would be *equivalent weights* up to 600, 600 to 800, and 800 to market. For yearlings, two groups would usually be adequate: 650 to 850, and 850 to market. It is recommended that the ration be formulated based on the lightest cattle in the group. Obviously protein supplementation will be the most efficient when the cattle are grouped carefully according to weight.

The lb/head/day of each ingredient can be estimated by multiplying the proportion given by the dry matter intake given in Tables 8-17 and 8-18. For example, a 500-lb steer fed ration 1 would be getting 11.7 × .15 = 1.75 lbs of supplement dry matter. The as fed lb/head/day can then be obtained by dividing the lb dry matter by the proportion of dry matter in the feed. If the supplement was 90 percent dry matter, then the as fed lb = 1.75/0.90 =

1.95. The percentage of supplement given assumes Rumensin is fed, as discussed previously. The percentage of supplement needed would be lower if Rumensin is not fed, due to the increased intake. As discussed previously, these levels of protein supplement may be 1.5-2.5 percentage units or 0.2-0.3 lb/head/day higher than needed, if current research underway clearly proves that Rumensin improves feed protein utilization.

8-8 NUTRIENT REQUIREMENTS OF BREEDING CATTLE

The goal of a cow-calf operation should be to produce a calf weighing a minimum of 450 pounds every 12 months from every cow. The key to success in developing a beef cow herd to reach this goal lies in becoming knowledgeable in nutrition, breeding, and health management of the beef herd. You must develop economical systems to harvest and utilize roughages for fall and winter feeding and economical pasture or drylot feeding systems for the critical lactating and breeding periods. The good cowman knows the nutrient value of the feeds available and knows how to use these to meet the requirements of the cow at the various stages of her reproductive cycle. Feed costs are over half of the total cost in producing a calf. This is the area where the greatest reduction in costs can probably be obtained. Underfeeding, however, is false economy.

Nutrient Requirements

Minimum nutrient requirements for various classes of cattle are presented in the tables that follow (Tables 8-20 to 8-24). Requirements are listed for energy (TDN or total digestible nutrients), crude or total protein, calcium, phosphorus, and vitamin A. Along with water and salt, these five nutrients are the ones that are of greatest practical concern to cattlemen. Salt (sodium chloride) and various trace minerals are certainly important, but their requirements are normally met by feeding trace mineralized salt. The vitamin D requirement is met by exposure to direct sunlight or by feeding sun-cured forages. Vitamin E deficiency is found only in young calves in the form of white muscle disease and is best prevented by a vitamin E—selenium injection at birth. Mature ruminants, including cattle, receive adequate amounts of B vitamins and vitamin K through bacterial synthesis in the rumen.

The nutrient requirements are presented in two ways: (1) in

Table 8-20
Requirements of Weaned Heifer Calves

Heifer Wt. (lb.)	Max. Daily DM (lb.)	TDN lb/day	TDN % of DM	CRUDE PROTEIN lb/day	CRUDE PROTEIN % of DM	CALCIUM g/day	CALCIUM % of DM	PHOSPHORUS g/day	PHOSPHORUS % of DM	VITAMIN A IU/day	VITAMIN A IU/lb.
				Average Daily Gain of 1.1 lb. per day							
330	9.9	5.7	61	1.00	11.0	14	.34	12	.29	9,000	1000
440	13.2	7.7	58	1.28	9.6	14	.23	13	.22	13,000	1000
550	14.3	8.6	58	1.37	9.5	14	.20	13	.20	14,000	1000
660	16.5	9.9	61	1.48	9.2	14	.19	14	.19	16,000	1000
770	18.3	11.2	61	1.61	8.7	15	.18	15	.18	18,000	1000
Avg., all wts.	15.0	8.6	60	1.35	9.6	14	.23	13	.22	14,000	1000
				Average Daily Gain of 1.5 lb. per Day							
330	9.9	6.2	69	1.10	12.4	18	.45	14	.35	9,000	1000
440	13.2	8.4	64	1.37	10.2	18	.30	16	.27	13,000	1000
550	14.3	9.1	72	1.37	10.5	17	.29	15	.26	14,000	1000
660	16.5	10.4	72	1.48	10.1	16	.24	15	.23	16,000	1000
770	18.3	11.9	69	1.61	9.2	15	.19	15	.19	18,000	1000
Avg., all wts.	15.0	9.2	69	1.40	10.5	17	.29	15	.26	14,000	1000
Overall Avg.	15.0	8.9	65	1.40	10.0	15	.26	14	.24	14,000	1000

Table 8-21
Requirements of Coming 2-Year-Old Heifers in the Last 3–4 Months of Pregnancy

Heifer Wt. (lb.)	Max. Daily DM (lb.)	TDN		CRUDE PROTEIN		CALCIUM		PHOSPHORUS		VITAMIN A	
		lb/day	% of DM	lb/day	% of DM	g/day	% of DM	g/day	% of DM	IU/day	IU/lb.
Average Daily Gain of 0.9 lb. per Day											
715	20.7	7.7	52	1.28	8.8	15	.23	15	.23	19,000	1275
770	22.0	8.1	52	1.35	8.8	15	.22	15	.22	19,000	1275
825	24.2	8.4	52	1.39	8.7	15	.21	15	.21	20,000	1275
880	25.6	8.7	52	1.43	8.7	16	.21	16	.21	21,000	1275
935	26.7	9.0	52	1.52	8.8	16	.20	16	.20	22,000	1275
935	26.7	9.0	52	1.52	8.8	16	.20	15	.21	20,000	1275
Avg., all wts.	23.8	8.4	52	1.40	8.8	15	.21				
Average Daily Gain of 1.3 lb. per Day											
715	20.7	9.9	52	1.65	8.8	18	.21	18	.21	23,000	1275
770	22.0	10.3	52	1.72	8.8	19	.21	19	.21	25,000	1275
825	24.2	10.8	52	1.78	8.7	19	.20	19	.20	26,000	1275
880	25.6	11.3	52	1.85	8.7	19	.20	19	.20	27,000	1275
935	26.7	10.8	52	1.80	8.8	19	.20	19	.20	26,000	1275
Avg., all wts.	23.8	10.6	52	1.80	8.8	19	.20	19	.20	26,000	1275
Overall Avg.	23.8	9.6	52	1.60	8.8	17	.21	.7	.21	23,000	1275

Table 8-22
Requirements of Dry Pregnant Mature Cows

Cow Wt. (lb.)	Max. Daily DM (lb.)	TDN lb/day	TDN % of DM	CRUDE PROTEIN lb/day	CRUDE PROTEIN % of DM	CALCIUM g/day	CALCIUM % of DM	PHOSPHORUS g/day	PHOSPHORUS % of DM	VITAMIN A IU/day	VITAMIN A IU/lb.
Middle ⅓ of Pregnancy, 0.0 lb. Average Daily Gain												
772	20	6.6	52	.71	5.9	10	.18	10	.18	15,000	1275
882	22	7.3	52	.79	5.9	11	.18	11	.18	17,000	1275
992	24	7.9	52	.86	12	12	.18	12	.18	19,000	1275
1002	26	8.6	52	.93	5.9	13	.18	13	.18	20,000	1275
1213	28	9.2	52	.99	5.9	14	.18	14	.18	22,000	1275
1323	30	9.8	52	1.08	5.9	15	.18	15	.18	23,000	1275
1433	32	10.4	52	1.15	5.9	16	.18	16	.18	25,000	1275
1545	34	11.0	52	1.21	5.9	17	.18	17	.18	27,000	1275
Avg., all wts.	27	8.9	52	1.00	5.9	14	.18	14	.18	21,000	1275
Last ⅓ of Pregnancy, 0.9 lb. Average Daily Gain												
772	20	8.0	52	.90	5.9	12	.18	12	.18	19,000	1275
882	22	8.7	52	.97	5.9	14	.18	14	.18	21,000	1275
992	24	9.4	52	1.06	5.9	15	.18	15	.18	23,000	1275
1102	26	10.0	52	1.12	5.9	15	.18	15	.18	24,000	1275
1213	28	10.7	52	1.19	5.9	16	.18	16	.18	26,000	1275
1323	30	11.2	52	1.26	5.9	17	.18	17	.18	27,000	1275
1433	32	11.9	52	1.32	5.9	18	.18	18	.18	29,000	1275
1545	34	12.6	52	1.39	5.9	19	.18	19	.18	30,000	1275
Avg., all wts.	27	10.3	52	1.20	5.9	16	.18	16	.18	25,000	1275
Overall Avg.	27	9.6	52	1.10	5.9	15	.18	15	.18	23,000	1275

Table 8-23
Requirements of Lactating Cows in the First 3–4 Months After Calving

Cow Wt. (lb.)	Max. Daily DM (lb.)	TDN lb/day	TDN % of DM	CRUDE PROTEIN lb/day	CRUDE PROTEIN % of DM	CALCIUM g/day	CALCIUM % of DM	PHOSPHORUS g/day	PHOSPHORUS % of DM	VITAMIN A IU/day	VITAMIN A IU/lb.
				Average Milking Ability (10-12 lb/day)							
770	25	9.7	52	1.65	9.2	24	.29	24	.29	19,000	1775
880	27	10.4	52	1.79	9.2	25	.28	25	.28	21,000	1775
990	29	11.0	52	1.90	9.2	26	.28	26	.28	23,000	1775
1100	31	11.7	52	1.98	.92	27	.28	27	.28	24,000	1775
1210	33	12.3	52	2.14	9.2	28	.27	28	.27	26,000	1775
1320	35	13.0	52	2.23	9.2	28	.25	28	.25	27,000	1775
1430	37	13.7	52	2.32	9.2	29	.25	29	.25	29,000	1775
1540	39	14.4	52	2.41	9.2	30	.25	30	.25	31,000	1775
Avg., all wts.	32	12.0	52	2.05	9.2	27	.27	27	.27	25,000	1775
				Superior Milking Ability (21-23 lb/day)							
770	30	12.8	55	2.45	10.9	45	.44	40	.39	32,000	1775
880	32	13.5	55	2.58	10.9	45	.42	41	.38	34,000	1775
990	34	14.1	55	2.71	10.9	45	.40	42	.37	36,000	1775
1100	36	14.8	55	2.84	10.9	46	.39	43	.36	38,000	1775
1210	38	15.4	55	2.98	10.9	46	.37	44	.35	41,000	1775
1320	40	16.1	55	3.11	10.9	46	.36	44	.34	43,000	1775
1430	42	16.8	55	3.22	10.9	47	.35	45	.33	45,000	1775
1540	44	17.5	55	3.33	10.9	48	.34	46	.32	47,000	1775
Avg., all wts.	37	15.1	55	2.90	10.9	46	.38	43	.35	40,000	1775
Overall Avg.	34	13.5	53.5	2.50	10.0	37	.33	35	.31	32,000	1775

Table 8-24
Requirements of Bulls
(Growth + Maintenance, Moderate Activity)

Bull Wt. (lb.)	Avg. Daily Gain (lb.)	Max. Daily DM (lb.)	TDN lb/day	TDN % of DM	CRUDE PROTEIN lb/day	CRUDE PROTEIN % of DM	CALCIUM g/day	CALCIUM % of DM	PHOSPHORUS g/day	PHOSPHORUS % of DM	VITAMIN A IU/day	VITAMIN A IU/lb.
660	2.4	19	13.2	77	2.16	12	29	.41	23	.32	34,000	1775
880	2.0	24	15.4	64	2.27	22	23	.21	23	.21	43,000	1775
1100	1.5	27	16.5	61	2.36	10	22	.18	22	.18	48,000	1775
1323	1.1	29	16.1	61	2.25	9	22	.18	22	.18	48,000	1775
1543	0.3	31	17.0	55	2.38	8.5	23	.18	23	.18	50,000	1775
1764	0	33	12.8	55	1.96	8.5	19	.18	19	.18	41,000	1775
1984	0	35	13.9	55	2.32	8.5	22	.18	22	.18	48,000	1775
2205	0	37	15.2	55	2.32	8.5	22	.18	22	.18	48,000	1775
2425	0	39	16.4	55	2.40	8.5	23	.18	23	.18	51,000	1775

pounds, grams or international units (IU) per day; and (2) in percentage of the ration dry matter (DM). In each of the tables, an estimate is made of the animal's maximum possible daily dry matter intake. As a guide to total daily DM consumption, most dry hays and grains contain 85 to 90 percent dry matter, whereas most silages contain only 30 to 50 percent. Maximum dry matter intake varies with the moisture content of the ration, season of the year, palatability of the ration, size and age of the animal, and whether or not the animal is lactating. For example, maximum intake is usually higher on dry feeds than on silages; higher in cold weather than in hot; increases with size and age; and is generally higher for the lactating cow than for the dry cow. As can be seen in the tables, the amount of a given nutrient required per day tends to increase as young cattle grow larger; however, the required concentration of that nutrient in the diet tends to decline with age and size.

Effect of cold weather on feed requirements. Research in Western Canada, Kansas, and elsewhere has shown that the stress of extremely cold weather increases the energy requirements of cattle. This can be an important consideration in the wintering of brood cows in the Northern states. The requirements listed in the tables here are valid for a temperature range of 30° to 80° F, which is normally considered the comfort zone for most cattle. Energy requirements increase when the temperature goes above, or below, this range. The increase is especially dramatic for cattle in extremely cold weather with no shelter. Wind, together with cold stress, further increases the need for additional energy to maintain body temperature and body weight. Wind chill factors for beef cattle are given in the following chart.

8-8 Nutrient Requirements of Breeding Cattle

Wind speed mph	\multicolumn{7}{c}{Temperature (°F)}						
	0	5	10	15	20	25	30
0	0	5	10	15	20	25	30
5	-5	1	5	10	15	20	25
10	-8	-6	-4	4	9	14	19
15	-16	-11	-6	-1	4	9	14
20	-20	-15	-10	-5	-1	3	8
25	-27	-22	-17	-1	-9	-2	3
30	-36	-31	-26	-21	-16	-11	-6
35	-50	-45	-40	-35	-30	-25	-20
40	-66	-62	-59	-53	-48	-43	-34

Generally speaking, an 1100-lb dry brood cow in good condition with a full coat of winter hair and no access to shelter will require 13 percent more energy or TDN for each 10° decline in the wind chill factor below 30°. For example, if the temperature were 0°F and the wind velocity were 20 mph, the wind chill factor would be -20°, or 50° below the critical temperature of 30°. This means that her maintenance requirement for energy would be increased by 65 percent. According to Table 8-22, a 1000-lb cow in midpregnancy needs 8.6 lbs TDN daily if she is in the comfort zone of 30° to 80°F; therefore, her TDN requirement would be 8.6 × 1.65 = 14.2 lbs, or an increase of 5.6 lbs of TDN. If the dry matter in the hay she receives averages 50 percent TDN, this would require the feeding of 28.4 lbs hay DM, which slightly exceeds her expected maximum daily DM intake of 26 lb. However, cattle consume more DM during cold weather so she would likely meet her requirement. If the wind chill factor were to drop significantly lower, a higher energy feed such as corn silage or grain would probably have to replace some of the hay in order to maintain her body weight. If the wind chill factor fell to -66°, her daily TDN requirement would be increased by 125 percent or 10.7 lb. This amount added to 8.6 lbs would come to a total of 19.3 lbs TDN or 38.6 lbs of hay dry matter. At this point, she could not consume enough hay to maintain her body weight.

An extremely thin cow with a poor hair coat is stressed even further by cold weather. Her energy requirement increases by about 30 percent with every 10° drop in wind chill factor below 30°F. On the other hand, cattle on feed are not stressed as much by low temperatures; their TDN needs are increased by about 8 percent for every 10° decline in wind chill factor below 30°F.

When using the requirements listed in the tables that follow, allowances should be made for cattle that are under extreme cold stress for extended periods of time with no access to shelter or windbreak. One or two days of cold stress are no cause for alarm, but long periods of below zero weather should be accounted for when feeding the cow herd.

Feeding the Cow Herd During a 12-Month Reproductive Cycle

Period 1. Midgestation (Spring Calving, Nov.-Jan.; Fall Calving, May-July). During this time, the nutrient requirements of the cow will be at a low point. From weaning up to 2-3 months before calving, the beef cow is fed primarily for maintenance. Grazing crop residues and diverted acres or medium to poor quality hay, straw, chaff, or other harvested crop residues can furnish much of the nutrients needed, when properly supplemented. Fat cows can and should lose some weight in early gestation. However, all cows should be maintaining their weight or gaining slightly (¼ to ½ lb/day) within 60 days of calving. After calving they should gain weight for at least 90-120 days or until the end of breeding season.

Period 2. 60-90 Days Before Calving (Spring Calving, Jan.-March; Fall Calving, July-Sept.). During this time, nutrients are needed for rapid fetal growth, in addition to those needed for maintenance. The nutritional level needed in the ration will depend primarily on the general condition of the cows. Additional silage or some grain may be needed if the cows are too thin. We do not want the cows too fat at calving time, however, as calving difficulties may result. Feeding for fat gain is too expensive. In addition we want the cow in a gaining condition between calving and rebreeding for best conception. It's difficult to flush a fat cow.

The effect of body condition at calving and heat after calving (Table 8-25) indicates poor return to heat of cows underfed prior to calving. Note that at the end of a normal 60-day breeding season, 46, 61, and 91 percent of the cows had shown heat in the thin, moderate, and good condition groups respectively. Cows that are thin at or near calving don't show heat during the breeding season, or only show heat late in the breeding season.

Tables 8-25 and 8-26 reflect the effect of cow condition and calving date. Notice the decrease in calf crop as cows calve later. In thin cows, calf crop decreased to 68 percent, a drop of 10 percent when cows calved January 10 to January 30, and then to 56 percent, a drop of 12 percent when cows calved from January 31 to February 19. In cows in good condition, the decrease in later calving cows was much less, only 1 percent in cows calving January 10 to January 30, and then 6 percent in cows calving January 31 to February 19.

Table 8-25
Body Condition at Calving and Heat After Calving

Body Condition at Calving	No. Cows	Proportion in heat at various days post-calving					
		40	50	60	70	80	90
		(percent in heat)					
Thin	272	19	34	46	55	62	66
Moderate	364	21	45	61	79	88	92
Good	50	31	42	91	96	98	100

Wiltbank, 1975.

Table 8-26
Effect of Cow Condition and Weight Gain and Heat After Calving and Calving Date on Pregnancy During a 60-Day Breeding Season (March 10—May 10)

Calving Time	Weight Change After Calving	Cow Condition at Calving		
		Thin %	Moderate %	Good %
Dec. 20 to Jan. 9	Gain	78	93	94
	Loss	64	87	90
Jan. 10 to Jan. 30	Gain	68	90	93
	Loss	64	83	90
Jan. 31 to Feb. 19	Gain	56	78	87
	Loss	64	77	88

Wiltbank, 1975.

This occurs because body condition causes cows to start cycling sooner after calving. This is especially true for late-calving cows.

Period 3. Calving Through Rebreeding (Spring Calving, Mar.-July; Fall Calving, Sept.-Jan.). This is the period of greatest nutritional needs. The cow loses about 125 pounds at calving and this weight should be regained in 90 to 120 days after calving, with most of it recovered by the start of breeding. In addition, she has to produce milk for a calf and get her reproduction tract in shape for rebreeding and conception besides meeting her maintenance requirements. Proper feeding is important to get the cows rebred quickly to avoid a strung out calf crop, which results in a lower average weaning weight and some cows not getting rebred in time to stay within a 12-month calving interval. The bulls should be removed after 60-90 days to prevent late calves next year. Then, pregnancy check and cull those not pregnant.

Period 4. End of Breeding to Weaning (Spring Calving, July-Nov.; Fall Calving, Jan.-May). Nutrients for milk production as well as maintenance are still needed, but the critical feeding period is over after the cow is rebred. Also, the calves are consuming other feeds in addition to milk. Use whatever feeds are readily available, such as temporary or permanent pastures.

All of these phases, and guidelines for ration needs, are summarized in Figure 8-4.

Feeding Replacement Heifers

The objectives here are to have replacement heifers calve as two-year-olds and then calve at the same time as mature cows the following year. This requires having them weigh 600 to 800 pounds at 14-15 months of age when first bred, and then feeding first and second calf heifers separately and at a higher nutritional level than the mature cows. The level of feeding needed from weaning to first breeding depends on their weaning weight and breed. If we want them to weigh 600 to 800 pounds at the start of breeding, which should be 20 to 30 days ahead of the mature cows, they will usually need to gain 200 to 250 pounds in 180-210 days, requiring a gain of 1 to 1½ pounds per day from weaning to first breeding (Tables 8-27 and 8-28).

During breeding season (14 to 16 mos) heifers should gain about 1.3 lbs per day. After breeding season, up to 120 days prior to calving (16 to 20 mos), they can afford to gain as little as ½ lb per day. During the last 120 days of gestation (20 to 24 mos), they should be fed to gain 0.9 to 1.3 lbs per day. After calving, they should continue to gain weight until they are bred.

In order for heifers to obtain the level of feed needed to gain properly, they should ideally be fed separately from the rest of the herd during their first and second winters. If not, the mature cows may consume more than their share of the feed, and the heifers are apt to suffer. This especially is true in larger herds of cattle and

8-8 Nutrient Requirements of Breeding Cattle 243

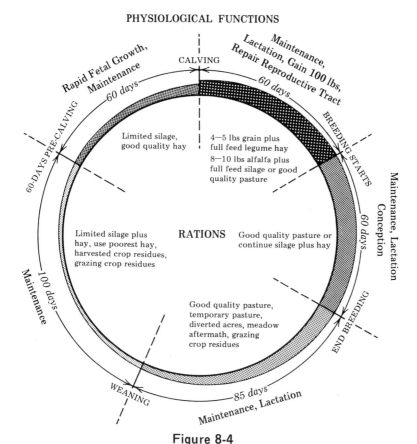

Figure 8-4
Beef Cow Reproductive and Feeding Cycle to Obtain One Calf per Cow Every Twelve Months

in herds where feeding space is limited. It is also a good idea to winter the coming 3-year-olds separately during their third winter if they are extremely thin from raising their first calf. In fact, many good producers feed their first and second-calf heifers and their old, thin cows all together as one nutritional management group.

Fall Calving vs Spring Calving

Some producers prefer fall calving since it allows them to wean calves in the spring when feeder cattle prices are often at their peak. They may also use these calves to utilize summer pasture

Table 8-27
Relationship of Weight and Breed to the Percentage of Heifers Showing Heat

Breeds	Proportion Cycling		
	85-95%	65-70%	50%
	Pounds		
Angus	650	600	550
Hereford	725	700	675
Shorthorn	600	550	535
Charolais	750	725	700
Angus x Hereford cross	675	760	625
Simmental x English breeds	750	700	675
Limousin x English breeds	750	700	675
Brahman x English breeds	750	725	700

Wiltbank, 1975.

Table 8-28
Effect of Feeding Charolais Heifers so They are Different Weights at Start of Breeding

Weight at start of breeding	Expect in heat after		Expected pregnant after	
	20 days	40 days	20 days	40 days
	of breeding		of breeding	
750	95	100	57	83
700	70	95	42	74
650	40	70	24	52

Wiltbank, 1975.

and then sell them in the fall as yearlings, resulting in more pounds of calf being marketed per cow every 12 months. This also avoids calving during the busy spring planting season.

This system requires more intensive management of the cow during the winter in the northern United States, as harvested feeds must be fed during nearly all of the critical lactation and breeding periods. Good quality spring pasture meets requirements in a spring calving system with little additional feed other than minerals. For most producers, spring calving is the preferred system.

9

Evaluation of Feedstuffs and Ration Formulation

9-1 THE USE OF FEED ANALYSIS

Nutrient values for individual feeds vary considerably from farm to farm and from year to year, as a result of differences in variety, soil fertility, weather, date of harvest, and harvesting and storage procedures.

Thus, it is obvious that feed analysis is important. The problem is to decide what it is practical to test for; when and how often to sample your own feeds; and how to get a representative sample. Nutrients of concern include energy, protein, calcium, phosphorus, Vitamin A, and some of the trace minerals. In deciding what to test for, the variability of the nutrient, the cost of analysis, and the cost of supplementation must be considered.

Energy

The primary source of energy in home grown feeds is carbohydrates. Carbohydrates include sugar, starch, cellulose, and hemicellulose. The cellulose and hemicellulose are contained in plant cell walls in a complex with lignin, an indigestible compound, while the sugars and starches are found in the cell contents. The cell walls are the lowest in digestibility, and cell contents highest. Thus, the higher the proportion of cell walls, the lower the TDN and net energy value of the feed.

The following laboratory procedures are used to estimate the usable energy content of a feed sample sent in for analysis.

1. Crude fiber. This is the standard analysis for the fibrous parts of the plant. Its development dates back to 1864. Digestibility, and thus energy value, typically decrease as the crude fiber percentage increases. However, it is only a rough guide to differences in fiber among feeds. The fiber of immature plants contains mostly cellulose and little lignin; the cellulose is highly digestible. In contrast, the fiber of mature plants contains large amounts of lignin, which is indigestible. Another shortcoming is that the crude fiber procedure does not measure accurately all of the lignin.

2. Nitrogen free extract (NFE). NFE represents the energy in the highly digestible cell contents, or sugars and starches. NFE, however, is not directly determined. It is estimated by subtracting the amount of moisture, crude fiber, fat, protein, and minerals—items directly determined by analysis—from the total weight (on an as-fed basis) of the sample. It contains, as a consequence, the measurement errors of estimating each item, including the lignin missed in the crude fiber determination. Thus, NFE is only a rough estimate of sugar and starch content.

3. Fat. Fat has 2.25 times as much energy as carbohydrates and is a valuable nutrient. However, most home grown feeds contain less than 5 percent fat. It is measured as ether extract, a reasonably accurate procedure.

4. Total digestible nutrients (TDN). TDN represents the proportion of the energy sources—fiber, protein, sugars, starches, and fat—that are digestible. TDN is calculated by determining the moisture, crude fiber, NFE, protein, fat, and minerals in the feed, multiplying each item by its digestibility, and adding up the percentages. Usually, average digestibilities from many experiments are used. The fat value is multiplied by 2.25 in the calculations to adjust for its higher energy value.

The example that follows shows how this is done:

Component	% in Feed	Average Digestibility	% of Digestible Components of Feed	
Water	10	—		
Crude fiber	10	50	50 × 10% =	5.00
Ether extract (fat)	3	90	3 × 90% × 2.25 =	6.08
Protein	10	75	10 × 75% =	7.50
Minerals	2	—		
NFE	65	90	65 × 90% =	58.50
Total	100			77.08

For this example, TDN is 77.08 percent. Since the feed contains 10 percent moisture (or is 90 percent dry matter) the TDN content of the dry matter is 85.7. That's 77.08 ÷ 0.9.

If you have a total feed analysis of a feed sample, the estimated TDN value would be calculated in this manner. Because of the errors in this system it is only an estimate of the energy content. Feeds high in TDN will also be high in net energy, however. TDN values can be used to rank feeds in order of their net energy content.

5. *New methods.* New methods potentially more accurate, are being tested. Acid detergent fiber (ADF) appears to be the most promising. This procedure accurately measures the amount of poorly digestible cell wall components, primarily lignin; formulas are under development that can be used to estimate net energy content of a feed from an analysis for ADF.

6. *Other means of estimating energy content.* There are methods of estimating energy content of the feed that probably are more useful in identifying the energy content of a feed than chemical analysis.

   a. *Plant maturity*—Early cut, immature plants are much higher in energy than late cut, mature plants due to a lower fiber (cell wall) content. Therefore, comparing the maturity of your forage with those listed in Table 9-1, "Feed Composition Values," provides a reasonable estimate of the energy content.

   b. *Weather damage*—Weather-damaged hay, even though early cut, is lower in energy because rain leaches out some of the cell contents, primarily soluble sugars. As a result, weather-

Table 9-1
Feed Nutrient Values

Feed Specification Code	(1) Feed Code	(2) TDN %	(2) NE$_m$ Mcal/lb DM	(3) NE$_g$ Mcal/lb DM	(4) Total Protein % of DM	(5) Net Protein % of DM	(6) Calcium % of DM	(7) Phos. % of DM	(8) Pot. % of DM	(11) Dry Matter Usual %	(12) Feed Type Code
CONCENTRATES:											
Alfalfa dehy. protein @ 15%	001	61	.59	.31	16.3	4.9	1.32	.24	2.50	93	1
@ 17%	002	61	.59	.31	19.2	5.8	1.43	.26	2.68	93	1
Barley, 48 to 52 lbs/bushel,											
dry	006	83	.97	.64	12.0	4.6	.09	.47	.63	89	2
high moisture	007	83	.97	.64	13.0	3.9	.09	.47	.63	75	2
Barley, light-dry	008	78	.80	.53	13.4	4.7	.07	.37	.63	89	2
high moisture	009	78	.80	.53	13.4	4.0	.07	.37	.63	75	2
Barley screenings	010	80	.84	.56	13.4	4.0	.07	.37	.63	90	2
Beans, navy, cull	011	83	.90	.60	25.4	8.9	.17	.63	1.89	90	2
Beans, soy (whole beans)	012	94	1.09	.69	42.1	14.7	.26	.64	2.21	90	2
Beet pulp	013	72	.73	.43	10.0	3.5	.74	.11	.23	91	2
Brewer's grain, dry	015	83	.91	.64	28.1	9.8	.29	.54	.09	93	2
wet	016	83	.91	.64	28.1	9.8	.29	.54	.09	20	2
Corn, shelled, 56 lbs/bu dry	017	91	1.02	.67	10.0	3.5	.03	.40	.46	85	2
high moisture	018	91	1.02	.67	10.0	3.0	.03	.40	.46	70	2
45 lbs/bu, dry	019	88	.98	.64	11.0	3.9	.03	.45	.58	85	2
high moisture	020	88	.98	.64	11.0	3.3	.03	.45	.58	70	2
35 lbs/bu, dry	021	85	.95	.62	12.0	4.2	.03	.47	.90	85	2
high moisture	022	85	.95	.62	12.0	3.6	.03	.47	.90	70	2
Corn, ear, 56 lbs/bu as											
grain, dry	023	82	.90	.56	8.9	3.1	.05	.33	.61	86	2
high moisture	024	82	.90	.56	8.9	2.9	.05	.33	.61	70	2
45 lbs/bu as grain											
dry	025	77	.82	.49	9.8	3.4	.05	.38	.82	86	2
high moisture	026	77	.82	.49	9.8	2.9	.05	.38	.82	70	2

Feed	Code							
34 lbs/bu as grain								
dry	027	74	.75	.46	10.7	3.7	.05	.39	1.61	86	2
high moisture	028	74	.75	.46	10.7	3.2	.05	.39	1.61	70	2
Cottonseed meal	032	76	.78	.50	44.8	15.7	.16	1.21	1.36	93	1
Distillers solubles	033	88	.98	.66	28.9	10.1	.38	1.38	1.87	98	1
Linseed meal, solvent	038	76	.78	.52	38.5	13.5	.44	.91	1.52	91	1
Malt sprouts	039	69	.70	.45	28.2	9.9	.24	.78	.23	93	1
Millet	042	77	1.02	.67	13.3	4.7	.06	.31	.40	90	2
Milo, fine ground or rolled	043	83	.89	.59	12.4	4.34	.04	.37	.88	89	2
ensiled or steam flaked	044	90	.96	.64	12.4	3.72	.04	.33	.88	70	2
head chop	045	75	.78	.52	11.1	3.33	.13	.23	.44	32	2
Molasses beet	048	91	.93	.61	8.7	2.61	.21	.04	6.20	88	1
cane	049	89	.91	.58	4.3	1.29	1.07	.11	3.17	83	1
Oats, 38 lbs/bu	050	80	.83	.57	12.2	4.3	.10	.43	.42	90	2
32 lbs/bu	051	75	.68	.52	13.2	4.6	.10	.43	.42	90	2
Potatoes	054	79	.80	.59	9.0	2.7	.04	.22	2.13	25	2
Rye	055	86	.93	.62	13.4	4.7	.07	.38	.52	90	2
Soy 44	056	81	.88	.59	50.8	17.8	.36	.75	2.21	90	1
Soy 49	057	81	.88	.59	54.8	19.2	.26	.72	2.21	90	1
Speltz	058	75	.78	.52	12.2	4.3	.10	.33	.42	88	2
Tallow	070		2.07	1.89						99.5	2
Triticale	080	83	.97	.64	13.0	4.6	.06	.40	.48	89	2
Urea, 45% nitrogen	081	0			281	98.4				100	1
42% nitrogen	082	0			262	91.7				100	1
Anhydrous ammonia	083	0			512					512	1
Wheat, hard	084	88	.98	.64	14.6	5.1	.06	.44	.46	89	2
soft	985	88	.98	.64	12.3	4.3	.10	.33	.33	89	2
Wheat, middlings	089	83	.89	.59	17.2	6.0	.16	1.00	1.08	90	1
screenings	091	77	.80	.53	16.9	5.9	.09	.40	.56	90	1
Whey, acid, dry	092	85	.90	.60	11.0	3.3	2.00	1.00	3.20	93	1
wet	093	85	.90	.60	11.0	3.3	2.00	1.00	3.20	6	1

[a] Only useful as long as requirements and ability to use the degraded protein are not exceeded (see protein adjustments). Values for forages marked by * should be estimated by using the multipliers given in the previous chart.

Feed Specification Code		(2)	(3)	(4)	(5)	(6)	(7)	(8)	(11)	(12)
		NE$_m$	NE$_g$	Total Pro- tein	Net Pro- tein	Cal- cium	Phos.	Pot	Dry Matter	Feed Type
Feed Description	Feed Code	TDN %	Mcal/ lb DM	Mcal/ lb DM	% of DM	% of DM	% of DM	% of DM	% of DM	Usual %	Code
MINERALS:											
Bone meal	123						30.5	14.3		100	1
Calcium sulfate	124						20.3			100	1
Dicalcium phosphate	125						23.0	18.0		100	1
Deflor, rock phosphate	127						32.0	18.0		100	1
Limestone (cal, carbonate)	129						38.0			100	1
Monosodium phosphate	131							22.5		100	1
Potassium chloride	132								52.3	100	1
Potassium sulfate	133								41.0	100	1
Salt	135										
ROUGHAGES:*											
Alfalfa-prebloom, hay	150	63	.71	.40	19.4	*	1.25	.30	2.10	90	3
haylage	151	63	.71	.40	19.4	*	1.25	.30	2.10	60	3
haylage	152	63	.71	.40	19.4	*	1.25	.30	2.10	50	3
silage	153	63	.71	.40	19.4	*	1.25	.30	2.10	30	3
Alfalfa, early bloom, hay	154	61	.57	.27	18.4	*	1.25	.23	2.10	90	3
haylage	155	61	.57	.27	18.4	*	1.25	.23	2.10	60	3
haylage	156	61	.57	.27	18.4	*	1.25	.23	2.10	50	3
silage	157	61	.57	.27	18.4	*	1.25	.23	2.10	30	3
Alfalfa, medium, hay	158	58	.51	.18	17.1	*	1.25	.22	1.35	90	3
haylage	159	58	.51	.18	17.1	*	1.25	.22	1.35	60	3
haylage	160	58	.51	.18	17.1	*	1.25	.22	1.35	50	3
silage	161	58	.51	.18	17.1	*	1.25	.22	1.35	30	3
Alfalfa, mature, hay	162	50	.45	.15	13.6	*	.56	.20	1.55	90	3
haylage	163	50	.45	.15	13.6	*	.56	.20	1.55	60	3
haylage	164	50	.45	.15	13.6	*	.56	.20	1.55	50	3
silage	165	50	.45	.15	13.6	*	.56	.20	1.55	30	3
Alfalfa, weather damaged, hay	166	55	.51	.18	10.0	*	.90	.21	1.25	90	3
haylage	167	55	.51	.18	10.0	*	.90	.21	1.25	60	3
Alfalfa, pasture	168	61	.60	.33	19.4	*	1.25	.30	2.10	20	3

Alfalfa-brome, early bloom,									
hay	169	58	.57	.25	16.0	*	1.30	.36	3.86	90	3
haylage	170	58	.57	.25	16.0	*	1.30	.36	3.86	60	3
haylage	171	58	.57	.25	16.0	*	1.30	.36	3.86	50	3
silage	172	58	.57	.25	16.0	*	1.30	.36	3.86	30	3
Alfalfa-brome, medium, hay	173	55	.54	.23	13.0	*	1.00	.23	1.50	90	3
haylage	174	55	.54	.23	13.0	*	1.00	.23	1.50	60	3
haylage	175	55	.54	.23	13.0	*	1.00	.23	1.50	50	3
silage	176	55	.54	.23	13.0	*	1.00	.23	1.50	30	3
Alfalfa-brome, mature, hay	177	50	.45	.15	10.0	*	.56	.20	1.00	90	3
haylage	178	50	.45	.15	10.0	*	.56	.20	1.00	60	3
haylage	179	50	.45	.15	10.0	*	.56	.20	1.00	50	3
silage	180	50	.45	.15	10.0	*	.56	.20	1.00	30	3
Alfalfa-brome, pasture	181	61	.61	.34	19.5	*	1.52	.36	3.86	20	3
Apple pumace	186	69	.70	.46	5.4	*	.13	.12	.47	30	3
Barley straw	187	41	.45	.05	4.1	*	.34	.09	2.28	90	3
Beet top silage	188	47	.47	.20	12.0	*	1.00	.22	1.80	30	3
Birdsfoot trefoil hay	190	61	.59	.31	15.5	*	1.75	.21	2.20	90	3
Bluegrass, early cut, hay	191	65	.64	.37	11.6	*	.30	.29	1.59	90	3
Bluegrass pasture	192	70	.70	.45	16.0	*	.50	.39	2.01	20	3
Brome, early cut, hay	193	57	.57	.25	11.8	*	.40	.30	2.20	90	3
late cut, hay	194	50	.45	.15	6.3	*	.31	.14	2.00	90	3
pasture	195	68	.69	.43	20.3	*	.59	.37	4.30	20	3
Clover hay	196	58	.56	.27	14.8	*	1.61	.22	1.75	90	3
Corn cobs	199	47	.49	.11	2.8	*	.12	.04	.84	90	3
Corn silage, 6.7 bu corn/ton										
without NPN	200	75	.75	.48	8.0	1.6	.28	.21	.95	32	3
with NPN	201	75	.75	.48	13.0	3.4	.28	.21	.95	32	3
with NPN + minerals	202	75	.75	.48	13.0	3.4	.50	.30	.95	32	3
Corn silage, 5.4 bu corn/ton										
without NPN	203	70	.71	.45	8.0	1.6	.28	.21	.95	32	3
with NPN	204	70	.71	.45	13.0	3.4	.28	.21	.95	32	3
with NPN + minerals	205	70	.71	.45	13.0	3.4	.50	.30	.95	32	3

Feed Specification Code	(1) Feed Code	(2) TDN %	(2) NE_m Mcal/lb DM	(3) NE_g Mcal/lb DM	(4) Total Protein % of DM	(5) Net Protein % of DM	(6) Calcium % of DM	(7) Phos. % of DM	(8) Pot. % of DM	(11) Dry Matter Usual %	(12) Feed Type Code
Feed Description											
Corn silage, 3.5 bu corn/ton											
without NPN	206	66	.65	.41	8.0	1.6	.28	.21	.95	32	3
with NPN	207	66	.65	.41	13.0	3.4	.28	.21	.95	32	3
with NPN + minerals	208	66	.65	.41	13.0	3.4	.50	.50	.95	32	3
Corn stalks, grazing	209	66	.65	.40	7.0	2.1	.30	.12	1.43	50	3
Corn husklage	210	55	.55	.23	3.7	1.1	.16	.08	1.43	80	3
Corn stalklage	211	45	.45	.18	4.2	1.3	.27	.12	1.43	50	3
Fescue, early, hay	213	62	.60	.33	10.5	*	.50	.36	1.87	90	3
medium hay	214	55	.50	.20	8.0	*	.40	.25	1.87	90	3
mature hay	215	50	.45	.15	6.0	1.8	.30	.15	1.87	90	3
Milo pasture	217	66	.65	.40	11.6	*	.50	.36	1.87	20	3
Milo residue	218	45	.45	.18	4.7	1.4	.49	.13	1.41	70	3
Oat hay	219	59	.52	.18	9.9	*	.23	.21	.97	90	3
straw	220	52	.50	.16	4.4	1.3	.33	.10	2.24	90	3
silage (dough)	222	59	.58	.29	9.7	1.9	.37	.30	3.41	32	3
Prairie hay	223	45	.45	.13	7.8	*	.51	.17	.97	90	3

Orchard grass, early, hay	224	57	.57	.25	11.0	*	.45	.37	2.10	90	3
medium, hay	225	50	.50	.20	9.7	*	.40	.30	2.10	90	3
mature hay	226	45	.45	.15	6.0	1.8	.30	.15	2.00	90	3
pasture	227	65	.68	.41	15.2	*	.42	.47	3.38	20	3
Quackgrass hay	228	45	.45	.18	7.7	*	.40	.30	2.00	90	3
Reed canarygrass hay	229	50	.50	.20	8.5	*	.33	.16	2.35	90	3
Rye grass pasture	230	71	.70	.45	16.3	*	.58	.56	3.40	20	3
Rye straw	231	31	.34	.00	3.0	.9	.28	.10	.97	90	3
Sorghum silage, grain type	232	59	.57	.30	7.9	1.6	.34	.21	.44	32	3
forage type	233	59	.57	.29	10.2	2.0	.64	.23	3.00	32	3
Sorghum-sudan, silage	234	59	.57	.29	12.7	*	.55	.23	3.00	32	3
hay	235	59	.57	.29	12.7	*	.63	.23	3.00	90	3
pasture	236	63	.61	.35	12.7	*	.66	.23	3.00	20	3
Timothy, early, hay	237	58	.57	.28	8.7	*	.66	.34	.42	90	3
medium, hay	238	55	.50	.20	7.0	*	.52	.26	.42	90	3
mature hay	239	50	.45	.15	6.3	1.9	.38	.18	.42	90	3
pasture	240	66	.66	.40	9.3	*	.50	.35	2.40	20	3
Wheat straw	242	45	.40	.09	3.5	1.1	.17	.08	1.11	90	3
pasture	243	62	.60	.32	5.3	1.6	.35	.15	2.00	20	3

damaged early cut hay becomes like undamaged hay cut later in energy content. The same is true of corn stalks or grass left for winter grazing. They will have the highest energy content early in the season.

c. *Grain content of silage*—Recent research shows that corn silage with a high grain content, such as short stalked, high grain yielding varieties is higher in energy than tall corn varieties with a similar grain yield. Although the latter would likely give more energy per acre, the energy content/pound dry matter will be lower. The Feed Composition Table (9-1) gives an estimate of the energy value of corn silage with various grain contents, based on recent research. To estimate the grain content of your silage, divide the bushels of #2 corn yield/acre by the tons of 32 percent dry matter silage/acre.

Protein

The (crude) total protein content of a feed sample can be accurately determined by laboratory analysis. It is determined by measuring the nitrogen in the feed, and converting it to protein by multiplying by 6.25. The basis for this is protein contains 16 percent nitrogen, or 1 part of nitrogen/6.25 parts protein. Thus if a shelled corn sample was found to have 1.61 percent nitrogen in the dry matter, it would be estimated to have $1.61 \times 6.25 = 10.06$ percent total protein. However, it does not tell the amount of protein that will be available to the animal for maintenance, growth, or milk production because part of the protein will not be digestible or will be lost as ammonia when it is degraded in the rumen, as discussed earlier.

In spite of the variation in quality, analysis for total protein is the most important because it is one of the most expensive nutrients and because the protein content of corn and corn silage is inadequate to meet the needs of growing and finishing cattle. If accurate nutrient values are obtained by feed analysis the safety-factor used in determining the amount of protein supplement to feed can be reduced. Even though analysis does not give us the usable protein, it does tell us whether our silage is 4.4 or 10.8 percent protein. Conversion of crude protein values to net protein to adjust for protein quantity will be discussed in the feed composition section.

Minerals

Accurate methods are available for mineral analysis, and these are valuable as most rations need some supplemental minerals. The most important one, however, is the phosphorus content of silage and hay since it is the most expensive to supplement and most rations, particularly high forage rations, are deficient in phosphorus. However, many forages will be adequate in phosphorus, and less expensive supplementation programs can be developed.

Calcium can be cheaply supplemented with calcium carbonate (ground limestone), and it is therefore of no great concern to obtain an analysis in order to save supplemental calcium.

Most feeds are adequate, or nearly adequate in trace minerals. Although trace mineral analyses are helpful in pinpointing specific deficiencies, feeding trace mineral salt will usually provide enough safety factor to cover any trace mineral deficiencies that might exist in the home-grown feed.

Vitamins

The only vitamins normally of concern are A and D, and sometimes E. Vitamin A content can be determined through an analysis for carotene content, as cattle convert carotene to vitamin A. However, it is not of great concern, because typical management practices provide adequate quantities of Vitamin A. New feeder cattle are usually injected with 1-2 million IU of Vitamin A to aid in combating stress. Also many rations are high in corn silage, which is usually high in carotene content. Further, the cost of providing supplemental Vitamin A is minimal. The guides given in the Mineral-Vitamin Requirement section and in the Feed Composition table can be used to figure supplemental Vitamin A, D, and E needs.

Sampling Feeds for Analysis

Inaccurate sampling can lead to greater errors than using average values from feed composition tables. The sample must be representative of all the feed in question. The important factors to consider are when and how to sample feeds for analysis. In most states, your local county agent has sample bags and information on obtaining feed analysis. Also, most reputable feed companies will analyze your feed. There are also private laboratories that offer this service. After obtaining sample bags, the following guides can be used to obtain a representative sample.

High Moisture Corn, Corn Silage, and Haylage

Sampling at harvest. Collect 3 to 5 handfuls of silage or grain from one or two loads each day. For example, you might sample the last load before lunch and the last load of the day. Place the sample in a plastic bag and put in a freezer immediately. Then mix samples together after the silo is filled, and place about one quart in a plastic bag and send it in for analysis. Keep separate samples from different silos if they are filled with silage from different varieties and/or at different maturities. When they are fed, some adjustments in supplement can then be made if they are different.

Sampling from the silo. (Silage needs to be sampled after filling if treated with NPN.)

1. *Upright silo*—Collect a one-to-two-quart sample from the discharge of the silo unloader when you are about through feeding. Be sure two or three feet have been removed before sampling to avoid spoiled or exceptionally dry material.
2. *Bunker silos or piles*—Take 15 or more handfuls from all over the face of the silo after it is opened and you are into well-packed, good quality silage. Mix the samples in a clean pail, then place about one quart in a plastic bag and either freeze it or send immediately for analysis. It is desirable to sample several times during the feeding period, particularly if there is any great variation in plant maturity, variety, or soil type.

Dry grain sampling. Take a minimum of 5 grain samples, with a grain probe if possible, from various places in the bin or truck. Mix them in a clean pail, then place about one pint in a plastic bag, seal, and send in for analysis.

Hay sampling. To sample loose or chopped hay, take samples from various locations in the pile or stack, using a core forage sampler. To sample baled hay, take core samples from the end of a dozen or more bales taken from various places in the mow or stack. Mix samples together, then send in about one quart in a sealed plastic bag for analysis.

9-2 FEED COMPOSITION VALUES

This section summarizes the average nutrient content of feeds normally of concern to beef producers in balancing rations.

Average nutrient content values are typically used in the absence of laboratory analysis. The values work reasonably well in most cases, but should be adjusted when actual values are avail-

9-2 Feed Composition Values

able. Nutrient values for an individual feed may vary considerably. For example, total protein as a percent of dry matter for untreated corn silage samples runs from 4.4 percent to 11 percent. Feeds—especially corn silage, treated corn silage, and hays—should be tested for protein and moisture. Forages are described by factors such as time of harvest and percent grass in alfalfa-grass mixtures to obtain better estimates of nutrient content. The bushel corn/ton of silage and treatment methods are used to categorize corn silages. Grains are described according to test weight.

Energy, protein, calcium, phosphorus, and potassium values for feeds are given in Table 9-1. Typical dry matter percent values are given. All nutrient values are on a dry matter basis. For example, if 32 percent dry matter corn silage has a total protein value of 2.56 percent, then its protein value on a dry matter basis is $(2.56 \div .32) = 8$ percent. Energy values are given for two commonly used systems. The total digestible nutrient (TDN) values are provided for those who prefer to use that system. The TDN system works reasonably well for beef cow ration analysis and formulation, and will be used for balancing beef herd rations. The Net Energy System is widely used in most of the cattle feeding areas of the United States for ration evaluation and formulation for growing and finishing cattle. It will be used in the ration balancing section for growing and finishing cattle.

Protein values are stated on both a total (crude) and net basis. Individuals must be careful in use of total protein values since protein quality varies widely among feedstuffs, particularly between corn silage and corn grain. Also, within a particular feedstuff, the method of harvesting and storage has an influence.

The Vitamin A content of general classes of feeds is given in Table 9-2.

Feeds are coded to facilitate computerized ration formulation used in some states. For example, 56 lb test weight corn is Code Number 017 when it is 15 percent moisture, and Code Number 018 when it is 30 percent moisture. Also, each nutrient is given a code for reference in computerized ration formulation.

Adjustments to Feed Composition Values

A number of conditions affect the values for energy and protein given in these tables. Adjustment factors will be given that can be used to correct for these conditions.

Energy values. Cattle condition and certain feed additives affect the energy digestibility of a feed. Thin cattle utilize energy

Table 9-2
Estimated Vitamin A Content of Feeds

	100% DM Basis	
	Mg. Carotene/lb.	Vitamin A, IU/lb.
Fresh green legumes and grasses, immature	75-200	30,000-80,000
Mature pasture	5-10	2,000-4,000
Dehydrated alfalfa meal, fresh very bright green color	120-150	48,000-60,000
After considerable storage time, bright green	50-80	20,000-32,000
Legume hays; cured rapidly with minimum exposure, bright green and leafy	20-30	8,000-12,000
Grass hays, well cured, good green color	10-15	4,000-6,000
Average quality, bleached, some green color	5-10	2,000-4,000
Legume silage	60-100	24,000-40,000
Corn silage, medium to good green color	5-30	2,000-12,000
Grains, mill feeds, protein concentrates by-products	.01-.2	4-80
Yellow corn and its by-products	1.5-2	600-800
Crop residues	1-2	400-800
Mature hay	2-4	800-1,600

more efficiently; fleshy cattle will utilize the same feed less efficiently. Multiply the feed NE value by the following factors to adjust for these conditions.

Body condition	Multiplier for feed	
	NE_m	NE_g
Very fleshy	.95	.90
Average	1.00	1.00
Very thin	1.05	1.10

Feed additives such as Rumensin and antibiotics increase the digestibility of the feed. When these are used, the feed energy value can be adjusted as follows:

Feed additive	Multiplier for feed	
	NE_m	NE_g
Antibiotics	1.04	1.04
Rumensin	1.10	1.10

Protein values. Feed crude protein values can be used in balancing many rations to get a reasonable estimate of supplemental protein needs. Use of the crude protein requirement tables and feed crude protein values will give similar results to those obtained by using NRC recommendations. However, protein will be overfed in high grain rations and underfed in high forage rations, especially those containing silage. Currently complex protein systems are under development that will pinpoint the metabolizable or net protein value of a feed, and how much NPN can be fed with various combinations of feeds. Many of the values presented to date, however, need further evaluation before it will clearly be of value to use a complex system to balance beef rations. The guidelines given in Chapter 8 can be used to estimate how much NPN can be used in the ration being formulated. Then crude or net protein values can be used to balance for protein. Although not fully tested, using the net protein values given here, especially with corn-corn silage rations, will give ration protein levels that have been optimum in recent experiments.

Multipliers to convert feed CP to NP values were estimated from feeding and metabolic trials. They were used to calculate the NP values in Table 9-2. They are given here to be used for estimating NP values from feed CP values obtained from feed analysis. These values will be refined as more data becomes available.

Feed	Crude protein multiplier when feed used in the following diet:		
	High grain	50% grain	Alone
Dry grain	.35	.35	.35
High moisture grain	.30	.30	.30
High protein feed, when used to balance for protein	.35	.35	.35
Grass hay — early	.30	.30	.20
— late	.30	.30	.30
Legume hay or undamaged silages	.30	.25	.17
Grain crop silage	.20	.20	.20
Heat damaged haylage —			
Slightly brown	.16	.16	.16
Brown to dark brown	.12	.12	.12

Current research suggests that if Rumensin is fed under certain conditions, the feed net protein value may be increased 5–10 percent. These conditions have not been clearly defined to date, however. Be sure to decrease feed intake by 10 percent when balancing the ration, if the feed energy and protein values are increased to correct for feeding Rumensin.

Table 9-3
Estimated TDN and Net Energy Levels in Various Complete Protein Supplements
(As Fed Basis)

Total Protein %	Physical form	Total Protein from NPN,%	Estimated TDN %	Estimated NE_m MCal/lb.	Estimated NE_g MCal/lb.
32	dry	0	70	.78	.46
32	dry	6	60	.60	.34
32	liquid	30	50	.50	.32
40	dry	0	70	.78	.46
40	dry	14	65	.65	.38
40	dry	33	60	.60	.35
40	liquid	38	47	.47	.30
60	dry	55	39	.40	.25
60	liquid	62	37	.37	.23
80	dry	73	30	.34	.20

9-3 THE IMPACT OF FEED PROCESSING ON FEEDSTUFF UTILIZATION

Feed is processed primarily to improve digestibility. This is accomplished by reducing the particle size, increasing the surface area, or modifying the starch in grains to make them more susceptible to microbial action in the rumen. Under conditions where cattle thoroughly chew the grain or forage, processing is not necessary for efficient digestion, and may even reduce performance. Chewing not only reduces particle size, but also increases secretion of buffering compounds. The response to grain processing thus primarily depends on the density of the kernel, hardness of the seed coat, and form of the starch. Roughage is processed to improve ease of handling and mixing and to reduce waste. Other factors affecting the value of processing include age of the cattle, dryness of the grain, and level of roughage in the ration.

The importance of reducing feed requirements and improving rate of gain were discussed in the ration formulation section. However, the cost of processing often exceeds the value of the improvement in performance. This section will begin by outlining the various processing methods and the expected improvement by processing various grains. Then an evaluation of expected returns vs cost of processing will be given that can be used to determine the value of processing grain under various conditions.

Grinding. The grain is ground to varying degrees of fineness with a hammermill. Screen size and type of grain are the major factors affecting fineness of grind. The fineness of grind desired depends on many factors. In general, coarse grinding with a minimum of fine particles is preferred when the grain is the major ration ingredient. Fine grinding is preferred when several ingredients are being combined, and when small amounts of grain are being fed with high roughage rations.

Hay is ground primarily for ease of handling, to reduce wastage, and to allow complete mixing of ration ingredients. Cattle normally chew long hay well enough that it does not need to be processed to improve digestibility.

Rolling. The grain is crushed between rollers, which are usually grooved. The degree of fineness depends primarily on roller spacing. Grain moisture content and rate of grain flow also influence fineness. In some operations, the grain is first held in a steam chamber for one to eight minutes to soften the grain. This process gives similar performance, however, to dry rolling or grinding.

Pelleting. Grinding followed by steam and pressure treatment is used to compact the ingredients into a pellet. This method is used primarily to combine ingredients into a complete feed or a protein supplement for convenience of feeding and to prevent separation. It is also used to condense dehydrated alfalfa meal, for ease of handling and transportation. Feed intake may improve on all pelleted rations, especially with forages.

Steam Flaking. The grain is subjected to steam in a chamber for 15-30 minutes, then rolled through closely spaced rolls to reduce the kernel to a thin flake. This process causes the complex starch molecule to rupture (gelatinize). This process improves the digestibility of most grains under most conditions. The moisture is increased to 16-20 percent.

In some operations, the grain is *pressure cooked* for one to two minutes at temperatures up to 300°F. It is then cooled to below 200°F before flaking. This method may further increase gelatinization of the starch, and may help keep the flake intact during mixing.

Extruding. Dry whole grain is forced through an orifice by a tapered screw, producing flakes. The friction increases temperature, with the net effect being somewhat like steam flaking.

Roasting. The grain is heated to about 300°F. The starch is gelatinized or expanded and, in addition, part of the protein may be made less soluble, resulting in more bypassing to the small intestine before being degraded.

Popping. Grain is heated to 700-800°F for 15-30 seconds. The starch is gelatinized and expanded. The moisture content is reduced to about 3 percent. Then the grain may be rolled and moisture may be added.

Exploding. Steam is applied to grain in chambers, increasing the temperature and pressure (about 250 psi). When the grain is released after about 20 seconds, it swells, looking like puffed breakfast cereal.

Micronizing. Dry grain is heated with gas fired infrared generators as it passes along an oscillating steel plate, then is dropped on knurling rolls. The grain has an intact, flakelike appearance; the effect on the starch is similar to other heat treatments.

Storing grain at a high moisture content. Grain is harvested at a high moisture content at physiological maturity but before it dries. It is placed in silos and is fermented; the acids produced preserve the grain. Feed efficiency is improved by alteration of the starch, or the fermentation in the silo being more efficient than fermentation in the rumen. The grain should have 25 to 30 percent moisture for best results. All types of structures (oxygen limiting, conventional tower, or bunker silos) are used. High-moisture shelled corn can be stored whole in tight tower silos. Sorghum grain is stored whole in oxygen limiting towers. All grains must be ground for storage in bunker silos.

The major advantages of harvesting and storing grain at a high moisture content are reduced field losses, elimination of drying and handling costs and losses, and spreading out the harvesting period.

Whole high-moisture grain can be stored in wood bins, on cement floors, or in trench silos by treating it with acid. Usually propionic, or acetic plus propionic acids are used, the amount depending on the moisture content of the grain. Performance of the cattle is not affected.

Water may be added to dry grain to reconstitute it to 25 to 30 percent moisture. The grain is stored whole in a silo for a minimum of 3 weeks and rolled before feeding. Effects are similar to storing grain harvested at a high moisture content.

Expected improvement in performance by processing grain. Table 9-4 summarizes the effect of various methods of corn and milo processing on dry matter requirements/100 lb gain. Some may improve rate of gain; however, the major economic benefit is due to reduced feed requirements. Only one of the grains is listed for several of the methods, either because the method is only useful for that grain or little or no data is available for the grain not listed. All methods are compared to fine grinding or dry rolling.

This data shows that processing is much more beneficial for milo. Some form of processing is necessary for milo, so the problem is to determine which one is best. In the case of corn, however, many methods may not be profitable unless the operation is large enough for the benefits to offset the cost. Grinding or rolling does not improve performance in high concentrate rations using corn as the grain. Actually, grinding or rolling may be detrimental in all concentrate corn rations. Extensive rumen wall damage and liver abscesses may occur. Corn should be ground or rolled, however, when fed in rations containing 40 percent or more roughage, especially when fed to older cattle. Also, extremely dry corn (less than 10 percent moisture) should be processed.

Data on processing small grains (wheat, barley and oats) is

Table 9-4
Effect of Grain Processing Method of Feed Efficiency[1]

	Corn	Milo
	-- % improvement --	
High moisture harvested		
Ensiled whole	5	10
Ensiled ground	3	–
Re-constituted	3	9
Steam flaked	7	7
Roasted	10	–
Micronized	–	15
Popped	–	8
Extruded	7	12

[1] Based on data summarized by S. D. Farlin, University of Nebraska.

limited. Some form of processing is likely to be profitable. It would appear that coarse rolling or grinding may be the most profitable for them. Larger feedlots that are equipped with one of the other more extensive processing methods described, however, will likely obtain additional improvement in feed efficiency over grinding or rolling small grains.

Economic Advantage of Processing

The best processing method is one that gives the greatest returns above costs. Profits from processing depend on the cost compared to the benefits. Table 9-5 illustrates the importance of feedlot size in deciding how much can be spent on feed processing equipment.

Small feeders cannot afford to spend much for feed processing. Even medium sized feedlots (1,000-4,000 head) must have an improvement of over 5 percent in feed efficiency before more than $40,000 can be spent for processing equipment. Large feeders, however, can justify large investments in processing equipment, even if improvement in feed efficiency is only 3-5 percent.

The most practical solution for small feeders is to grind or roll small grains, and to use high-moisture storage for corn or milo, and

Table 9-5
Effect of Equipment Cost and Feedlot Size on Profits from Processing[1]

Number fed/year	Feed Savings of 5% Original equipment cost				Feed savings of 10% Original equipment cost			
	10,000	20,000	40,000	80,000	10,000	20,000	40,000	80,000
	-- Profits from processing --				-- Profits from processing --			
500	480	-1,520	-5,520	-13,520	4,560	2,560	-1,440	-9,440
1000	2,960	960	-3,040	-11,040	11,120	9,120	5,120	2,880
2000	7,920	5,920	1,920	-6,080	24,240	22,240	18,240	10,240
4000	17,840	15,840	11,840	3,840	50,480	48,480	44,480	36,480
8000	37,680	35,680	31,680	23,680	102,960	100,966	96,960	88,960

[1]Equipment cost assumed; 20% annual fixed cost, plus $2/ton operating cost. Based on a ration cost of $5/cwt., 400 lb gain and a feed conversion of 7.6 lb feed/lb gain at 5% feed savings and 7.2 lb feed/gain at 10% feed savings.

roll before feeding. Most small feeders are farmer feeders, and thus get the additional benefits of early harvest, ease of handling, elimination of drying costs, etc. Part of the cost of high moisture storage can be charged to the farming operation, as the grain must be stored somewhere. Large feedlots, however, should use the best method available.

9-4 DETERMINING DAILY DRY MATTER INTAKES OF GROWING AND FINISHING CATTLE

Feed intake is one of the key factors influencing feedlot performance. Perhaps no other factor has such an overriding impact in determining rate of gain and, ultimately, the profits derived from the feeding operation. The larger the animal's daily intake, the smaller the percentage of the energy consumed that goes for maintenance and the larger the percentage that is available for growth. This section summarizes normal daily dry matter intakes for alternative frame sizes of cattle for feeders started as calves and as yearlings. The impact on performance of above and below normal intakes is summarized.

This section is used in ration formulation and feed use projection. But, one of its most important uses is as a diagnostic tool. Cattle not eating normally are often evidence of a management error, something wrong in the feeding program, or in animal

266   Evaluation of Feedstuffs

health. Tables 9-6 to 9-8 give expected dry matter intakes for steer and heifer calves and for yearling steers. These tables are based upon research conducted at a number of experiment stations with adjustments based upon observations in farm and commercial feedlots.

Feeder cattle gradually eat less dry matter, as a percent of body weight, as they increase in weight. The rate of reduction becomes larger as the animals approach a fatness of high good to low choice. Typically, absolute daily matter intake plateaus, or decreases, once the choice grade is reached.

Table 9-6

Expected Daily Dry Matter Intake (lbs) for Steers Started on Feed as Calves and Carried to Slaughter*

Weight (lbs.)	Frame Size		
	Small (grade @ 840 lbs.)	Average (grade @ 1050 lbs.)	Large (grade @ 1260 lbs.)
400	10.91	10.91	10.91
450	11.91	11.91	11.91
500	12.89	12.89	12.89
550	13.85	13.85	13.85
600	14.78	14.78	14.78
650	15.63	15.70	15.70
700	16.18	16.59	16.95
750	16.67	17.48	17.48
800	16.97	18.34	18.34
850	17.12	18.87	19.20
900	17.12	19.37	20.04
950		19.82	20.87
1000		20.06	21.39
1050		20.24	21.88
1100		20.24	22.33
1150			22.68
1200			22.90
1250			23.07
1300			23.07

*Intakes will average 10% less when grain is over 80% of the ration dry matter.

Table 9-7
Expected Daily Dry Matter Intake (lbs) for Heifers Started on Feed as Calves and Carried to Slaughter*

	Frame Size		
Weight (lbs.)	Small (grade @ 690 lbs.)	Average (grade @ 840 lbs.)	Large (grade @ 990 lbs.)
400	10.91	10.91	10.91
450	11.91	11.91	11.91
500	12.89	12.89	12.89
550	13.57	13.85	13.85
600	14.10	14.78	14.78
650	14.40	15.63	15.70
700	14.48	16.18	16.59
750	14.48	16.67	17.48
800		16.97	18.10
850		17.12	18.60
900		17.12	19.07
950			19.33
1000			19.52
1050			19.52

*Intakes will average 10% less when grain is over 80% of the ration dry matter.

Table 9-8
Expected Daily Dry Matter Intake (lbs) for Steers Started on Feed as Yearlings (Approximately 18 Months of Age) Off Pasture and Carried to Slaughter*

	Frame Size		
Weight (lbs.)	Small (grade @ 880 lbs.)	Average (grade @ 1080 lbs.)	Large (grade @ 1290 lbs.)
550	13.85		
600	14.78	16.26	
650	15.70	17.27	17.27
700	16.59	18.25	18.25
750	17.48	19.22	19.22
800	18.34	20.18	20.18
850	19.20	20.99	21.11
900	19.62	21.57	22.04
950	19.62	22.12	22.95
1000		22.55	23.78
1050		22.83	24.35
1100		22.95	24.89
1150		22.95	25.39
1200			25.76
1250			26.02
1300			26.13

*Intakes will average 10% less if grain is over 80% of the ration dry matter.

268 Evaluation of Feedstuffs

Yearling cattle typically eat about 10 percent more than calves when compared at the same weight. In adjusting for age, we have assumed yearling cattle are 30 lb heavier at the same *equivalent* body composition than an animal of comparable frame started on feed as a calf. A 60-lb adjustment is used for two-year-old steers. These adjustments are based primarily on experience in feedlots since there is little data where cattle of the same breeding and previous nutritional history were compared in experiments.

These tables should provide reasonable benchmarks, but experience may call for adjustments in your situation. Reasons for adjustment include feedlot condition and types of feed used, such as degree of fermentation or date at which hay is cut.

Expected daily gains for four levels of grain feeding using corn-

Table 9-9
Expected Daily Gain for an All Corn Silage Ration, Properly Supplemented*

Lbs. daily dry matter intake	Body weight, lb.					
	450	650	850	1050	1250	
	Expected Daily Gain, Lbs.					
8	0.87	0.17				
9	1.21	0.45				
10	1.55	0.73	0.22			
11	1.87	0.99	0.45			
12	2.18	1.25	0.67	0.28		
13	2.48	1.50	0.89	0.47		
14	2.77	1.75	1.11	0.66	0.33	
15	3.06	1.99	1.32	0.85	0.50	
16	3.34	2.22	1.52	1.03	0.67	
17		2.45	1.72	1.21	0.84	
18		2.68	1.92	1.39	1.00	
19		2.90	2.11	1.57	1.16	
20		3.11	2.30	1.74	1.31	Steer
21		3.32	2.49	1.91	1.47	Calf
22			2.67	2.07	1.62	Yearling
23			2.85	2.23	1.77	Steer
24			3.03	2.39	1.91	
25			3.21	2.55	2.06	
26				2.71	2.20	
27				2.86	2.34	

*Intake needed for the indicated rate of gain will be 10% less if Rumensin is fed.

corn silage rations are summarized in Tables 9-9 to 9-12. These projections assume average frame steers in average flesh given a growth stimulant and that the feedlot is stress-free. Adjustment factors for other conditions were discussed earlier. A line is drawn through the expected gains at average intake for a particular weight. Thus, those below the line would be below average and those above the line would be above average in performance.

Adjustment for Feeding Rumensin

Rumensin is a feed additive for beef cattle that improves feed efficiency by increasing the energy available from a given amount

Table 9-10

Expected Daily Gain for Shelled Corn (Dry Corn Equivalent) at 0.6 to 0.7 lb/cwt of Body Weight/Day Plus a Full Feed of Corn Silage*

Lbs. daily dry matter intake	Body weight, lb.					
	450	650	850	1050	1250	
	Expected Daily Gain, Lbs.					
8	1.01	0.31				
9	1.36	0.59				
10	1.69	0.87	0.36			
11	2.01	1.13	0.59	0.21		
12	2.33	1.39	0.82	0.42		
13	2.63	1.65	1.04	0.61		
14	2.93	1.89	1.25	0.80	0.47	
15	3.22	2.14	1.46	0.99	0.64	
16	3.50	2.37	1.67	1.18	0.81	
17		2.60	1.87	1.36	0.98	
18		2.83	2.07	1.54	1.14	
19		3.05	2.26	1.71	1.30	
20		3.27	2.45	1.88	1.46	Steer
21		3.48	2.64	2.05	1.61	Calf
22				2.22	1.76	Yearling
23				2.38	1.91	Steer
24				2.55	2.06	
25					2.21	
26					2.35	
27					2.50	

*Intake needed for the indicated rate of gain will be 10% less if Rumensin is fed.

of ration. Daily dry matter intake will average approximately 10 percent lower when Rumensin is fed, with a drop in intake of 10-30 percent when it is first fed to a drop of only 5-10 percent when the cattle become adjusted to it. Rate of gain is about the same, with or without Rumensin.

Thus cattle gain the same on about 10 percent less intake. *The expected gains in Tables 9-9 to 9-12 for the various weights and types of cattle would not be changed by feeding Rumensin, but the intake required to obtain the gains given in these tables would be about 10 percent less.*

Table 9-11

Expected Daily Gain for Shelled Corn (Dry Corn Equivalent) at 1.4 to 1.5 lbs/cwt of Body Weight/Day*

Lbs. daily dry matter intake	Body weight, lb.					
	450	650	850	1050	1250	
	Expected Daily Gain, Lbs.					
8	1.33	0.57				
9	1.70	0.88				
10	2.06	1.17	0.62			
11	2.41	1.46	0.88			
12	2.75	1.74	1.12	0.69		
13	3.08	2.02	1.36	0.90		
14	3.40	2.28	1.59	1.11	0.75	
15	3.71	2.54	1.82	1.31	0.93	
16	4.01	2.80	2.04	1.51	1.12	
17		3.05	2.26	1.71	1.29	
18		3.29	2.47	1.90	1.47	
19		3.53	2.68	2.09	1.64	
20		3.76	2.89	2.27	1.81	Steer
21		3.99	3.09	2.45	1.98	Calf
22			3.29	2.63	2.14	Yearling
23			3.48	2.81	2.31	Steer
24			3.67	2.98	2.47	
25			3.86	3.15	2.62	
26				3.32	2.78	
27				3.49	2.93	

*Intake needed for the indicated rate of gain will be 10% less if Rumensin is fed.

Factors Causing Poor Intake

The daily dry matter intakes presented in Tables 9-6 to 9-8 are only averages. Actual intakes vary considerably as a result of the following factors:

1. *Sickness.* If intakes are below normal, the cattle may not be healthy. A reduction in daily dry matter intake is typically one of the first signs of illness. When cattle are recovering, however, they will begin to increase intake back to normal rates.

2. *Weather.* Just prior to a change in weather, cattle will increase intake. During the change, intake will decrease and then

Table 9-12

Expected Daily Gain for Shelled Corn (Dry Corn Equivalent) at 1.8 to 1.9 lbs/cwt of Body Weight/Day*

Lbs. daily dry matter intake	Body weight, lb.				
	650	850	1050	1250	
	Expected Daily Gain, Lbs.				
8	0.80	0.30	−0.05	−0.31	
9	1.13	0.59	0.21	−0.08	
10	1.45	0.86	0.45	0.15	
11	1.77	1.13	0.69	0.37	
12	2.07	1.40	0.93	0.58	
13	2.36	1.65	1.16	0.79	
14	2.64	1.90	1.38	0.99	
15	2.92	2.14	1.60	1.20	
16	3.19	2.38	1.82	1.39	
17	3.46	2.62	2.03	1.58	
18	3.72	2.84	2.23	1.77	Steer
19	3.97	3.07	2.43	1.96	Calf
20		3.29	2.63	2.14	Yearling
21		3.50	2.83	2.32	Steer
22		3.71	3.02	2.49	
23		3.92	3.21	2.67	
24			3.39	2.84	
25			3.57	3.01	

*Intake needed for the indicated rate of gain will be 10% less if Rumensin is fed.

return to normal after the change. Often, cattle will eat more at night during hot weather and more during the day during cold weather. Anticipating these changes and making appropriate adjustments in feed offered can help avoid a stale feed problem or running out of feed. If underfed and hungry, cattle will often overeat when fed and have digestive upsets. Thus, following feeding they will tend to go off feed.

3. Stale feed. Cattle tend to reduce intake when feed is not fresh. It is better to have clean feedbunks for a short period of time prior to feeding than to place fresh feed on top of stale feed, hoping that the cattle will eventually clean up the leftover feed. This is usually the biggest problem during periods of weather change, after a rain, or when cattle go off feed for some reason.

4. Mud and lot conditions. The greatest problem with mud occurs right behind the bunk apron where the largest amount of travel occurs. Cattle will avoid wading through mud to reach the bunk until extremely hungry. Then they tend to overeat when they finally go to the feedbunk, causing digestive upsets. Roughness of frozen ground will cause similar problems. Typically, ground will be muddy and rough as a result of rain or snow prior to freezing. Then it will freeze and stay rough until the ground thaws or until it is gradually worked down by the cattle.

5. Amount of concentrate in the ration. The rate of grain feeding has an impact on daily dry matter intake, particularly when high rates of grain are fed. As a working rule, dry matter intakes are similar across grain feeding rates until the rate of corn fed/day reaches 60 percent to 70 percent of ration dry matter. That is corn at the rate of 1.4 percent to 1.5 percent of body weight (dry corn equivalent) in the steer-calf feeding program. Above that level, as the rate of grain feeding is increased, the animal will typically not gain more, but will reduce feed intake. Thus, on a ration containing 10 percent to 15 percent corn silage in ration dry matter, daily intake will likely be 90 percent to 92 percent of what it is when 30 percent to 40 percent of the ration is composed of corn silage. Thus, your gains may be good on the high grain ration, even though they are eating less.

6. Roughage quality. For cattle on haylage rations, the quality of the forage can have substantial impact on intake and economy of gain. For example, at the same rate of grain feeding per day, daily dry matter intake may be as much as 20 percent to 30 percent higher for high quality haylage vs late cut weather-damaged haylage. Thus, the problem of lower energy in the poor quality forage is compounded by a reduced intake.

7. Protein level of the ration. Cattle fed a ration deficient in protein will eat 10 percent to 20 percent less dry matter than those fed properly supplemented rations. As a result, gains will be considerably less than potential, and than expected. This is due to the fact that they both eat less and, since the protein level is below requirements, their efficiency of energy utilization is reduced.

8. Feed quality. Weather-damaged or moldy feed can reduce intake, and feeding a high proportion of these types of feedstuffs will cause intake problems. A small amount of spoilage well mixed in the ration will not likely reduce intake; it is not clear what the maximum proportion of spoiled silage is that can be fed before intake is reduced.

9. Types of feed. Cattle will eat some kinds of feed better than others. For example, a ration of ground hay and shelled corn will likely be consumed more readily than one based on corn silage, especially in new feeder cattle. However, the ration should be based primarily on feeds grown on the farm, and those that are least-cost sources of energy and protein. Palatability, if feedstuffs are properly harvested, stored, and supplemented, is not a major problem with high quality corn-corn silage rations.

10. Use of growth stimulant. Part of the impact of growth stimulants is to increase intake. Thus, if a growth stimulant is not used or cattle are not reimplanted at the proper intervals (about 100 days), intake may be below normal.

11. Body condition. As cattle approach a fatness of low choice, it will be increasingly difficult to maintain intake. This is a result, in part, of time on feed; other factors related to body condition are important influences also. The net result is that as cattle approach the fatness of choice, the cost/unit of gain increases at faster rates as a result of lower relative intake and more energy required/lb of gain as a result of the amount of fat relative to protein being deposited.

9-5 CORRECTING FOR MOISTURE CONTENT OF FEEDS

Variation in the moisture content of the feeds making up a ration is the largest single source of serious variation in rations using silage, haylage, green chop, or high-moisture grains. Moisture variations are almost always greater than variations caused by chemical components such as energy, protein, minerals, or fiber. There is no way to ensure a successful feeding program using any feed that can vary in moisture content without formulating on a standard moisture basis and adjusting for moisture as often as necessary.

Problems Due to Moisture Variation

Grains vary in moisture content from 12 to 30 percent and silage from 50 to 75 percent or more which greatly affects the as fed nutrient composition. For example, 10 lbs of corn silage at 50 percent moisture contains 5 lbs of dry matter, 3½ lbs of TDN and 0.42 lb of protein. Ten lbs of corn silage at 70 percent moisture, however, contains only 3 lbs of dry matter, 2.1 lbs TDN and 0.25 lb of protein. A steer eating 20 lbs of corn at 20 percent moisture would only have to eat about 18 lbs of corn at 10 percent moisture to get the same nutrients. If the price per pound was the same for both moisture levels, buying the 10 percent moisture corn would result in getting more than 11 percent more dry matter for the same money.

Also, it is necessary to correct for moisture in order to properly balance the ration. For example, an 800-lb steer requires 2.19 lbs of total protein for a 2.6 lb per day gain. The requirement for energy and protein could be met by feeding 5 lbs alfalfa hay at 12 percent moisture and 15 lbs shelled corn at 14 percent moisture. If alfalfa silage at 60 percent moisture was substituted for the alfalfa hay, however, 11 lbs of the alfalfa silage would be needed to balance the ration.

A sudden drop in roughage dry matter in a high concentration ration of 1 to 2 lbs (a 5 to 15 percent roughage change in most high grain rations) may result in digestive disorders and cattle going off feed, and even relatively small changes in moisture content of feedstuffs can result in an important change in the amount of roughage fed.

Ways to Avoid Problems Due to Moisture Variation

1. Formulate on a standard moisture basis. The simplest way to avoid errors in ration formulation is to formulate on a 100 percent dry matter or 90 percent dry matter basis and then correct for moisture after the ration is properly balanced.

Table 9-13 gives conversion factors from a dry matter basis. Tables 9-14, 9-15, and 9-16 show how rations are corrected for moisture.

Caution: Many feed composition and nutrient requirement values are given as a percentage composition of the feed. If, for example, 11 percent protein is given as adequate for a ration containing 90 percent dry matter, the protein content for the same ration calculated on a 100 percent dry matter basis must be 11/0.9 = 12.2 percent.

To convert nutrient requirements or feed composition values from 100 percent dry matter basis to an as fed basis, multiply the dry matter value times the percent dry matter of the feed. For example, if No. 2 shelled corn has 10 percent protein on a dry basis and a feeder has 30 percent moisture corn, 10×70 percent dry matter = 7 percent protein on an as fed basis.

2. *Adjust for day to day variation in moisture content.*

• When feeding on a per head basis: If a grain intake is controlled but the silage is fed free choice as in a growing ration, the cattle will tend to eat more pounds of silage as the moisture content increases and therefore will adjust for variation in the moisture content of the silage. The total pounds of grain fed will have to be increased as the grain increases in moisture, however. The reverse would be true for a finishing ration where the grain is being fed to appetite but the silage is limit-fed. In this case the pounds of silage fed will need to be increased as the silage increases in moisture to reach the desired roughage dry matter intake, but the cattle will tend to adjust the grain intake as its moisture varies.

• When feeding on a percentage or proportional basis: New proportions for the various ingredients need to be calculated each time a major ingredient changes significantly in moisture content. Ingredients expected to vary more than 3 to 5 percent in moisture during the feeding period should be checked periodically, followed by calculation of new feed formulas or proportions as major changes in moisture occur. Small changes in moisture content of just one major ingredient will alter the amount of energy or other nutrients consumed daily, which would be expected to affect performance. Many times when ingredients are proportioned out rather than weighed out, quality control is more satisfactory as pounds of dry matter per unit of volume may not vary greatly with small changes in moisture. Weighing feed ingredients daily aids in management and recordkeeping, but adjustments need to be made for moisture variation to achieve good quality control.

Table 9-13
Correcting for Moisture Content of Feedstuffs

1. To determine lb. as fed from lb. dry matter, multiply factor in column next to moisture content times lb. dry matter. For example, 10 lb. corn silage dry matter = 25 lb. corn silage at 60% moisture (10 x 2.50).
2. To determine lb. dry matter from lb. as fed, multiply factor in second column next to moisture content times lb. as fed. For example, 25 lb. corn silage at 60% moisture = 10 lb. corn silage dry matter (25 x .4).

% Moisture	Dry to as Fed	As Fed to Dry	% Moisture	Dry to as Fed	As Fed to Dry	% Moisture	Dry to as Fed	As Fed to Dry
10	1.11	.90	38	1.61	.62	66	2.94	.34
12	1.13	.88	40	1.66	.60	68	3.12	.32
14	1.16	.86	42	1.72	.58	70	3.33	.30
16	1.19	.84	44	1.78	.56	72	3.57	.28
18	1.22	.82	46	1.85	.54	74	3.84	.26
20	1.25	.80	48	1.92	.52	76	4.17	.24
22	1.28	.78	50	2.00	.50	78	4.54	.22
24	1.31	.76	52	2.08	.48	80	5.00	.20
26	1.35	.74	54	2.17	.46	82	5.55	.18
28	1.39	.72	56	2.27	.44	84	6.25	.16
30	1.43	.70	58	2.38	.42	86	7.14	.14
32	1.47	.68	60	2.50	.40	88	8.33	.12
34	1.51	.66	62	2.63	.38	90	10.00	.10
36	1.56	.64	64	2.78	.36			

9-5 Correcting for Moisture Content of Feeds

Table 9-14
Conversion of a Feeding Formula from a Dry Matter Basis to an As Fed Basis

Feed	% In Ration DM	% Moisture	Conversion Factor	As fed lb. needed per 100 lb. of dry matter	As Fed Composition
Corn	5	20	1.25	6.25[A]	2.13[D]
Silage	90	68	3.12	281.25[B]	95.97[E]
Supplement	5	10	1.11	5.56[C]	1.90[F]
Total	100			293.06	100.00

A 5 x 1.25 = 6.25
B 90 x 3.12 = 281.25
C 5 x 1.11 = 5.56

D 6.25/293.06 = 2.13%
E 281.25/293.06 = 95.97%
F 5.56/293.06 = 1.90%

Table 9-15
Conversion of Lb. Dry Matter per Head Daily to Lb. As Fed per Head

Feed	Lb. DM	Moisture	Conversion Factor	Lb. as fed
Corn	1	20	1.25	1.25
Silage	18	68	3.12	56.16
Supplement	1	10	1.11	1.11
Total	20			58.52

Table 9-16
Conversion of Lb. per Head Daily to Lb. Dry Matter

Feed	Lb. as fed	% Moisture	Conversion Factor	Lb. Dry Matter
Corn	1.25	20	.80	1
Silage	56.16	68	.32	18
Supplement	1.11	10	.90	1
Total	58.52			20

278 Evaluation of Feedstuffs

3. Adjust for moisture when buying feeds. The multipliers in Table 9-13 may be used to determine the price per unit of dry matter simply by multiplying price times the appropriate factor for the indicated moisture. For example, shelled corn at 30 percent moisture and costing $60.00 per ton costs 60 × 1.43 = $85.94 per ton on a dry basis. Another source of corn costs $65 per ton at 25 percent moisture, or 65 × 1.33 = $88.85. In this case the 30 percent moisture corn is the better buy.

Table 9-17 gives the correction factors to correct shelled corn of different moisture contents to a 15.5 percent or standard No. 2 basis. For example, if No. 2 corn (standard 15.5 percent moisture) is priced at $80.00 per ton, 30 percent moisture corn is worth $80 × 0.8284 = $66.27 per ton. If the feeder was receiving 19 percent moisture corn and paying for 15.5 percent moisture, he would receive only 95.86 percent of the dry matter he paid for. If corn is delivered with 7 percent moisture, while paying on a 15.5 percent moisture basis, the feeder would receive 110.06 percent of corn he paid for. If 15.5 percent moisture corn is the purchase basis, it will require 1.1834 units of purchase base corn to make 1 unit of 100 percent dry matter base corn.

When evaluating commodity purchases, a feeder should never lose sight of how much water he is forced to buy. If the feeder, for example, assumed that corn and wheat had equal nutritional characteristics per unit of dry matter, the trading basis (15.5 percent moisture for U.S. No. 2 Corn and usually about 8–10 percent moisture for wheat) is probably much more significant than any nutritional difference found in the two grains.

In any area there are so called "norms" in terms of how much moisture should be in feed commodities. In general, livestock feeders are very lax in observing moisture standards until the feeds become so wet as to cause storage or handling problems. Reputable suppliers usually observe the standard very closely, being sure that they do not supply more or less moisture than the standard calls for.

A large elevator could lose $1 million a year simply by selling grain containing 2 percent or 3 percent less moisture than the standard allows. In some cases there is no standard or established moisture level for commodities. In these cases commodities are usually sold using protein, fat, and fiber guarantees. In the case of oil meals, as the moisture content rises the protein content usually goes down. Feeders should always remember that "as is" feeds are similar to a pie where moisture is like the first slice removed from the pie. The larger the first slice the less is left. There are many

Table 9-17
Relative Value of U.S. No. 2 Corn (15.5% Moisture) as Affected by Changes in Moisture

Moisture %	Multiplier	Moisture %	Multiplier
0	1.1834		
1	1.1716	19	.9586
2	1.1598	20	.9467
3	1.1479	21	.9349
4	1.1361	22	.9231
5	1.1243	23	.9112
6	1.1124	24	.8994
7	1.1006	25	.8876
8	1.0888	26	.8757
9	1.0769	27	.8639
10	1.0651	28	.8521
11	1.0533	29	.8402
12	1.0414	30	.8284
13	1.0296	31	.8166
14	1.0178	32	.8047
15	1.0059	33	.7929
16	.9941	34	.7811
17	.9822	35	.7691
18	.9704	36	.7574

implications to the pie concept. If, for example, one sample of soybean meal had 44 percent protein and 12 percent moisture, and another sample had 44 percent protein and 7 percent moisture, the prospective purchaser would need to look critically at the difference in the two samples. By removing the water, the prospective purchaser can see that the first sample is 50 percent protein on a moisture-free basis (100 −12 = 88; 44 ÷ 0.88 = 50%) and the second sample is 47 percent (100 - 7 = 93; 44 ÷ 0.93 = 47%). The first sample is a higher protein meal on a moisture-free basis, and quite possibly a better quality meal even though it contains more water.

Testing for Moisture

Proper sampling and access to a moisture tester are essential to determine feed moisture levels. The most useful moisture tests are taken on the farm, due to losses in moisture and changes in the feed when sent to a testing laboratory and time lag between when the sample is sent and when an analysis is received. An inexpensive relatively accurate moisture tester can be purchased from your local Dairy Herd Improvement Association milk tester or from Koster Crop Tester (4716 Warrensville Center Road, North Randall, Ohio 44128). Many county agents also have moisture testers in their offices. Many local elevators have testers that can test for moisture in grains.

The best recommendation, however, is to purchase your own tester. It costs less than $150 and can also be useful in determining the best point to harvest hay, haylage, silage, or grains. A moisture tester is a relatively small investment compared to the large investment in silos and feeding and harvesting equipment and the losses that can occur due to harvesting at the wrong time or errors in buying and selling due to not knowing the moisture level.

If the feed is being sampled for nutrient analysis also, your local county agent can give you instructions as to cost and how to prepare the sample for sending to the laboratory.

9-6 RATION FORMULATION

There are several commonly accepted methods of balancing rations. All are useful, depending on the situation. The goal of all systems is to provide adequate supplemental nutrients to allow most efficient utilization of energy and protein in common feedstuffs. Most cattle feeders, especially farmer feeders, will not consider all possible rations. Most simply wish to supply missing nutrients in homegrown grain, silage, and hay, and to develop the best feeding system to utilize their own feeds most efficiently. Developing least cost-maximum profit feeding programs will be discussed in the next chapter.

In general, high-silage programs are more profitable when grain prices are high (over $2.50–$3.50/bu for corn). However, other factors are usually just as important such as amount of roughage that can be handled and stored (in commercial feedlots) and a feeding system that will best market through cattle the crops grown (i.e., late planted or drought, frost-damaged corn is harvested for silage; harvesting part of the crop as silage extends the harvest season, etc.).

Let's assume, for all examples in this section, a cattle feeder purchases 500-lb average-frame steer calves. He wants to average, over the entire feeding period, about 50 percent shelled corn and 50 percent silage, on a dry matter basis. He is equipped to weigh all of the feed, and can purchase a commercial protein supplement or can mix his own. Using this example, the different methods of formulating the ration and developing the best overall feeding system will be outlined.

Step 1. Determine the weight groups for the cattle. Experience indicates that calves can be sorted, fed, and managed efficiently in about three weight categories, considering variation in the cattle, labor and facilities, and efficient ration balancing. It would be practical to formulate rations for these three groups: to 600 lbs, 600–800 lbs, and 800 lbs to market.

Step 2. Determine the timing of grain feeding. This will be discussed later in detail. The best system is to feed a high forage (the energy equivalent of an all corn silage ration) followed by a high grain ration when overall 30 to 60 percent of the feed is to be forage (two-phase feeding). The high grain ration should contain 10–15 percent roughage. Many prefer, however, to feed a constant proportion of grain for ease in management. Overall feed efficiency will be reduced about 3–8 percent, but ability to manage the program (type of storage and feeding equipment available, labor, etc.) may be more critical. Budgets will be given later that can be used to estimate feed requirements on different systems. For this feeder, let's assume an all silage ration to 800 lbs, then a gradual switch to a high grain, minimum roughage ration (15 percent corn silage, 85 percent grain, plus supplement in the ration dry matter).

Step 3. Convert crude protein values from feed analysis and the supplement tag to net protein values, if the net protein system is to be used to balance the ration. In this case the corn silage tested 8.7 percent crude protein (CP) and the high moisture corn tested 10.5 percent CP. Using the factors from section 9-2, the estimated % net protein is $8.7 \times 0.2 = 1.7$. For the corn, $10.8 \times 0.3 = 3.2$ percent. His 40 percent protein supplement is $40 \times 0.35 = 14$ percent estimated NP.

Methods for Balancing the Ration

1. Guideline rations. Prebalanced guideline rations given in the next chapter that are close to his can be used. These are the most useful to the small feeder. Table 10-8 shows that he needs

1.9 lbs of a 40 percent supplement to 600 pounds, then 1.7 - 1.4 lbs from 600 to 800 pounds. When he switches to the high grain ration at 800 pounds, only 0.7 lb of supplement is needed. His feed was analyzed, however, and the feed nutrient values are different than those in Table 9-1; thus, the ration should be evaluated.

2. Ration evaluation. This method is used to simply check the ration he is now feeding for adequacy in protein and minerals. Ration Evaluation Worksheets 9-1 and 9-2 demonstrate this method. Expected gains and nutrient requirements are used from Table 8-17. He is feeding corn silage with 1 lb of a typical commercial 40 percent protein supplement to the 450-lb calves. The tag reads "Contains 40% protein, with not more than 30% protein from non-protein nitrogen." It also contains 2 percent calcium, 1 percent phosphorus, and 3 percent salt. Immediately you should see that it exceeds the guideline of "not more than 1/4 to 1/3 of the supplemental protein from urea for cattle under 600 lbs equivalent weight." In his supplement, 30/40 = 3/4 of the supplemental protein from urea. Thus, a supplement higher in natural protein should be substituted.

First calculations indicated that the ration was deficient in protein, calcium, phosphorus, and salt (Worksheet 9-1). When the minerals from the additional protein supplement needed were added, however, the mineral needs were nearly met.

An evaluation for the cattle in the second group (600-800 lbs) indicated that the supplement could be reduced to 1.4 lbs/day for 600-lb cattle and 1.1 lb/day when most averaged about 700 pounds. The ration is still adequate in minerals. However, 0.022 lb more salt/head daily is needed.

For the 800 lb to market group, a different supplement is needed. Only 0.8 lb/day of a 40 percent protein supplement is needed to meet the protein needs. However, the supplement would have to contain 7 percent calcium and 7 percent salt to meet mineral needs; 0.09 lb limestone and 0.03 lb salt must be fed/head daily if the same supplement is fed.

The expected rate of gain and thus protein needs were based on averages given in Table 8-17. However, there are many combinations of feeds used to finish cattle. Ration Evaluation Worksheet 9-2 shows how the rate and cost of gain and protein needs could have been calculated on the finishing ration. Several combinations can be tested for cost of gain to find the optimum combination. The vitamin and feed additive level should be checked when the ration is balanced. A level of 20,000-30,000 IU Vitamin A should

9-6 Ration Formulation

be adequate. Rumensin level should be between 200 and 300 mg/head daily.

3. Ration Formulation. This method is used when all ration ingredients are to be weighed, and/or a protein supplement is to be formulated. All formulation calculations are made on a dry matter basis; then the ration is converted to an as fed basis.

- Step 1: Determine the level of NE_g percentage of grain or roughage desired. In the case of the example finishing ration, 15 percent corn silage in the dry matter is desired.
- Step 2: Balance the ration for protein, using a simple algebraic equation.

Example: Let x = parts urea needed; the silage = 15% or 0.15 parts. Leave 0.01 parts for minerals. Then the parts of grain = 0.84 - x. The net protein needed = 3.51% (Tables 8-17 to 8-19); this is obtained by multiplying the parts of each ingredient times its % net protein, as follows:

$$3.51 = 0.15 \, (1.7) + x \, (98.4) + 0.84 - x \, (3.2)$$

$$95.2 \, x = 0.567 \quad x = 0.0060, \text{ or } 0.60\% \text{ urea}$$

- Step 3: Balance for minerals, and add feed additives. Add 0.28% salt, 0.05% Vitamin A–D premix (the one being used contains 2 million IU A and 200,000 IU D/lb), and 0.02% Rumensin Premix (60 gm/lb).

Ingredient	% in ration DM	% Ca	lb Ca	% P	lb P	% K	lb K
Corn silage	15.00	.28	.042	.21	.032	.95	.143
Shelled corn	83.60	.03	.025	.40	.336	.46	.386
Urea	.60	—	—	—	—	—	—
Salt	.28	—	—	—	—	—	—
Limestone	.66	—	—	—	—	—	—
Potassium chloride	.14	—	—	—	—	—	—
Vitamin A premix	.05	—	—	—	—	—	—
Rumensin premix	.02	—	—	—	—	—	—
	100.35	x	.067	x	.368	x	.529

$$\text{Limestone} = \frac{.31 - .06}{.38} = 0.66\%$$

$$\text{Potassium chloride} = \frac{.60 - .529}{.523} = 0.14\%$$

Worksheet 9-1
Ration Evaluation

Ration No. 1 Expected DM intake 11.7 Expected daily gain 1.9 (See Tables 8-17 to 8-19.)

Table 1. Amount in Present Ration

	Col. 1 As fed lb Actual	Col. 2 DM % Actual	Col. 3 DM lb Col. 1 x 2	Col. 4 Protein % Table 9-1 or actual	Col. 5 Protein lb Col. 3 x 4	Col. 6 Ca % Table 9-1	Col. 7 Ca lb Col. 3 x 6	Col. 8 P % Table 9-1	Col. 9 P lb Col. 3 x 8	Col. 10 Salt % Actual	Col. 11 Salt lb Col. 6 x 10
Grain and Roughages											
Corn Silage	31	35	10.8	1.7	.183	.28	.030	.21	.023	--	--
Protein supplement**				% protein x .35	Col. 1 x 4	% calcium from tag	Col. 1 x 6	% phosphorus from tag	Col. 1 x 8	% salt from tag	Col. 1 x 10
"Hi Gain"	1	90	.9	40 x .35 = 14	.140	2	.020	1	.010	3	.03
Totals of Columns ***			11.7		.323		.050		.033		.030

*Divide grain fed in Col. 3 by expected total dry matter intake to get % gain in ration. Then get expected daily gain from Table 9-9.
**To get amount of nutrient fed from supplement, multiply actual lb fed times % of nutrient shown on tag of the supplement. To estimate supplement dry matter intake, multiply actual lb of a dry supplement x .9 and actual lb of a liquid supplement x .6.
***To estimate silage dry matter intake, subtract DM intake of other ingredients from expected total intake in Tables 8-17 to 8-19.

Table 2. Requirements and Deficiencies

	Col. 12 DM lb	Col. 13 Protein %	Col. 14 Protein lb	Col. 15 Ca %	Col. 16 Ca lb	Col. 17 P %	Col. 18 P lb	Col. 19 Salt %	Col. 20 Salt lb
Amount needed (Tables 8-17 to 8-19)	11.7	3.67	.430	.50	.059	.37	.043	.28	.033
Amount of nutrient in ration (Col. 3, 5, 7, 9, 11)			.323		.050		.033		.020
Deficiencies (amt. needed - amt. in ration)			.107		.009		.010		.013

Amount of Supplement needed

Protein Supplement: .107/.14 = .75 lb/day more =
Phosphorus Supplement None .015 .008 .022
Calcium Carbonate (Limestone) None
Salt None

Methods used for Worksheet 9-1

Step 1. Calculate amount in ration.
A. Put down in Column 1 what you are feeding. If you only know the grain and supplement consumption, estimate the silage DM intake by subtracting the grain and supplement DM intake from the expected total DM intake from Tables 8-17 and 8-18.
B. Calculate the DM intake for each ingredient. Get the protein, % calcium (Ca) and % phosphorus (p) in the feed from Table 9-1 or from your own feed analysis.
C. Multiply the % of each nutrient in the DM times the DM intake.
D. Add up the amount in each column to get the total for each nutrient.

Step 2. Determine the Requirement
A. Find the ration in Tables 8-17 to 8-19 that is similar to yours.
B. Put the expected DM intake, % protein, % Ca and % P across from amount needed.
C. Multiply the expected DM intake times the % required of the nutrient and enter the result in the indicated column.

Step 3. Determine the deficiencies
A. Enter the amount of the nutrient in the ration in Table 2 from Table 1.
B. Subtract from the requirement to get the deficiency.

Step 4. Determine the amount of supplement needed.
A. First see how much protein supplement is needed.
Lb protein = Lb of deficiency
supp. needed Lb protein in 1 lb of supplement
B. Then multiply this amount times the % Ca, P, and salt in the supplement and add to amount in the ration.
C. If still short on phosphorus, determine the amount of phosphorus supplement needed.

Lb phosphorus = Lb of deficiency
supplement needed Lb phosphorus in 1 lb of phosphorus supplement
D. Multiply the lb of phosphorus supplement needed times its % calcium and salt, and subtract from the calcium and salt deficiencies.
E. If still short on calcium, determine the amount of limestone needed.
Lb limestone needed = lb of deficiency
                        0.38
F. The amount of salt deficiency left is the lb of salt needed/head/day.

Worksheet 9-2
Ration Formulation

Ration for: Sex steer; Weight 800 lb. Ration type Finishing
Ration specifications: Daily expected 100% dry matter consumption 14.8 lb. (Table 8-17 or 8-18)

FEEDSTUFFS*	Col. 1 Dly.lb. 100% DM	Col. 2 NE_m per lb. (Find in Table 9-1)	Col. 3 Total Mcal NE_m (Multiply Col. 1 × Col. 2)	Col. 4 NE_g per lb. (Find in Table 9-1)	Col. 5 Total Mcal NE_g (Multiply Col. 1 × Col. 4)	Col. 6 percent Net Protein (Find in Table 9-1)	Col. 7 lb. Net Protein (Multiply Col. 1 × Col. 6)	Col. 8 percent moisture (From feed analysis)	Col. 9 moisture multiplier (Table 9-13)	Col. 10 lb. as fed (Multiply Col. 1 × Col. 9)	Cost 1 Day
Shelled Corn	11.9	1.12	13.30	.73	8.69	3.2	.380	30	1.43	17.0	62.7
Corn Silage	2.5	.78	1.95	.50	1.27	1.7	.042	68	3.12	7.8	7.8
Supplement*	0.4	.71	.28	.42	.17	14.0	.056 used as is			0.7	6.0
Totals of Columns	14.8		15.53		10.13		.479				76.5

*Feed net energy values from Table 9-1 were increased 10% to adjust for the effect of rumensin.
**%net protein of the supplement = % crude protein × .35.

METHODS

A – Predicting rate of gain
A1. NE_m per lb. of ration = 15.53/14.8 = 1.05 (Col. 3 total)
                                                  (Col. 1 total)

A2. Daily NE_m required = 6.47 megcal (from Tables 8-10 to 8-12)

A3. Lb. needed for maintenance = 6.16 $\frac{(Daily\ NE_m\ required)}{(Ration\ NE_m\ per\ lb.)}$ (from A1 and A2)

A4. Lb. left for gain = 14.8 − 6.16 = 8.63 (Col. 1 total-lb. needed for NE_m) (from A3)

A5. NE_g per lb. of ration = 10.13/14.8 = .68 $\left(\dfrac{\text{Col. 5 total}}{\text{Col. 1 total}}\right)$

A6. Mcal NE left for gain = 8.63 × .68=5.87 (lb. left for gain from A5)

A7. Expected rate of gain = 2.6 (This if ound in Tables 8-10 to 8-12 for the weight and sex of the cattle and across from the value that corresponds to the Mcal NE left for gain as calculated above).

B - Meeting the protein requirements
B1. Lb. protein furnished by ration = .48 (Sum of Col. 7)

B2. Lb. protein required daily = .52 (from Table 8-16)

B3. Lb. protein still needed = .04 (Lb. required - sum of Col. 7) (lb. protein still needed)

B4. Lb. protein supplement needed = .04/.14=.3 (percent protein of supplement) × 100
.3 + .4= .7

C. - Minerals: (calculate in the same way the protein needs were calculated)

D - Cost of ration
D1. Total cost of ration =76.5 (lb. of each ingredient to be fed x per lb. cost of ingredient, then total up cost of all ingredients).

D2. Feed cost per lb. of gain = 76.5/2.6=29.4 $\dfrac{\text{(total cost of ration)}}{\text{(expected gain, from A7)}}$

D3. Total cost per lb. of gain = 76.5 + .30/2.6 = 41¢ $\left(\dfrac{\text{(total cost of ration 30 cents)}}{\text{expected gain}}\right)$, or D1 + 30 cents from (A7)

E - Break-even price on cattle
E1. Total cost of gain = 600 × .41 = $246 (Total lb. gain of cattle x total cost per lb. of gain)
E2. Break-even price = 337 + 246/1050 = .56¢ $\dfrac{\text{(Cost of feeder cattle per hd. + total cost of gain)}}{\text{expected sale wt. of cattle}}$

288 Evaluation of Feedstuffs

• Step 4: Adjust to 100% by removing or adding corn. In this example, the corn is adjusted to 83.25%.

• Step 5: Formulate the protein supplement. Place all supplemental ingredients into a supplement; put enough corn in supplement so that it is at least half of the supplement, or to give the percentage of supplement in the ration desired. If a fixed lb/head/day is wanted, divide that amount by the expected dry matter intake. In this case, assume 1/2 lb/head/day, or 3% of ration DM.

Ingredient	% in total ration	% in supplement	Moisture factor	As fed lb in 100 lb supplement DM	Lb as fed/ 100 lb supplement*
Urea	.60	20.00	1.00	20.00	19.00
Salt	.28	9.33	1.00	9.33	8.86
Limestone	.66	22.00	1.00	22.00	20.90
Potassium chloride	.14	4.67	1.00	4.67	4.44
Vitamin A premix	.05	1.67	1.11	1.85	1.76
Rumensin premix	.02	.67	1.11	.74	.70
Sub-total	1.75	58.34			55.66
% shelled corn	1.25	41.66	1.12	46.66	44.34
Total	3.00	100.00		105.25	100.00

*Lb ingredient in column 4 ÷ column 4 total.

• Step 6: Compute the as fed formula for the total ration.

	lb/100 lb ration DM	Moisture factor	As fed lb in 100 lb ration DM	lb/100 lb. as fed ration*
Corn silage	15.00	3.12	46.80	27.98
Shelled corn	82.00	1.43	117.26	70.10
Supplement	3.00	1.07	3.21	1.92
	100.00		167.27	100.00

*Lb ingredient in column 3 ÷ column 3 total.

10

Simple Guidelines for Feeding Beef Cattle

Nutrient requirements, feed composition values, and ration formulation were discussed in Chapters 8 and 9. Using the values and methods discussed there, simple guideline rations for feeding cattle were developed for this chapter.

10-1 GUIDELINE RATIONS FOR THE BEEF HERD

In the paragraphs that follow, suggested rations are given for various ages and classes of breeding cattle. These rations are based on the requirements listed in Chapter 8. Dr. Harlan Ritchie, Michigan State University, contributed to the development of the guidelines for feeding the breeding herd.

In addition to the rations presented below, a salt-mineral mix should be offered free-choice at all times. Several possible mixes are listed later in this section. Vitamin A should also be added to the diet or injected intramuscularly if the forage is of low quality and apt to be deficient in this vitamin. If injected, a dose of 1 to 3 million IU is recommended. The injected dose will last for 90 to 100 days.

Rations for Weaned Heifer Calves

The goal in feeding open heifers is to achieve enough gain in weight so that they may be bred one heat period prior to the main cow herd at about 14 months of age, as mentioned before. Their daily gain from weaning to breeding should average 1.0 to 1.5 lbs per day. The following rations are possibilities.

1. High quality pasture + 5# grain
2. 12# hay (full-feed) + 5# grain
3. 40# corn silage (30% DM) + 1# soybean meal or equivalent
4. 30# wet haylage (35% DM) + 5# grain
5. 20# dry haylage (50% DM) + 5# grain
6. 5# hay + 30# corn silage + ½# soybean meal or equivalent
7. 10# hay + 20# corn silage
8. 40# forage sorghum silage (30% DM) + 2½# grain + 1# soybean meal or equivalent
9. 40# oat silage (30% DM) + 2½# grain + ¼# soybean meal or equivalent

Rations for Coming 2-Year-Old Pregnant Heifers

The goal in feeding pregnant heifers is to achieve about 1-lb average daily gain for 120 days prior to calving; for example, from 800 lbs to 920 lbs. Underfeeding pregnant heifers can be disastrous because they are still growing as well as developing fetus and preparing for the stress of their first lactation. Overfeeding, however, can lead to too much internal fat, resulting in a higher incidence of calving difficulty.

1. High quality pasture
2. 20 to 25# hay (full-feed)
3. 45# corn silage (30% DM) + 1¼# soybean meal or equivalent
4. 55# wet haylage (35% DM)
5. 40# dry haylage (50% DM)
6. 5# hay + 35# corn silage + ¾# soybean meal or equivalent
7. 10# hay + 25# corn silage + ¼# soybean meal or equivalent
8. 15# hay + 15# corn silage
9. 55# forage sorghum silage (30% DM) + 1# soybean meal or equivalent
10. 55# oat silage (30% DM)

Rations for Dry 1100-lb Mature Cow, Middle 1/3 of Pregnancy

The goal here is to maintain the body weight of pregnant mature cows in good condition after their calves have been weaned.

1. Low to medium quality pasture
2. 17 to 25# hay
3. 7# hay + 15# straw
4. 45# corn silage (30% DM)
5. 50# wet haylage (35% DM)
6. 35# dry haylage (50% DM)
7. 5# hay or 7# straw, + 30# corn silage
8. 10# hay or 13# straw, + 20# corn silage
9. 15# hay or 21# straw, + 10# corn silage
10. 50# forage sorghum silage (30% DM)
11. 50# oat silage (30% DM)
12. 1 to 2 acres cornstalks per cow + hay or supplement as needed
13. Full-feed dry corn plant refuse (13# DM) + 7# hay
14. Full-feed corn plant refuse silage (15# DM) + 3# corn + ½# soybean meal or equivalent

Rations for Dry 1100-lb Mature Cow, Last 1/3 of Pregnancy

The goal during the last 90 to 120 days of pregnancy is to achieve an average daily gain of 0.5 to 1.0 lbs per day. Ideally, cows should be on a rising plane of nutrition prior to and after calving so as to be in proper condition for the start of breeding season.

1. Medium to high quality pasture
2. 20 to 22# hay
3. 50# corn silage (30% DM)
4. 60# wet haylage (35% DM)
5. 40# dry haylage (50% DM)
6. 5# hay + 35# corn silage
7. 10# hay + 25# corn silage
8. 15# hay + 15# corn silage
9. 60# forage sorghum silage (30% DM)
10. 60# oat silage (30% DM)

Rations for 1100-lb Lactating Cow (Average Milking Ability)

The goal here is to keep the cow in a positive nutritional status so she will conceive by 80 days after calving and average 10 to 12 lbs of milk per day during the first 3 to 4 months of lactation. This level of milk production would be typical of most British beef cows.

1. High quality pasture
2. 25 to 30# hay (full-feed)
3. 55# corn silage (30% DM) + 1# soybean meal or equivalent
4. 65# wet haylage (35% DM), full-feed
5. 45# dry haylage (50% DM), full-feed
6. 8# hay + 40# corn silage
7. 12# hay + 30# corn silage
8. 16# hay + 20# corn silage
9. 7# forage sorghum silage (30% DM), full-feed + 1# soybean meal or equivalent
10. 70# oat silage (30% DM), full-feed

Rations for 1100-lb Lactating Cow (Heavy Milker)

The goal is the same as for the average milking cow except that milk production is 20 to 24 lbs per day, which is typical of dairy x beef crossbred females and some dual-purpose exotics. It is difficult for females of this type to consume enough energy to get back in shape for breeding season so as to conceive on schedule.

1. High quality pasture + grain if necessary
2. 30# hay (full-feed) + grain if necessary
3. 65# corn silage (30% DM), full-feed + 2¼# soybean meal or equivalent
4. 85# wet haylage (35% DM), full-feed + grain if necessary
5. 60# dry haylage (50% DM), full-feed + grain if necessary
6. 80# forage sorghum silage (30% DM), full-feed + 2# soybean meal or equivalent
7. 85# oat silage (30% DM), full feed + ¾# soybean meal or equivalent

10-1 Guideline Rations for the Beef Herd

Rations for Mature Herd Sires

The goal is to maintain the weight of mature bulls in good condition and to put weight on thin bulls.

1. High quality pasture + grain to condition
2. 30# hay + grain to condition
3. 70# corn silage (30% DM) + 1½# soybean meal or equivalent
4. 85# wet haylage (35% DM) + grain to condition
5. 60# dry haylage (50% DM) + grain to condition
6. 90# forage sorghum silage (30% DM) + 1# soybean meal or equivalent
7. 85# oat silage (30% DM)

Rations for Young Herd Sires (Yearlings and 2-Year-Olds)

The goal is to provide adequate nutrition to support an average daily gain of 1.5 lbs on yearling bulls and 0.7 lb on 2-year-old bulls.

1. High quality pasture + 12# grain
2. 20# mixed hay + 12# grain
3. 80# corn silage (30% DM) + 2.0# soybean meal or equivalent
4. 50# wet haylage (35% DM) + 12# grain
5. 35# dry haylage (50% DM) + 12# grain
6. 70# forage sorghum silage (30% DM) + 6# grain mix + 1¾ soybean meal or equivalent
7. 80# oat silage + 3# grain + ½# soybean meal or equivalent

Winter Feed Budgets

Tables 10-1, 10-2, and 10-3 attempt to illustrate the total winter feed requirements for a 50-cow beef herd, utilizing various combinations of feed stuffs. In Table 10-1, hay is the only roughage source, whereas in Table 10-2 corn silage is the only roughage. In Table 10-3, about 50 percent of the roughage dry matter is furnished by hay and 50 percent by corn silage. In developing these budgets, minimum nutrient requirements for 1100-lb mature cows were used, and no allowance was made for cold stress. Furthermore, it is assumed that any mineral deficiencies would be offset

Table 10-1
Winter Feed Budget for a 50-Cow Beef Herd Using Hay as Roughage

Class of Cattle	No. Head	No. Days	HAY			GRAIN		
			lb per Head per Day	Herd Total for Winter (T)	Total per cow Unit (T)	lb. per Head per Day	Herd Total for Winter (lb)	Total per cow Unit (lb)
Pregnant mature cows	40	120	20	48.0	0.96	---	---	---
Lactating mature cows	40	60	27	32.4	0.65	---	---	---
Pregnant 2-yr. heifers	10	120	20	12.0	0.24	---	---	---
Lactating 2-yr. heifers	10	60	27	8.1	0.16	---	---	---
Open yrlg. heifers	13	180	11	12.9	0.26	5	11,700	292
Mature herd sire	1	180	30	2.7	0.05	---	---	---
Young herd sire	1	180	20	1.8	0.04	8	1,440	36
Total	---	---	---	117.9	2.92	---	13,140	263

As shown in Table 10-1 it takes about 2.4 T. of hay and 260 lb of grain per producing female to winter a herd consisting of 50 breeding age females, 13 open yearling replacement heifers and 2 herd sires for 180 days. In addition, it would take a total of about 25 lb of salt-mineral mix per cow unit, or a total for the entire herd of approximately 1,625 lb over the 180-day period.

Table 10-2
Winter Feed Budget for a 50-Cow Beef Herd Using Corn Silage as Roughage

Class of Cattle	No. Head	No. Days	30% DM CORN SILAGE			40% PROTEIN SUPPLEMENT		
			lb per Head per Day	Herd Total for Winter (T)	Total per Cow Unit (T)	lb per Herd per Day	Herd Total for Winter (lb)	Total per Cow Unit (lb)
Pregnant mature cows	40	120	45	108.0	2.16	---	---	---
Lactating mature cows	40	60	65	78.0	1.56	2.0	4,800	96.0
Pregnant 2-yr. heifers	10	120	45	27.0	0.54	1.3	1,560	31.2
Lactating 2-yr. heifers	10	60	65	19.5	0.39	2.0	1,200	24.0
Open yrlg. heifers	13	180	40	46.8	0.94	1.1	2,574	51.5
Mature herd sire	1	180	70	5.4	0.11	1.4	252	5.0
Young herd sire	1	180	75	6.8	0.14	1.5	270	5.4
Total	---	---	---	290.7	5.81	---	10,656	213.1

Table 10-2 shows that it takes slightly over 5 3/4 T of corn silage and about 210 lb of soybean meal equivalent per producing cow to winter a 50-cow herd for 180 days. In many instances, NPN compounds such as urea would be a more economical source of supplemental crude protein than soybean meal, as mentioned previously.

Table 10-3
Winter Feed Budget for a 50-Cow Beef Herd Using Hay and Corn Silage

Class of Cattle	No. Head	No. Days	HAY lb per Herd per Day	HAY Herd Total for Winter (T)	HAY Total per Cow Unit (T)	30% DM CORN SILAGE lb per Herd per Day	30% DM CORN SILAGE Herd Total for Winter (T)	30% DM CORN SILAGE Total per Cow Unit (T)
Pregnant mature cows	40	120	5	12.0	0.24	35	84.0	1.68
Lactating mature cows	40	60	10	12.0	0.24	45	54.0	1.08
Pregnant 2-yr. heifers	10	120	15	9.0	0.18	15	9.0	0.18
Lactating 2-yr. heifers	10	60	10	3.0	0.06	40	12.5	0.24
Open yrlg. heifers	13	180	10	11.7	0.23	20	23.4	0.47
Mature herd sire	1	180	15	1.4	0.03	30	2.7	0.05
Young herd sire	1	180	10	0.9	0.02	55	5.4	0.10
Total	--	--	--	50.0	1.00	--	190.1	3.80

As shown in Table 10-3, about 1 T of hay and 3.8 T of corn silage per producing cow are required to winter a 50-cow herd for 180 days. With hay at 90% DM and corn silage at 30% DM, approximately half of the total dry matter is supplied by each of these feeds.

by free-choice feeding of a salt-mineral mix. In addition, the following assumptions were made: (1) a total winter feeding period of 180 days; (2) on an average, the herd is lactating during the last 60 days of the winter feeding period; (3) half of the cows are average milkers, half are heavy milkers; (4) 13 open yearlings are kept as herd replacements; (5) the pregnant herd consists of 40 mature cows and 10 coming 2-year-olds; (6) the mixed hay contains 50 percent TDN and 10 percent crude protein.

Free-Choice Mineral Mixtures

Mixture 1. For the cow herd during breeding season to provide extra phosphorus:

	% of Mix	% Ca	% P
Trace mineralized salt	33	—	—
Bonemeal or dicalcium phosphate	67	22–27	13–18
Total in mix	100	14.7–18.0	8.7–12.8

Mixture 2. For the cow herd before and after breeding season:

	% of Mix	% Ca	% P
Trace mineralized salt	50	—	—
Bonemeal or dicalcium phosphate	50	22–27	13–19
Total in mix	100	11.0–13.5	6.5–9.5

Mixture 3. For cattle in drylot on grain or other feedstuffs low in calcium content:

	% of Mix	% Ca	% P
Trace mineralized salt	33.3	—	—
Bonemeal or dicalcium phosphate	33.3	22–27	13–19
Ground limestone	33.3	38	—
Total in mix	100.0	20.0–21.7	4.3–6.3

Mixture 4. For feeding to herds during late winter and early spring in areas where grass tetany (magnesium deficiency) is a

problem (No other salt or mineral mixture should be offered, or daily magnesium intake may be too low.):

	% of Mix	% Ca	% P	% Mg
Magnesium oxide	25	—	—	60
Trace mineralized salt	25	—	—	—
Bonemeal or dicalcium phosphate	25	22–27	13–19	—
Ground corn	25	—	.35	—
Total in mix	100	5.5–6.8	3.3–4.8	15.0

Commercial mixtures. Salt-mineral mixtures comparable to those listed above may be purchased commercially. Beware of mineral blocks that are extremely hard and dense because it is very difficult or impossible for cattle to obtain their daily mineral requirements from such blocks.

Feeding salt and mineral separately. To ensure adequate intake of salt and all other mineral elements, it is often considered preferable to feed Mixture 1, 2, or 3 in one feeder and straight trace mineralized salt in another feeder.

How to feed mineral mixes. All salt or mineral mixes should be fed under cover to keep out rain and/or snow. When fed outside, weather-vane type feeders that rotate with the wind are the most desirable. They may be constructed at home or purchased commercially at a cost of approximately $40 to $125. Mineral feeders should be located in sites where cattle have daily contact.

How to budget mineral consumption. When fed free-choice, cattle will consume approximately 0.1 to 0.2 lb of salt-mineral mix per head per day. A figure of 0.15 lb per day or 55 lb per cow per year would be a rough average.

Adding vitamin A to mineral mixes. Adding a Vitamin A premix to the mineral mix is a convenient method of providing this vitamin. However, vitamin A loses its potency with time, so these mixes should not be stored for extended periods of time. Enough vitamin A should be added to the mineral mix so that each animal receives its requirement (10,000 to 50,000 IU) in 0.1 to 0.2 lb of total mix.

10-2 GUIDELINES FOR FEEDING BULL CALVES

Bull calves fed a growing ration containing 60 to 65 percent TDN (50 to 60 percent concentrate) and 12.5 to 13.0 percent crude protein will gain about 2½ lbs per day. This feeding program will usually result in bulls capable of breeding 15 to 20 cows at 14 to 15 months of age. Mature breeding bulls will usually maintain satisfactory condition on the same kind of pasture and wintering management provided for the cow herd.

If bulls have not regained their weight during the winter, they should be given an extra feed for 1½ to 2 months before the breeding season begins to get them into a strong and vigorous condition. Bulls that are still growing (between 2-4 years of age) need higher energy and protein rations through the winter than the cow herd. A 1000 to 1300 lb bull should gain 1.5 to 2.0 lbs per day on a full feed of good quality legume hay plus 10 lbs of grain, or a full feed of corn silage plus 2 lbs of 40 percent protein supplement.

Daily rations for growing bull calves for 2.3 to 2.5 lbs daily gain are given below.

Ration #1	*Ration #2*
Full feed corn silage	Free-choice legume hay or hay silage
1 lb shelled corn per 100 lbs body wt	1½ lb shelled corn per 100 lbs body wt
3 lbs of 40% protein supplement	
40,000 IU vitamin A	40,000 IU vitamin A
Free-choice mineral mix	Free-choice mineral mix

Feeding a ration that contains 65 to 70 percent TDN (70 to 80 percent concentrate) to bull calves for 140 to 150 days after weaning to measure their growth potential probably will not reduce their breeding ability when placed into service if energy intake is reduced following the test period. Good quality pasture or a ration that is about 50 percent concentrates should be satisfactory following the test period.

Guideline Rations for 140-Day Feed Test

The following summary is a result of a survey of bull testing stations and performance herds conducted by G. L. Minish. In reviewing the rations presented here, it will become apparent to the reader that a wide variety of feeding regimes are in use today. For this and other reasons, it is virtually impossible to compare the absolute performance of bulls tested in different stations and/or herds. In most instances, the weight ratio, which compares a bull with his test contemporaries, is a more reliable indicator of his ability to gain. The following rations are suggestions and representatives of the stations surveyed throughout the United States:

Alabama, Auburn University, Auburn, Ala.

66.2% ground snapped corn
10.2% molasses
10.0% hay
  9.0% soybean meal   TDN = 72.3 as fed, 84.0 dry
  3.3% alfalfa meal   C.P. = 10.9 as fed, 12.7 dry
  1.0% salt
  0.5% dicalcium phosphate
100.2%

Arkansas, University of, Fayetteville, Ark.

900# cracked corn, #2 yellow
700# prime cottonseed hulls
200# crimped oats
150# cottonseed meal (41%)
150# soybean meal (44%)   TDN = 63.2 as fed, 71.0 dry
 21# limestone   C.P. = 12.1 as fed, 13.6 dry
2121#

Colorado State University, Fort Collins, Colo.

20.0% steamed rolled oats
20.0% chopped hay (mostly grass)
18.5% cottonseed hulls
10.4% cottonseed meal
 8.0% alfalfa pellets (sun-cured)
 8.0% steamed rolled corn
 8.0% steamed rolled wheat   TDN = 59.2 as fed, 66.5 dry
 5.6% molasses   C.P. = 12.1 as fed, 13.6 dry
 1.0% salt
 0.5% mineral (high phosphorus)
   + 10,000 IU vit. A/ton
100.0%

Georgia Agricultural Experiment Station, Tifton, Ga.

900# shelled corn
400# peanut hulls
280# cottonseed meal (42%)
200# ground oats
150# cottonseed hulls   TDN = 59.1 as fed, 66.5 dry
150# molasses   C.P. = 12.2 as fed, 13.7 dry
 10# limestone
 10# trace mineralized salt
  0.2# vit. A supplement*   *supplies 1000 IU per lb of feed
—————
2100.2#

Iowa Beef Improvement Association, Ames, Iowa

	Days on Test			
	1–28	28–56	56–84	84–140
Corn silage (35% DM), lb	18	15	12	12
Shelled corn (85% DM basis), lb	9	12	15	Full fed (ap. 18)
Liquid supp. (60% CP), lb	.5	.6	.8	1.0
Linseed meal, lb	1	1	.75	.75
Calcium carbonate, lb	.1	.1	.1	.1
Total feed, lb	28.6	28.7	28.65	31.85
Total DM, lb	15.8	17.5	19.08	21.93
% DM	55.17	61.0	66.6	68.9
% TDN, as fed	43.9	49.8	55.4	57.7
% TDN, DM basis	79.6	81.6	83.2	83.8
% CP, as fed	6.85	7.64	8.38	8.73
% CP, DM basis	12.42	12.52	12.58	12.68

Louisiana Agricultural Experiment Station, Baton Rouge, La.

800# ground ear corn
400# oats
400# alfalfa hay
200# cottonseed meal
200# molasses   TDN = 68.1 as fed, 77.4 dry
 10# oystershell flour   C.P. = 12.8 as fed, 14.6 dry
 10# salt
  0.5# Vit. ADE premix
—————
2020.5#

Michigan State University #1 Pellet, East Lansing, Michigan*

    47.05% corn screenings
      15% soybean meal (44%)
      10% corn cobs (Andersons #4 fines)
      10% dehydrated alfalfa meal (17%)
      10% wheat
       5% cane molasses
       1% soybean oil
     0.5% trace mineral salt
     0.5% limestone   TDN = 72.3 as fed, 80.3 dry
     0.5% biophos   C.P. = 14.2 as fed, 15.8 dry
    0.25% Vit. ADE premix
    0.20% calcium propionate *to be full-fed with hay, free choice
  100.00%

*Michigan State University Conventional Low-Moisture, East Lansing, Michigan**

    38.0% crimped oats
    25.0% cracked corn
    10.0% crimped barley
     9.5% protein supplement (32%)
     8.0% wheat bran
     8.0% corn silage   TDN = 67.1 as fed, 78.9 dry
     0.5% trace mineral salt   C.P. = 12.2 as fed, 14.4 dry
     0.5% limestone
     0.5% Vit. A&D supplement *to be fed with hay, free choice
  100.0%

Montana–Midland Performance Testing Station, Billings, Mont.

    36% hay pellets
    20% winter speltz, oats, and middlings pellet
    19% beet pulp pellets
     9% protein supplement (20%)
  6.75% corn silage   TDN = 58.9 as fed, 68.9 dry
  6.75% upland hay   C.P. = 10.3 as fed, 12.1 dry
  2.50% molasses
  100.0%

Nebraska Beef Improvement Association

    1200# steamed flaked corn
     400# whole oats

200# sun-cured alfalfa pellets
100# molasses
  80# protein supplement (32%)
  10# antibiotic feed   TDN = 73.9 as fed, 83.0 dry
   5# salt   C.P. = 10.7 as fed, 12.0 dry
   5# vitamin premix
―――
2000#

Oklahoma—Noble Foundation, Ardmore, Okla.*

800# corn
500# cottonseed hulls**
300# oats   TDN = 65.0 as fed, 73.0 dry
300# cottonseed meal**   C.P. = 14.9 as fed, 16.8 dry
200# dehydrated alfalfa   *minerals and hay feed free-
100# wheat bran   choice
100# soybean meal   **During the last 50 to 60 days
  + Aurofac-10   of the test, the ration
―――   changes to 250 lbs cotton-
2300#   seed hulls and 100 lbs cot-
                               tonseet meal; all other in-
                               gredients remain the same.

Virginia—Virginia Polytechnic Institute & State University—Red House Bull Test Station

1300# corn silage
 540# shelled corn
 140# soybean meal   TDN = 44.0 as fed, 81.0 dry
  10# trace mineral salt   C.P. = 7.1 as fed, 13.2 dry
  10# limestone
―――
2000#

Virginia—Virginia Polytechnic Institute & State University—Culpeper Bull Test Station

1043# ear corn
 450# cottonseed hulls
 290# soybean meal (44%)
 100# alfalfa meal
 100# molasses   TDN = 66.0 as fed, 75.0 dry
  10# trace mineral salt   C.P. = 12.2 as fed, 13.85 dry
   7# limestone
―――
2000#

10-3 CREEP FEEDING CALVES

The primary reason for creep feeding calves is to supply additional energy. The milk a suckling calf receives is high in protein and minerals but relatively low in energy. Creep fed calves will usually weigh 50 to 60 pounds more than noncreep fed calves at weaning time. In addition, cows will weigh 20 to 30 pounds more if their calves receive creep versus those receiving no creep feed. On the average, 6 to 9 pounds of creep ration is required to produce an additional 1 pound of calf weight.

The feed used in a creep ration should be very palatable. Oats is a common feed used in creep rations for this reason. Whole shelled corn is also an ideal grain since it is high in energy and very palatable. If both corn and oats are produced on the farm, it is advisable to start feeding the calves at one month of age a mixture of two parts shelled corn and one part oats. Then change to a full feed of corn when the calves are two months of age and are accustomed to eating grain. Table 10-4 provides guideline rations for creep feeding calves.

Table 10-4
Guideline Rations for Creep Feeding Calves

Feeds	RATIONS				
	1	2	3	4	5
Shelled corn, lb.	100	65	90	60	85
Whole oats, lb.	--	35	--	30	--
40% protein supplement, lb.	--	--	10	10	15
	100	100	100	100	100

Number 1, 2 and 3 rations are for calves receiving normal milk on adequate pastures. Number 4 and 5 rations are designed to meet protein, vitamin and mineral requirements if pastures are poor and/or calves are not receiving a normal milk supply.

It is extremely important to use fresh high quality feed for creep rations. In addition, at least five days supply should be provided and the creep feeder never allowed to become empty. It is very easy to cause overeating disorders to the extent of founder in calves where this happens. Creep feeders should be located in an easily accessible place where cows and calves gather regularly, such as near water or mineral mix locations. A creep feeder should allow one linear foot of feeder space for every three calves. When calves are near weaning, they will be consuming 7 to 8 pounds of creep feed per head daily.

In summary, the cattle producer should consider creep feeding if:

1. Heavy weaning weights and extra bloom are desirable from a sales standpoint.
2. Creep feed is relatively cheap in relation to feeder calf prices.
3. There are many first calf heifers in the herd.
4. The pastures are poor.
5. Fall-calving cows.
6. Cows are maintained in drylot.

Don't consider creep feeding if:

1. The pasture is abundant and the cows are excellent milkers.
2. The calves are to be pastured or wintered on a high roughage ration after weaning.
3. Creep feeding costs are not favorable to feeder calf prices.
4. Weaning weight and milking potential data are major criteria in your performance testing program. Creep feeding will reduce the accuracy of these comparisons.

10-4 FEEDING GRAIN ON PASTURE

Grain feeding to growing calves or yearlings on pasture has a number of advantages. First, it is a means of finishing cattle with lower labor and equipment cost compared to feedlot finishing. There is no bedding and manure handling cost. Secondly, it provides a means of extending the grazing season, especially during dry weather. Grain feeding is most useful when the pasture begins to decline in mid to late summer. The grain and pasture are most efficiently utilized when the grain intake is limited to about 1 lb/100 lb body weight. This can be accomplished by hand feeding or by adding salt or stabilized fat to grain. The following guidelines will be useful in managing this program.

1. Begin feeding grain when the pasture begins to decline, usually midsummer, *or* when the cattle are first placed on pasture before it becomes too lush.
2. Mix 10% plain white salt or 10% stabilized fat with grain to limit intake to 1 lb/100 lbs body weight to self feed, or hand feed in bunks. Plans for pasture bunks and self-feeders can be obtained from your local county extension agent.
3. Be sure an adequate water supply is readily available.
4. Implant with a growth stimulant.
5. Check fecal samples for worm eggs, and worm if necessary. (Your veterinarian can test fecal sample for worms.)
6. If the pasture becomes "grazed out," move to a full feed by eliminating the salt or place in feedlot. A complete balanced ration will then be needed, as outlined in the ration formulation section.
7. Before turning to a self-feeder full of grain, hand feed until you are sure all are eating. Start with 1/2 lb grain/100 lbs body weight, then when all are eating move to 1 lb/100 lbs body weight over a period of a few days.
8. If the salt is removed, empty the self feeder and hand feed in bunks. Increase grain 1/2 lb/head daily until they do not clean it up, then place on the self-feeder. Once on self feed, don't let feeder become empty.
9. Watch for moldy feed or packed fines in the trough; clean as needed.
10. Be sure pastures are stocked adequately. If forage is wasted, the advantage of grain feeding will be lost.

10-5 GUIDELINE RATIONS FOR GROWING AND FINISHING BEEF

High Roughage Rations for Growing Beef

Many farmers feed cattle as a means of marketing home-grown feedstuffs, primarily roughages. Also, many cattle feeders prefer to grow lightweight cattle for a period of time before placing them on a finishing ration to obtain maximum utilization of the feedstuffs available and economy of gain.

It is usually advisable to have the cattle gain a minimum of about 1.8 to 2.2 lbs per day in the feedlot in order to obtain economical gains and still utilize large amounts of roughages. The rations given here are designed to allow cattle to gain at these rates under most conditions. These rations will not fit all situations, however, because of differences in composition of feedstuffs and the amounts and kinds of feedstuffs that are available, types of cattle fed, and rates of gain that will be the most economical under specific conditions.

High Grain Rations for Finishing Beef

Most cattle feeders who finish cattle to normal slaughter weights prefer to have the finishing ration that will help prevent management problems such as going off feed, bloat, and founder and still allow the cattle to gain as rapidly as possible. The level of roughage that is desirable varies considerably, depending on feeds available and levels that have been found to be satisfactory under specific conditions. The rations given here were designed to contain 10 to 15 percent roughage on a 10 percent moisture basis, and to allow steers to gain 2.7 to 3 lbs per head daily and heifers to gain 2.5 to 2.8 lbs per head daily under most conditions, and when the cattle are slaughtered as soon as they reach the choice grade. If feed consumption is above average, then gains may be higher than this, but gains would be expected to be lower if feed consumption is reduced. Gains may also be reduced during extremely hot or cold weather or from poor lot conditions, when lower quality or improperly processed grain is used, and if one of the growth stimulating compounds such as MGA, Ralgro, or Synovex H or S is not used.

These rations will not fit all situations because of differences in composition of feedstuffs and the amount and kind of feedstuffs available, types of cattle fed, and levels of roughage and rates of gain desired for economical gains under specific conditions. Also, the ability to ensure all cattle get the same amount of roughage daily and from day to day must be taken into account. Those who do not have space for all cattle to eat at once or do not mix the roughage with the grain may have fewer problems when twice the recommended level of roughage is fed.

Table 10-6 gives guideline rations for other than corn-corn silage rations. Tables 10-7 and 10-8 give specific recommendations for corn-corn silage rations. This separation was made because protein supplementation is more critical in rations containing corn

Table 10-5

Adjustment Factors for the Protein-Mineral Supplement Feeding Rate for Frame Size, Sex, and Age

Age	Sex	Frame	Adjustment Factor, % of Average frame steer calf feeding rate
Calf	Heifer	Small	72
		Average	84
		Large	96
	Steer	Small	84
		Average	100
		Large	113
Yearling	Heifer	Small	80
		Average	93
		Large	105
	Steer	Small	93
		Average	110
		Large	124

Table 10-6

Guideline Rations for Other Than Corn-Corn Silage Combinations

Grain - lb/100 lb bodyweight daily	Roughage	Protein supplement	Mineral supplement
--------	Growing cattle - weaning to 750 lb	--------	
Corn - 1 lb/cwt.	Legume hay, free choice	None	Dicalcium phosphate + salt
Oats - 1 3/4 lb/cwt.	Legume hay, free choice	None	None except salt
Corn, wheat or barley - 1/2 lb/cwt.	3 parts corn silage, 1 part legume hay	None	Dicalcium phosphate + salt
Corn, wheat or barley - 1 lb/cwt.	Oat silage, free choice	1 lb 40% if corn, none if wheat or barley	Limestone + salt
Corn - 1 1/2 lb/cwt.	grass hay, free choice	1 lb 40%	Dicalcium phosphate + salt
Ground ear corn - 1 1/2 lb/cwt.	Legume hay, free choice	None	Dicalcium phosphate + salt
------------	Finishing cattle - 750 lb to slaughter	--------	
Barley or corn, free choice	2 lb hay	NOne	High calcium + salt
Equal parts corn and wheat, free choice	2 lb hay	None	" " " "
Equal parts ear and shelled corn	None	3/4 lb 40%	" " " "
Ground ear corn, free choice	None	1 lb 40%	" " " "

Table 10-7
Supplements for Corn-Corn Silage Rations

	A1	A2	A3	A4	A5	A6
	------------Lb/1000 lb batch------------					
Soybean meal, 44%	900	870	-	-	-	-
Ground shelled corn	-	-	705	560	470	-
Urea, 45% N	-	-	120	125	125	-
Dicalcium phosphate	30	-	70	40	-	-
Limestone	40	100	-	185	260	750
Calcium sulfate	-	-	25	25	25	-
Trace mineral salt	30	30	80	70	120	250
	Vitamins, million IU/1000 lb batch					
Vitamin A	20	20	30	30	40	100
Vitamin D	2	2	3	3	4	10
	% of Supplement					
Total protein	40	40	40	40	40	-
Calcium	2.4	4.0	1.5	4.5	10.0	28.5
Phosphorus	1.2	-	1.5	1.0	-	-
Salt	3.0	3.0	8.0	7.0	12.0	25.0

silage. Recommendations for supplementing NPN-treated corn silage are discussed in Section 10-6.

These rations were formulated by using average feed composition values. If the feed is known to be different in composition from average, then the ration should be balanced as discussed previously.

Adjusting for Age, Sex, and Frame Size

Table 10-5 gives the adjustment factor for amount of supplement fed/head/day for age, sex, and frame size. All amounts/head/day in subsequent sections are given for an average-frame steer started on feed as a calf. But, a 780-lb large-framed steer has *equivalent*

Table 10-8
Supplement Feeding Rates for Corn-Corn Silage Rations*

	None		30% high moisture corn, lb/cwt. body wt. daily					
			0.8		1.6		2.4	
	Supplement		Supplement		Supplement		Supplement	
	Lb	No.	Lb	No.	Lb	No.	Lb	No.
To 600	1.9	A1	2.2	A1	2.4	A1	1.7	A2
600–700	1.7	A1	1.8	A1	2.2	A1	1.6	A2
700–800	1.4	A3	1.5	A4	1.5	A4	1.3	A4
800–900	1.1	A3	1.0	A4	1.0	A4	0.7	A5
900–market	0.8	A3	0.7	A5	0.4	A6	0.4	A6

*Rumensin was not included in the supplement, due to the wide variation in feeding rates. For cattle under 700 lb, feed 1/2 day of a mix containing 0.6 Rumensin premix (60gm/lb) and 99.4% ground shelled corn. Cattle over 700 lb should receive 0.7 lb/head of this mix daily.

nutrient requirements, as a % of ration dry matter, to an average-framed steer weighing 650 lbs. Thus protein-mineral supplement, as a % of ration dry matter, is the same for the 650-lb average-framed steer and the 780-lb large framed steer. At 780 lbs, however, the large-framed steer eats 13 percent more dry matter/day than the average-framed steer and thus would need more pounds of supplement a day.

Feed Additives and Mineral Supplements

Rumensin is recommended for all beef growing and finishing rations. If a protein supplement is not needed, Rumensin® can be provided by feeding 1/2 lb/head daily of a grain-Rumensin mix. Many feed companies make such a special supplement.

Vitamin A should be fed with all rations. Provide a minimum of 20,000 IU/head daily. If a protein supplement is not needed, inject 1 million IU every 90 days or feed 600,000 IU/head once every month. This can be accomplished by mixing it with ground grain, 600,000 IU/lb, then feed 1 lb/head.

The type of mineral mix needed is given. If not given, mixtures and amounts to feed can be determined in the mineral requirements Section 8-3.

10-6 TREATING CORN SILAGE WITH NPN FOR GROWING AND FINISHING BEEF CATTLE

This section outlines "treating" corn silage at ensiling with alternative nonprotein nitrogen (NPN) products including urea, anhydrous ammonia, and ammonia mineral suspension (Prosil)® to raise the silage's protein content. A 40 percent protein, all-natural supplement is used as a reference base. The section begins with a summary of research with treating corn silage with NPN and expected savings at different protein supplement prices, followed by recommendations for treating silage and supplementing treated corn silage.

Performance of Cattle Fed NPN Treated Silage

Research has been conducted on the addition of NPN products to corn silage at ensiling time at a number of midwestern experiment stations. The following conclusions can be drawn, compared to rations containing a high proportion of corn silage and soybean meal.

First, treating corn silage with an NPN product at ensiling typically gives better performance than when urea is included in the protein supplement at feeding time. In trials to date, 3 to 10 percent less feed was required per pound gain depending upon products compared, when NPN was added at ensiling vs at feeding. Possible reasons for this include:

1. For corn silages treated with NPN, more soluble sugars are converted to lactic acid in the silo; lactic acid is more efficiently utilized than soluble sugars in the rumen.
2. There may be less breakdown of plant protein during fermentation for NPN treated silages.
3. Some (or all) advantages of the NPN treatment can be lost if treatment protein completely replaces a natural protein supplement when the cattle are placed on feed, particularly for cattle with equivalent weights of less than 600 pounds.
4. The effect of Rumensin on utilization of NPN-treated silage is not well known. To date, only one experiment has been conducted to evaluate the performance of cattle fed treated corn silage with and without Rumensin added at feeding. Performance differences due to Rumensin were comparable to those that would have been expected under soybean meal supplementation; that is, similar daily gains and a 10 percent reduction in feed/gain with Rumensin.

5. Feedbunk life is improved and secondary fermentation on the face of bunker silos and storage losses are probably reduced by NPN-treatment.

A number of NPN products can be utilized for treating corn silage at ensiling to raise its protein content.

Urea. Urea has been effectively used for many years by cattle feeders and dairymen. It is simple to apply and retention rates (amt recovered/amt applied) are high, above 90 percent. Gains are similar to slightly lower than when cattle are supplemented with soybean meal; 3 percent less feed was required per pound gain, when treated with both urea and minerals. Feeder cattle were started on soybean meal and untreated silage for 3-4 weeks in these studies before being placed on treatment, however.

Ammonia-mineral suspension (Pro-Sil)®. Systems based upon Pro-Sil® treated silage have given similar gains and 0 percent to 10 percent better feed efficiency than systems based upon untreated silage and soybean meal with an average in the 4 to 5 percent range. Retention rates with this product are high.

Anhydrous ammonia (applied using the cold flow method). Research results with the application of anhydrous ammonia to corn silage at ensiling with the cold flow chamber are limited. Four feeding trials and one metabolic trial have been conducted with beef cattle at Michigan State University. The results to date suggest that it is an effective method of adding NPN to corn silage. Gains and feed efficiency of cattle started on feed as calves have been 3 to 8 percent lower, compared to untreated silage plus soybean meal fed to calves in an all silage ration.

Urea-based supplement at feeding. In Nebraska trials, cattle started on feed as calves fed high corn silage rations supplemented with urea had 5 percent poorer feed efficiency than cattle receiving soybean meal and 10 percent poorer than animals receiving Pro-Sil® treated silage. Comparable results were obtained in 1975 and 1976 at Purdue University. Data on the utilization of urea-based supplements, as contrasted to soybean meal-based supplements, must be interpreted with caution, however. A large number of experiments conducted on high grain finishing rations throughout the Corn Belt suggest that heavy cattle perform nearly as well or as well on urea-based supplements as upon soybean meal-based supplements. Most trials show that the cattle above 600 pounds equivalent weight receiving urea-based supplements do not do quite as well in the first 28 days, but performance is equivalent thereafter. Heavy cattle on high grain finishing rations supplemented with urea-based supplements will have feed efficiencies and daily gains within 2-5 percent of cattle receiving soybean meal-based supplements.

Recommendations for Treating Silage with NPN and Supplementing Treated Silage

Table 10-9 summarizes recommendations for treatment levels and supplemental soybean meal for various types of feeding programs. All recommendations assume the protein is at least 8 percent in the corn plant dry matter before it is treated. The treated silage should be analyzed to ensure adequate protein levels. The ration protein level in a high silage ration should never drop below 9.5-10.5 percent of the dry matter for heavy cattle, and needs to be 12 to 13 percent for the calves.

1. Systems 1 and 2 are for all silage rations for cattle started on feed as calves. System 1 is for the "full treat," while system 2 is for the "1/2 treat" level accompanied by soybean meal until the animals reach an equivalent weight of 700 pounds.
2. System 3 is for all silage rations for yearlings. A lower level of treatment, "1/2 treat" is indicated. This would be 5-6 lbs urea, 5-7 lbs anhydrous ammonia, or 20-25 lbs Pro-Sil/ton of 32 percent dry matter corn silage.
3. Systems 4 and 5 are for corn fed at 1 lb/100 lbs of body weight. A "full treat" level is indicated.
4. Systems 6 and 7 are for a two-phase feeding program where an all silage ration is fed up to an equivalent weight of 800 lbs followed by corn at 1.5 percent of body weight/head/day thereafter. A "full treat" level is indicated.

The systems for cattle started on feed as calves show supplemental soybean meal being fed initially even though the silage has been treated. Experimental data from a number of experiment stations suggest that cattle under 550 to 600 pounds equivalent weight do not perform as well when supplemental protein is provided by treating corn silage with NPN, as when it is supplied by soybean meal. For example, in starting on feed trials at Michigan State University in 1975-76, calves fed untreated corn silage and soybean meal gained 2.02 pounds/day but those receiving ammonia-treated corn silage gained only 1.35 pounds/day. Calves fed

Table 10-9

NPN Treatment Rates and Soybean Meal 44 Supplement Recommendations for Treated Corn Silage

System No.		Cattle type (equivalent weight basis)	% protein in Corn silage	Supplemental soybean meal
1	All silage	Calves, 450# to market	13	28 days @ 1.0#/day
2	All silage	Calves, 450# to market	11	28 days @ 1.20#/day 475 to 600# @ .75#/day 600 to 700# @ .40#/day
3	All silage	Yearlings, 675# to market	11-11.5	21 days @1.0#/day
4	Corn @ 1% of body weight/head/day	Calves, 450# to market	13	28 days @ 1.0#/day 475 to 600# @ .7#/day 600 to 700# @ .4#/day
5	Corn @ 1% of body weight/head/day	Yearlings, 675# to market	13	21 days @ 1.0#/day
6	Two-Phase: All silage to 800# Corn @ 1.5% of body weight/head/day	Calves, 450# to market	13	28 days @ 1.0#/day
7	Corn @ 1.5% of body weight/head/day	Yearlings, 675# to market	13	21 days @ 1.0#/day

ammonia-treated corn silage and supplemented with soybean meal at the levels suggested in Table 10-9 had gains similar to the calves receiving untreated corn silage and soybean meal at the levels suggested in Table 10-8. While the calves started on ammonia-treated silage will often exhibit compensatory growth later, the data suggest it is profitable to supplement with soybean meal (or an equivalent natural supplement) at the rates suggested in Table 10-9.

Economics

The "to treat" or "not to treat" corn silage decision should be based upon the relative cost of energy, protein, and minerals from alternative sources. The amount of NPN required for alternative nitrogen sources and the appropriate mineral mixes for alternative rations are given in the next section. In addition, the cattleman must take into account performance differences which result from alternative treatment systems. Performance differences described previously are used in the economic analysis.

Figure 10-1 depicts the *maximum amount* you can afford to pay to treat one ton of corn silage with NPN + minerals given alternative prices per ton of an all-natural 40 percent protein–mineral supplement. The maximum value was obtained by the following steps:

1. Calculate the amounts of corn, corn silage, and supplement(s) that would be required for the untreated silage, natural supplement system and the treated silage program for the specified system in Table 10-8. Then calculate the cost of the feed.
2. Calculate the nonfeed costs including yardage and interest on the feeder (relevant where daily gains differ between treated silage and untreated silage + natural supplement systems).
3. Calculate total cost (TC) feed plus nonfeed for the untreated silage, natural supplement system.
4. Value of treated silage/ton =

$$\frac{\text{TC} - (\text{Nonfeed cost} + \text{feed cost} + \text{supplement cost for treated silage system})}{\text{tons of treated silage}}$$

For example, for cattle fed according to Systems 4 through 7 depicted in Table 10-9, you can pay up to $4.80/ton corn silage with urea plus minerals when a 40 percent protein-mineral supplement sells for $175/ton. Since the energy value of Pro-Sil treated

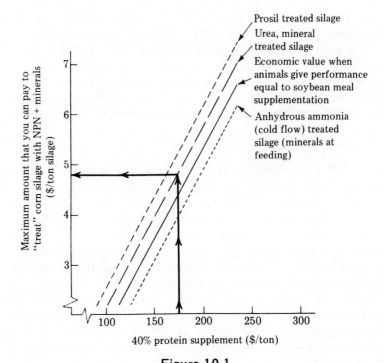

Figure 10-1
Estimated Economic Value of Treating Corn Silage with NPN and Minerals
(Relevant for Systems 4-7 in Table 10-9)

corn silage is higher, you can pay up to $0.30/ton corn silage more, or $5.10. For corn silage treated with anhydrous ammonia using the cold flow system and minerals the maximum amount is $1.00/ton corn silage less than with urea or $3.80. Economic analysis of the experimental data suggests you can afford to pay up to $1.30/ton of corn silage more to treat silage with Pro-Sil than with anhydrous ammonia plus minerals.

But, what is the net economic advantage, if any, for treating corn silage with NPN or NPN + minerals as contrasted to use of a natural 40 percent protein-mineral supplement? To complete these calculations we must subtract the treatment costs from maximum amount you can afford to pay to treat the silage (Figure 10-1).

Each cattle feeder will have to work out these calculations for himself based on his own system. Consider an example. Treating corn silage to raise its protein content from 8 to 13 percent requires 11.4 pounds of urea per ton of corn silage (Table 10-10). At

Table 10-10
Amounts of NPN Product Needed/Ton of Corn Silage to Raise Total Protein 5 Percentage Units

Corn Silage Dry Matter, %	Ammonia[1] Mineral Suspension	Aqueous[2] Ammonia	Anhydrous Ammonia Cold Flow[3]	Urea[4]
28	37	23	7.3	10.0
30	39	24	7.8	10.7
32	42	26	8.3	11.4
34	44	28	8.9	12.1
36	47	29	9.4	12.8

[1] Calculations are based upon an 85% protein content and a 10% loss during filling.

[2] Calculations are based upon a 137% protein (22% nitrogen) and 10% loss during filling.

[3] Calculations are based upon a 512% protein (82% nitrogen) and 25% loss during filling.

[4] Calculations are based upon a 281% protein (45% nitrogen) and no loss during filling.

$160/ton for urea, $0.08/pound, that would be 11.4 × 0.08 = $0.91/ton of corn silage. The mineral mix suggested in Table 10-11 for a 1 percent ration, systems 4 and 5, in Table 10-9, is fed at 0.15 pound/head/day. On the average, the animal will eat about 33 pounds of silage/day @ 750 pounds suggesting 9 pounds of mineral per ton of silage. The mineral mix costs $0.08/lb yielding a cost per ton of silage for minerals of $0.72. The range across systems is $0.40 to $0.80 depending upon the amount of dicalcium phosphate used. Thus, the total cost of treating corn silage with urea and minerals is ($0.91 for urea + $0.72 for minerals) or $1.63.

The cost of handling and equipment runs $0.40/ton resulting in a total cost per ton of treatment of $2.03/ton. Thus, at $175/ton for a 40 percent natural protein supplement, the advantage to treating corn silage would be $4.80 − $2.03, or $2.77/ton of corn silage fed. Alternatively, looking at Figure 10-1 we see that urea plus mineral treatment is profitable as long as the price of the 40 percent protein supplement is more than $120/ton.

The break-even prices for the "1/2 treat" systems, Systems 2 and 3, are about 65 percent of the values given in Figure 10-1. But, you are only adding 60 percent as much NPN in these systems, though the mineral levels are similar. The "break-even price" for full treating under the all corn silage to calves system, System 1, is approximately 90 percent of the value given in Figure 10-1, since relatively more protein is wasted during the later phases of the feeding program.

To summarize the steps:

1. Calculate the maximum value you can afford to pay to treat your silage with NPN plus minerals (Figure 10-1).
2. Calculate the amount of NPN product to be added and its cost (Table 10-9).
3. Include the cost for handling and equipment (typically $0.40/ton).
4. Calculate the value/ton of silage for minerals to be added (none in the case of Pro-Sil). Sixty cents is typically a "ballpark estimate."
5. Calculate the net economic advantage to treat (1) − (2) + (3) + (4).

NPN and Mineral Treatment Levels, and Supplementing Treated Silage

Table 10-10 gives the amounts of various NPN products needed to raise the total protein of corn silage by five percentage units. Corn silage averages 8 percent protein in the dry matter, so these levels will increase it to 13 percent. Keep in mind, however, that the actual level obtained will depend on the original silage protein (could be as low as 5 to 6 percent) and NPN losses during treatment. The amounts given are appropriate between 28 percent and 36 percent dry matter; above this range, NPN losses during treatment will likely increase. It may not be advisable to treat silage with less than 60 percent moisture. It is important to test the treated silage to be sure final protein level is adequate.

Table 10-11 gives mineral mixes for treating silage that can be added in addition to the urea or anhydrous ammonia. As discussed earlier, the addition of minerals at ensiling may improve feed efficiency an additional 3–5 percent over NPN addition alone.

Table 10-12 gives amounts needed/day, and Table 10-13 gives mineral mixes for supplementing corn silage treated with NPN alone. The mineral content of these supplements is given at the bottom of Table 10-13 to help the cattle feeder select appropriate commercial supplements.

10-6 Treating Corn Silage with NPN

Table 10-11
Mineral Mixes for Treating Corn Silage and Urea-Mineral Mix for High Moisture Snapped Ear Corn

1. **Mineral mix for treating corn silage** (NPN treatment level given in Table 9-10). To increase the calcium to .4%, phosphorus to .3%, and nitrogen:sulfur ratio to 15:1.

Ingredient	lb/1000 lb of mixture
Dicalcium phosphate	637
Calcium sulfate (gypsum)	363
	1000

    Amount of mix to add/ton silage

	Moisture of silage						
	72	70	68	66	64	62	60
Amt. of mix to add/ton	3.9	4.2	4.5	4.8	5.0	5.3	5.6

    An equivalent or superior mineral treatment is 10 lb calcium carbonate/ton (65% moisture basis), if a mineral supplement containing phosphorus and sulfur is given at feeding time.

2. Urea + mineral mix for treating snapped ear corn at ensiling

Ingredient	lb/1000 lb mixture
Urea	390
Calcium carbonate	546
Calcium sulfate (gypsum)	64
	1000

    This is added such that 100 lb treated snapped ear corn dry matter contains 1.9 lb urea-mineral mix and 98.1 lb snapped ear corn dry matter. The treated ear corn dry matter will contain .4% calcium, .3% phosphorus, nitrogen-sulfur ratio of 15:1, and 11% protein on the average.

    Amount to add/ton snapped ear corn at ensiling

	Moisture content				
	36	34	32	30	28
Lb of mix to add/ton	22.4	23.7	25.2	26.9	28.8

It is recommended that in addition a corn-Rumensin mix containing 8 lbs Rumensin premix (60 g/lb) and 992 lbs ground shelled corn be fed at the following rate.

Bodyweight, lb		
400-600	600-800	800-market
lb/head/day of corn	+	Rumensin mix
0.3	0.4	0.5

Table 10-12
Soybean Meal 44 and Mineral Supplement Recommendations for Varying Rates of Corn Fed/Day*

Weight (lbs.)	High moisture corn/head/day/cwt. body weight							
	None		0.8		1.6		2.4	
	---- Lbs./head/day ----							
	Soy	Min.	Soy	Min.	Soy	Min.	Soy	Min.
To 600	1.0	.20	1.2	.19	1.40	.18	X	X
600 to 700	-	.17	.3	.17	.80	.17	X	X
700 to 800	-	.15	-	.15	.40	.16	.9	.22
800 to 900	-	.15	-	.15	-	.16	.5	.22
900 to market	-	.15	-	.15	-	.16	-	.22

*The supplementation rates are given for an average frame steer started on feed as a calf. Table 10-5 gives the adjustment factors for other frame sizes, sexes, and ages.

Table 10-13
Mineral Supplements for Corn Silage Treated with NPN Only

Weight range (lbs)	Lbs corn/head/day/cwt. body weight		
	0 - .8	1.6	2.4
	400 to market	400 to market	400 to market
Ingredient	------ Lbs/1000 lb batch ------		
Dicalcium phosphate	390	-	-
Limestone	180	620	710
Calcium sulfate	140	90	-
T.M. salt	290	290	290
Vitamins	---- Million IU/1000 lb batch ----		
Vitamin A	130	130	130
Vitamin D	13	13	13
Nutrient composition of the mineral supplement	------ % of Supplement ------		
Calcium	18.6	25.0	27.0
Phosphorus	7.0	-	-
Salt	2.9	2.9	2.9
Sulfur	2.4	1.5	-

10-7 EXPECTED PERFORMANCE WITH VARIOUS LEVELS OF SILAGE FEEDING

Table 10-14 summarizes expected performance and feed requirements for various silage qualities and levels of grain feeding. These values assume that the ration is properly balanced for protein and minerals at each stage of growth. It is also assumed that a growth stimulant and Rumensin are used and the cattle are fed in a stress-free environment. Shelled corn was assumed to contain 30 percent moisture and corn silage is assumed to contain 68 percent moisture. Performance is given for average-frame cattle. Feed requirements/cwt gain will be similar for other frame sizes of cattle fed over similar stages of growth.

The pay to pay daily gains and feed requirements assume that 18 days are required to regain purchase weight and tissue shrink at sale is 0.5 percent. The feed requirements include provision for a 2 percent death loss with calves and 1 percent with yearlings.

Table 10-14
Expected Average Payweight-to-Payweight Daily Gain and Feed Requirements/100 lbs Gain for Corn-Corn Silage Rations[1]

	Corn Silage Quality (Bu. #2 Corn/ton 32% DM silage)								
	6.7			5.4			3.5		
	Added Corn[2]			Added Corn[2]			Added Corn[2]		
	0	.8	1.6	0	.8	1.6	0	.8	1.6
Steer Calf									
Daily Gain, lb.	1.90	2.19	2.49	1.75	2.06	2.43	1.46	1.85	2.27
Shelled corn, lb.	0	228	452	0	242	466	0	271	498
Corn Silage, tons	1.16	.757	.388	1.26	.802	.399	1.53	.899	.426
Soybean meal, lb.	39.3	29.3	21.0	42.1	31.0	21.7	50.3	34.0	22.8
Urea, lb.	6.50	7.17	7.17	7.17	7.67	7.33	8.67	8.50	7.83
Yearling Steer									
Daily gain, lb.	2.09	2.38	2.68	1.95	2.25	2.62	1.65	2.04	2.46
Shelled corn, lb.	0	285	523	0	303	541	0	333	577
Corn silage, tons	1.32	.860	.460	1.43	.912	.472	1.70	1.01	.503
Soybean meal, lb.	1.55	1.56	0	167	1.56	0	194	1.78	0
Urea, lb.	8.67	9.55	8.44	9.33	10.2	8.67	11.1	11.1	9.11
Heifer Calf									
Daily gain, lb.	1.61	1.85	2.11	1.48	1.74	2.05	1.23	1.56	1.92
Shelled corn, lb.	0	239	464	0	254	478	0	285	512
Corn Silage, tons	1.20	.781	.400	1.30	.826	.412	1.58	.926	.422
Soybean meal, lb.	34.2	24.4	17.6	36.7	25.3	17.8	43.1	27.6	18.7
Urea, lb.	6.89	7.55	7.11	7.56	8.00	7.33	9.33	9.11	8.00
Yearling heifer									
Daily gain, lb.	1.78	2.03	2.29	1.65	1.92	2.24	1.40	1.74	2.10
Shelled Corn, lb.	0	284	524	0	301	537	0	333	573
Corn silage, tons	1.32	.856	.458	1.43	.909	.470	1.70	1.01	.501
Soybean meal, lb.	1.48	1.32	0	157	1.58	0	183	1.58	0
Urea, lb.	8.68	9.47	8.16	9.21	10.0	8.42	11.1	11.1	8.95

[1] the values given in this table assume the following conditions.
A. 18 days required to recover shipping shrink. Feed requirements include provisions for a 2% death loss in calves and 1% death loss in yearlings.
B. A growth stimulant was used. Decrease daily gain 12% if a growth stimulant not fed or an implant is not given every 100 days
C. Rumensin was fed. Increase feed requirements 10% if rumensin not fed.
D. A stress free environment was used. Decrease daily gain 5-8% and increase feed requirements 8-12% for cemented outside lots or well drained outside dirt lots.
E. These cattle are assumed to be average frame British breed. Increase daily gain 6% for British x Exotic cross and 12% for large frame exotic breeds, but use same feed requirements/100 lb. gain. Increase daily gain 21% but also increase feed requirements 10%/100 lb. gain for Holsteins fed to low choice grade.

[2] Lb. 32% moisture shelled corn/100 lb. body weight daily.

11

Systems Analysis: Developing the Most Profitable Management System*

11-1 COMPUTERIZED RATION BALANCING AND PERFORMANCE SIMULATION FOR GROWING AND FINISHING CATTLE

Formulating rations is only part of developing a feeding system. Beef producers must first decide which feeding program will be the best for their total operation. Then the beef producer must develop feed budgets that can be used to determine how much of each feed he must be able to grow or buy and store, or how many cattle he can feed on the feedstuffs produced.

Computer programs are now available that can be used to formulate complete rations and supplements, to develop feed budgets, and to estimate profits on a given feeding program. Computer programs developed at Michigan State University (Telplan 44, beef ration formulation, and Telplan 56, beef feedlot gain simulation), Oklahoma State (beef feedmix programs), and University of Nebraska (AGNET) are being used in midwestern and plains states to formulate rations and predict performance, using

---

*Dr. J. R. Black, Michigan State University, collaborated with Dr. Fox in the development of the programs discussed in this chapter.

terminals in state and county extension offices. A few cattle feeders have their own terminals and connect directly to the computers used for these programs. The terminal looks much like a small portable typewriter, with the results being printed as the computer completes the calculations.

The cattle requirements, feed nutrient values, and factors affecting each discussed in Chapters 8 and 9 are stored in the computer, along with expected feed intakes. Costs of feeds available, feed analysis values, energy level desired, cattle type, combinations of feed additive and growth stimulants used, and environment are coded into the computer on the terminal, which is connected to the computer via telephone. The following example shows how the system works.

Assume a cattleman is purchasing 100 average-frame, British breed steer calves; they are in average flesh condition, with an average weight of 450 lbs. The steers will be fed in a slotted-floor, confinement barn. He wants a feeding program developed around corn fed at the rate of one percent of body weight/head/day along with all the corn silage they will eat. The following information is requested:

1. A protein-mineral supplementation program developed around soybean meal, dicalcium phosphate, limestone, and salt.
2. The ingredient composition of the protein-mineral supplement, and a description of the feed tag to permit consideration of a commercial supplement.
3. Expected feed disappearance, payweight to payweight basis.
4. Expected days on feed required for animals to grade low choice.
5. Expected net return per head purchased.

The input form shown demonstrates how the data is put into the computer. Performance desired is the first piece of data entered; it is keyed to energy level and equivalent weight, as outlined in Tables 8-17 to 8-19. In the example, ration "1" is the set of nutrient requirements associated with 0.45 Mcal NE_g/lb dry matter; ration "3" is the set associated with a 0.57 Mcal NE_g. The computer model permits the user to override nutrient requirement values stored in the computer data bank; thus, he can interpolate to obtain any set of values desired.

Feedstuffs under consideration are the second set of data entered. The feed number (Table 9-1) appropriate for the quality and storage method is entered; nutrient values can be modified in

Computer Input Form for Telplan Ration Evaluation and Formulation Model

I. Performance Desired
 a. Enter ration desired from Table 1 of Fact Sheet 1097.
 b. State code

Line No. 01.   Input Value: 002 | 21 |
                                  a   b

II. Feeds Available and Their Prices

(Enter "0" on the line following the last feed entered; the computer will skip to line 16. Feed codes are given in Fact Sheet 1102.)

Feed description	Line No.	Feed number	Price, $/cwt. as-fed basis	Mix code*
Soy 44	02.	056	10.50	1
Dical	03.	129	11.00	1
Limestone	04.	125	04.00	1
Salt	05.	135	04.00	1
Corn	06.	017	03.90	0
Corn Silage	07.	203	00.83	0
	08.	---	--.--	---
	09.	---	--.--	---

III. Commercial Protein Supplements

(Enter "0" in the entry following the last supplement; the computer will skip to line 20.)

Feed No. 098
 16a. Price, $/cwt.
  b. Crude protein, %
  c. Non-protein, nitrogen, %
  d. Mix code

16.   --.-- | --.-- | --.-- | 0
       a   b   c   d

 17a. Crude fiber, %
  b. Calcium, average %
  c. Phosphorous, %
  d. Salt, %

17.   --.-- | --.-- | --.-- | --
       a   b   c   d

IV. Commercial Mineral Supplements

(Enter "0" in the entry following the last supplement; the computer will skip to line 22.)

		Price, $/cwt.	Calcium, %	Phos. %	Salt, %	Mix code
Feed No. 138	20.	--.--	---	---	---	0
Feed No. 139	21.	--.--	---	---	---	---

V. Feed Additives
 22a. Percent of ration (100% dry basis)
  b. Cost, $/cwt.

22.   --.-- | --.--
       a   b   0

VI. Limits on Percentage of Ingredients Used

(Enter "0" on the line following the last limit; the computer will skip to line 33.)

Limit code:
 1 = Maximum (no more than)
 2 = Exact percentage
 3 = Minimum (no less than)

Type code:
 1 = % Protein-mineral supplement
 2 = % Concentrate
 3 = % Roughage
 4 = % Total

Line No.	Limit code	Type code	Feed number	Percent, 100% dry basis
23.	---	---	---	--.-- 0
24.	---	---	---	--.--
25.	---	---	---	--.--
26.	---	---	---	--.--
27.	---	---	---	--.--
28.	---	---	---	--.--
29.	---	---	---	--.--
30.	---	---	---	--.--
31.	---	---	---	--.--
32.	---	---	---	--.--

VII. Option to Calculate Nutrient Composition of Current Ration

(Enter "0" on the line following the last feed; the computer will skip to line 42.)

Feed description	Line No.	Feed No.	Lbs/head/day, lbs/batch mix, or percent of ration
	33.	---	--.-- 0
	34.	---	--.--
	35.	---	--.--

325

VIII. Modification of Nutrient Requirements

(Enter "0" on the line following the last modification; the computer will skip to line 53.)

Modification		Requirement no.	Requirement value
NEg	42.	03	00.54
Protein	43.	04	13.20
Calcium **	44.	06	00.55
Salt **	45.	09	00.50

...

IX. Modification of Feed Composition

(Enter "0" on the line following the last modification; the computer will skip to line 73.)

Modification		Feed no.	Nutrient code	Nutrient value
	53.	____	__	___.__0__
	54.	____	__	___.___
	55.	____	__	___.___

...

X. Animal Characteristics, Intake, and Batch Size

73. Animal characteristics   73. $\underline{05}_a \cdot \underline{3}_b | \underline{4}_c | \underline{5}_d | \underline{4}_e | \underline{7}_f | \underline{4}_g$
   a. Weight, cwt
   b. Age (1=calf, 2=yearling, 3=two years)
   c. Sex (1=steer, 2=heifer, 3=bull)
   d. Frame (1=very small, 5=average, 9=very large--a continuous code from 1-9)
   e. Condition (9=very thin, 5=average, 1=very fleshy--a continuous code from 1-9)
   f. Environment (1=outside lot, with frequent very deep muc; 3=outside lot without mounds; 5=outside lot with mounds and bedded during adverse weather; 7=partially or completely enclosed--free from mud, rain and wind, well ventilated--a continuous code from 1-7)
   g. Growth stimulant
      Without Rumensin: 1=none; 2=antibiotic; 3=DES(heifers); 4=Synovex H or MGA (heifers); DES, Synovex S, or Ralgro (steers)
      With Rumensin: 5=none, 6=antibiotic; 7=DES (heifers), 9=Synovex H or MGA (heifers); DES, Synovex S, or Ralgro (steers)

74. Expected daily dry matter intake, lbs.

(Enter "0" if the values given in Fact Sheet 1212 are acceptable.)

74.   ___.__0__

75. Batch sizes, cwt.   75.   __$\underline{0}_a$ | $\underline{0}\underline{0}_b$ | $\underline{0}\underline{0}_c$ | $\underline{0}$
   a. Premix (mix code "2")
   b. Concentrate (mix code "1")
   c. Bunk (mix code "0")

XI. Output Options

(Enter "1" if you want; "0" if you don't.)

76a. Ration?   76. $\underline{1}_a | \underline{1}_b | \underline{1}_c | \underline{1}_d | \underline{1}_e$
   b. Nutrient composition of ration?
   c. Prices at which feeds not cheap enough to be in the least-cost ration would enter?
   d. Prices at which ingredients currently in ration would be reduced or replaced with some other ingredient?
   e. Gain projection?

---

*Mix codes are: 1 = protein-mineral supplement; 0 = bunk.

Calcium and salt were "overspeced" for the "to 600 lb" ration. This is typically necessary if the same **protein-mineral supplement is going to be fed to all weight groups.

11-1 Computerized Ration Balancing and Performance Simulation

a subsequent section if warranted. The mix code permits selection of the manner in which feeds are to be mixed. For example, he can instruct the computer to mix all feeds together; or he can instruct the computer to mix the protein-mineral supplement separately and subsequently mix the protein-mineral supplement with the roughage and energy feeds.

Sections III and IV are for entering commercial protein-mineral and mineral supplements. The data requested is typically provided on the feed tag; equations in the computer program estimate NE_m and NE_g/lb dry matter based on the feed tag information. The user can override the computed values if warranted.

Section VI permits restriction of the level at which feedstuffs enter the ration based upon palatability and/or supply limitations. For example, a cattleman may be obtaining a fixed quantity of brewer's grain under contract. Or, he may want no more than 30 percent of the concentrate dry matter to come from wheat. A common use of Section VI is to balance a ration for the lightest pen weight, obtain the resultant protein-mineral supplement, and constrain the computer program to feed that supplement for all heavier weight groups at the level needed to meet protein requirements.

Section VII gives the user the option of calculating the nutrient content of an existing ration. This is a key section; it is important to know what the cattle feeder is currently doing before suggesting alternatives. If the current system is balanced and economical, changes probably are not warranted.

Section VIII deals with modification of the nutrient requirements stored in the computer data bank. The nutrient requirement code and preferred value are entered. Similarly, Section IX deals with modification of feed nutrient values; it is used primarily to adjust for feed nutrient values obtained from feed analysis.

Section X is for entry of animal characteristics, use of growth stimulants, and environment. Input 74 permits the user to enter his estimate of expected daily dry matter intake if it differs from that stored in the computer. The user can obtain the prices at which ingredients not currently in the ration would enter, and the lower and upper price bounds for feeds in the ration.

Results of Analysis

Ration formulation. Tables 11-1 and 11-2 depict the output for the case example. The output consists of five parts: (1) protein-mineral supplement(s); (2) ration ingredient composition on a

Table 11-1
Protein-Mineral Supplement

Ingredient	Lbs/1000 lbs	
	Dry basis	As-fed basis
Soybean meal - 44	883	893
Limestone	76	69
Salt	41	37
	1000	1000

Nutrient composition (as-fed basis):

Protein, %	40.8
Crude fiber, %	6.3
Calcium, %	3.0
Phosphorous, %	.7
Potassium	2.0
Salt, %	3.7

pounds per 1000 lbs DM and as-fed basis, and on a 100 head per day as-fed basis; (3) ration nutrient density; (4) dry matter intake and associated expected daily gain; and (5) cost projection. Output is typically divided into three or four weight groupings, depending upon the weight variation of animals in the pen. For the case example four groupings were used: (1) to 600 lbs; (2) 600 to 750 lbs; (3) 750 to 900 lbs; and (4) 900 lbs to market.

The nutrient density of the protein-mineral supplement is typically included, providing the cattleman with the option of finding a similar commercial supplement as well as formulating his own. Additional information on vitamin premixes and growth stimulants may be provided too.

This framework has proven sufficiently flexible for most applications. The information is detailed enough for an individual with a mixer truck; the output depicting the ingredients per 1000 lb batch can easily be used to develop feedsheets. If cattle are fed on a per head per day basis, the pounds of each feed per 100 head per day are directly available. Perhaps the most common strategy is to measure the amount of protein-mineral supplement fed per head carefully; to estimate the energy feed fed per head as accurately as feasible; and to feed the animals all of the roughage feed(s) they will eat. For the case example, steers in the 600 to 750 lb range would get 1.7 lbs of protein-mineral supplement per head per day along with 7.5 lbs of corn and all the corn silage they would eat.

The nutrient density breakdown provides a profile on the

Table 11-2
Balanced Rations, Gain, and Cost Prediction

Weight range, lbs.	450-600			600-750			750-900			900-1050		
		As-fed basis			As-fed basis			As-fed basis			As-fed basis	
	(DM) basis	Per 1000 lb batch	Per 100 head/day	(DM) basis	Per 1000 lb batch	Per 100 head/day	(DM) basis	Per 1000 lb batch	Per 100 head/day	(DM) basis	Per 1000 lb batch	Per 100 head/day
Ration												
Supplement	121	62	179	93	47	167	66	32	135	49	25	108
Corn	370	202	586	379	203	724	388	204	852	439	218	1025
Corn silage	509	736	2142	528	750	2677	546	764	3137	512	737	3179
	1000	1000	2908	1000	1000	2677	1000	1000	4124	1000	1000	4313
Nutrient density (DM basis)												
Dry matter, %		46.3			45.5			44.7			46.1	
NE$_m$, Mcal/lb		.83			.83			.83			.85	
NE$_g$, Mcal/lb		.54			.54			.54			.55	
Protein, %		13.2			12.2			11.2			10.7	
Crude fiber, %		14.9			15.3			15.6			14.8	
Calcium, %		.54			.46			.38			.32	
Phosphorus, %		.34			.32			.31			.32	
Potassium, %		.89			.86			.83			.78	
Salt, %		.50			.38			.27			.20	
Daily gain and intake projection												
Dry matter intake, lbs/day	12.47	13.47	14.47	15.24	16.24	17.24	17.67	18.67	19.67	18.87	19.87	20.87
Gain, lbs/day	2.22	2.51	2.79	2.27	2.51	2.74	2.27	2.47	2.68	2.18	2.37	2.55
Cost projection												
Per cwt dry matter, $		4.32			4.11			3.91			3.87	
Per cwt as-fed, $		2.00			1.87			1.75			1.78	
Per head/day @ average intake, $		.58			.67			.73			.77	

330   Systems Analysis

major nutrients and can be used in additional gain projection work including the use of Telplan 56 for gain, feed use, and economic projection.

Three daily dry matter intakes are considered in gain projection. The middle value is standard; however, the gain which would be expected if one pound more (or less) were eaten is calculated. This serves both control and educational functions. From a control standpoint, it provides the cattleman with a target; if his cattle are eating less than the target, there may be problems with health, bunk management, or other factors. Educational aspects key on the economic importance of dry matter intake. For the example, the difference for a 780-lb steer in daily gains between extremes is

Table 11-3
Feedsheet (for up to 600 lbs Ration)

Feed	% DM	Ration Composition	
		Dry basis	As-fed basis
Supplement	90	121	62
Corn	85	370	201
Corn silage	32	509	736
TOTAL		1000	1000

Dry Matter Pct   46.3

Load size, lbs					
Dry basis	As-fed basis		Protein supplement	Corn	Corn silage
1203	2600	Weight	162	524	1914
		Scale	162	686	2600
1249	2700	Weight	168	544	1988
		Scale	168	712	2700
1296	2800	Weight	174	564	2062
		Scale	174	738	2800
1342	2900	Weight	180	584	2135
		Scale	180	765	2900
1388	3000	Weight	187	604	2209
		Scale	187	791	3000
1434	3100	Weight	193	625	2282
		Scale	193	818	3100
1481	3200	Weight	199	645	2356
		Scale	199	844	3200
1527	3300	Weight	205	665	2430
		Scale	205	870	3300

11-1 Computerized Ration Balancing and Performance Simulation   331

nearly 0.5 lb/day; the resultant difference in feed cost/cwt gain is $1.75.

Feed sheet formulation. Table 11-3 depicts the feedsheet for a pen of 100 head on the "up to 600 lbs" ration. This information is typically placed on a clipboard in the feed truck, with the driver using it to set scale readings. For example, if he expected the cattle would "clean up" 3000 lbs of as-fed ration, he would set the scale reading at 187 lbs for the supplement, 791 lbs for the corn, and 3000 lbs for the corn silage. Expected daily dry matter intake for the pen, on the average, would be 1350 lbs. However, load sizes were run over a wider range of intakes to allow for variation in intake about "normal," and since the calves will be eating less when they start.

The composition of the ration is given in the table both on a dry and as-fed basis. Thus, the feed truck driver has an additional check if he has any question about the appropriate feedsheet.

Gain and feed use projection. Table 11-4 summarizes the projections for the case study. Information consists of four parts.

Table 11-4
Gain and Feed Use Projection[a]

Days on Feed	Shrunk weight (lbs)	Intake (lbs)	Period Daily gain (lbs)	Period Feed/gain	Pay to Pay Daily gain (lbs)	Pay to Pay Feed/gain	Feed Disappearance (as-fed basis) Supplement	Corn	Corn silage
38	500	12.85	2.50	5.13	0.99	11.30	57	185	676
58	550	13.80	2.50	5.51	1.49	7.95	92	299	1092
78	600	14.74	2.50	5.89	1.73	7.15	130	421	1538
98	650	15.66	2.50	6.25	1.88	6.87	161	554	2030
118	700	16.55	2.50	6.61	1.97	6.79	193	695	2552
138	750	17.44	2.50	6.97	2.04	6.80	228	844	3102
158	800	18.31	2.50	7.31	2.09	6.86	254	1004	3701
179	852	18.87	2.44	7.74	2.13	6.95	282	1179	4354
199	900	19.35	2.38	8.14	2.15	7.06	309	1350	4994
220	951	19.81	2.39	8.30	2.17	7.18	331	1558	5639
241	1000	20.05	2.29	8.75	2.18	7.30	354	1770	6295
264	1052	20.24	2.19	9.25	2.18	7.46	378	2004	7022
287	1101	20.24	2.07	9.79	2.17	7.62	403	2240	7752
312	1151	20.24	1.95	10.38	2.15	7.80	430	2496	8546
339	1202	20.24	1.84	11.02	2.13	8.01	459	2773	9403
367	1252	20.24	1.74	11.67	2.10	8.22	489	3060	10291
396	1301	20.24	1.64	12.34	2.07	8.45	521	3357	11212
428	1351	20.24	1.55	13.08	2.03	8.69	555	3685	12227

[a] Assumptions include 18 days to regain inshrink, 2% death loss, and a .5% tissue outshrink.

Expected daily dry matter intake provides a "target" variable which can be used to check if the cattle are eating normally. Expected daily gain and feed efficiency, at that weight, are given next followed by daily gain and feed efficiency on a payweight to payweight basis where adjustments have been made for death loss, time required to get animals on feed and regain tissue inshrink, and tissue outshrink. Last, feed disappearance to date is given on a payweight to payweight basis. For the case study, 3.5 tons of corn silage, 36 bushels of corn, and 3.8 cwt of protein supplement were required to take the animal from 450 to 1050 lbs. Much use is made of this segment of the model in feed budget development and other system comparisons. Cattlemen on two-phase corn-corn silage systems make alternative runs to assess the shift point from high silage to high corn rations that is consistent with their silage supplies.

Grade and sale price projections. Table 11-5 depicts estimated quality and yield grades, and sale price for various shrunk weights, where the grades are estimated according to the framework outlined in Table 8-1. The projected sale price is derived from information entered by the cattleman. He inputs the sale price for a low choice, yield grade 3 animal; the "normal" percentage differentials for cattle grading mid- and high-good as well as fatter cattle ranging through yield grade 5 are stored in the computer data bank. Differentials are available for market conditions when "green" cattle are selling above their normal relationship, and con-

Table 11-5

Grade and Sale Price Projections

Shrunk weight (lbs.)	Grade Quality[a]	Yield	Sale price ($/cwt)
902	9.1	2.2	37.80
950	9.4	2.5	37.80
1002	9.7	2.8	39.80
1052	10.0	3.1	42.00
1101	10.3	3.4	42.00
1150	10.5	3.6	41.67
1201	10.8	3.9	41.59
1250	11.1	4.2	39.27
1301	11.4	4.5	38.64

[a] 10 = low choice, 11 = mid choice

ditions under which yield grades 4s and 5s are selling above their normal relationship too. Alternatively, the user can input his estimate of the differentials based on his specialized knowledge.

Field testing is just beginning on this component of the model. As is well known, quality grade is sensitive to the type of cattle and to breeding within a particular breed. As a consequence, considerable judgment must be employed by the user in the implementation of this segment.

Economic projection. Table 11-6 summarizes the projections for the case study. There are three segments: cost of gain projection, net return per head projection, and information for making "when to sell" decisions. All economic data are presented on a payweight to payweight basis. The user inputs, in addition to information already described, data on the purchase and expected sale price of the animal, starting costs, and daily nonfeed variable costs.

Starting costs are costs that are incurred on a one-time basis and that are essentially independent of days on feed. Marketing, chute, and medical costs might fall into that category. The interpretation of what constitutes nonfeed variable costs will vary with application. For a typical farmer feeder, the costs associated with feeding such as interest on the feeder, fuel, wear and tear on equipment, and, perhaps, labor would be included. In contrast, for someone having cattle fed in a custom lot, the appropriate charge would be the daily charge of the custom operator plus interest on the feeder. Both the cost/cwt gain and net return projections must be interpreted carefully in the context of the assumptions made about cost structure.

The model can easily be used to budget the economics of a wide variety of alternatives including feeder systems, housing systems, growth stimulants, and different types of feeder cattle. The net return projection is useful to the cattle feeder in assessing the probable profitability of a particular system under consideration. For example, for the case study, the projected "net" return at low choice is $29/head which must cover costs of housing and machinery.

Four pieces of information are provided which may be of assistance in formulating saleweight decision. First, the added net return per head per day is calculated. For example, we see that at 1050 lbs and low choice the daily net return is increasing by $0.01/day; but at 1100 lbs, $0.05 is lost per day. In principal, in a

Table 11-6
Economic Projections[a]

Days on feed	Shrunk weight (lbs)	Variable cost/cwt Gain ($)			Net return to fixed resources ($)				Increase in sale price/cwt of Ch⁻ cattle required to give same net income/head as @ Ch⁻
					Current turn		Next turn		
		Feed	Nonfeed	Total	/head	Added /head /day	Avg /head /day	Avg /head /day	
199	900	28.6	11.9	40.5	-12.5	0.012	-0.063	-0.047	
209	925	28.7	11.5	40.3	-12.2	0.026	-0.058	-0.042	
220	951	28.9	11.2	40.1	-12.0	0.004	-0.055	-0.038	
230	975	29.1	11.0	40.0	-3.1	0.939	-0.014	0.004	
241	1000	29.3	10.7	40.0	7.2	0.943	0.030	0.048	
252	1025	29.5	10.5	40.0	17.6	0.945	0.070	0.089	
263	1050	29.7	10.3	40.1	29.2	0.006	0.111	0.131	
275	1075	30.0	10.1	40.2	29.1	-0.020	0.106	0.125	0.01
287	1101	30.1	10.1	40.3	28.6	-0.046	0.100	0.119	0.05
299	1125	30.6	9.9	40.5	26.4	-0.241	0.088	0.108	0.11
312	1151	30.9	9.8	40.7	23.1	-0.267	0.074	0.093	0.21
325	1176	31.3	9.7	40.9	20.8	-0.124	0.064	0.083	0.32
338	1200	31.6	9.6	41.2	19.0	-0.145	0.056	0.075	0.46
352	1225	32.0	9.5	41.5	-2.4	-0.814	-0.007	0.011	0.62
366	1250	32.4	9.4	41.8	-14.0	-0.841	-0.038	-0.021	0.81
381	1276	32.8	9.4	42.2	-25.4	-0.256	-0.067	-0.050	1.02
396	1301	33.2	9.3	42.6	-29.4	-0.274	-0.074	-0.058	1.25
411	1325	33.7	9.3	43.0	-33.6	-0.291	-0.082	-0.066	1.50
427	1350	34.1	9.3	43.4	-38.4	-0.308	-0.090	-0.074	1.77
443	1374	14.6	9.3	43.8	-43.5	-0.325	-0.098	-0.082	2.06

[a]The projections are based upon 85% DM corn @ $2.20/bushel, 32% DM corn silage @ $16.60/ton, and protein-mineral supplement @ $184/ton. Nonfeed costs are starting costs @ $25/head, with other nonfeed variable costs including interest on the feeder @ $.13/head/day. The purchase and sale (Ch⁻) prices are $38 and $42/cwt, respectively.

11-1 Computerized Ration Balancing and Performance Simulation

one-turn lot, the cattleman should keep his animal on feed as long as the added return from holding the animal another day is positive. In the case example the animal should be sold between 1050 and 1075 lbs. In contrast, for a continuous-turn lot, the animal currently in the lot must be able to bid space away from a "new" animal that could take his space. The net return for the "new" animal is his projected average net return per head per day. Clearly, given our price estimation skills, only a rough estimate can be ascertained. But, knowledge of cattle cycle characteristics gives us evidence as to whether the profit on the next turn of cattle is likely to be lower, the same, or higher than the current turn of cattle—at least on the average. In the case example, the maximum average net return per day on the next turn is $0.13 at 1050 lbs. Thus, the current animal must earn at least that much if he is to remain in the lot; he should be sold any time his net income added per day falls below $0.13.

Many cattle feeders ask an alternative question. "How much does the market for choice cattle have to rise to make it profitable to hold the animals currently in the lot to heavier weights?" In the case example, to carry an animal to 1350 lbs from 1050 lbs, the low choice, yield grade 3 sale price would have to rise by $1.80 for the cattleman just to earn the same net income per head as he would at 1050 lbs. These calculations do not include the opportunity cost of placing a new animal in the lot. Addition of the opportunity cost would increase the price rise needed. Further, the feeder cattle market tends to rise with the fat cattle market.

Using this basic system, feed budgets, optimum feeding systems, and bid sheets for purchasing feeder cattle can be developed. Tables 11-7 and 11-8 show the series of rations and supplements developed for an actual cattle feeder. As the ration number increases from 1 to 5, the energy level increases. Within each ration number, there are several alternatives for supplement type or grain. Rations 1 and 2 are used basically for light cattle. Thus, supplements with none, 1/2, or all of the supplemental protein from NPN are available, depending on the weight of the cattle. The grain used is high-moisture ear corn or dry shelled corn, depending on what is available. Table 11-9 shows feed requirements and projected profits from several possible cattle types and ration combinations developed to determine the optimum feeding system. These were also used to develop a bid sheet (Table 11-10), which he used to make decisions while buying cattle in auction sales.

Table 11-7
Willow Lane Feedlot Rations

	1	1A	1B	2	2A	3	3A	3B	4	4A	4B	5	5A
					Lb/100 lb Dry Matter								
Corn silage	863	923	943	460	660	-	400	200	-	300	150	-	150
HM snapped ear corn	-	-	-	483	-	953	-	473	540	-	270	290	-
Shelled corn	-	-	-	-	283	-	543	280	413	648	523	663	803
Rumensin Spl.	017	017	017	017	017	017	017	017	017	017	017	017	017
Supplement 1	120	-	-	-	-	-	-	-	-	-	-	-	-
Supplement 1A	-	060	-	-	-	-	-	-	-	-	-	-	-
Supplement 1B	-	-	040	-	-	-	-	-	-	-	-	-	-
Supplement 2	-	-	-	040	040	-	-	-	-	-	-	-	-
Supplement 3	120	-	-	-	-	030	040	030	030	030	040	030	030
NE_m, Mcal/lb	.72	.72	.72	.80	.80	.90	.90	.90	.94	.91	.91	.97	.97
NE_g, Mcal/lb	.48	.48	.48	.51	.51	.57	.57	.57	.60	.60	.60	.63	.63
CP, %	12.2	11.5	10.8	11.5	11.5	10.5	10.5	10.5	10.5	10.5	10.5	10.5	10.5

Table 11-8
Willow Lane Feedlot Supplements

1. Rumensin Premix = Rumensin (60 gm/lb) $\frac{11}{989}$ As fed
   ground shelled corn (dry) $\frac{989}{1000}$
   contains 660 gm/1000 lb or 660 mg/lb
   gives 25 grams/ton ration DM or 22 gm/ton @ 90% DM

2. Protein - Mineral Supplements

	Supplement 1	1A	1B	Supplement 2	Supplement 3
Soybean meal	900	737	-	-	-
Ground shelled corn	-	-	514	476	487
Urea	-	130	273	305	170
Dicalcium phosphate	50	68	91	-	-
Calcium sulfate	-	23	57	54	30
Limestone	20	-	-	95	228
Trace mineral salt	$\frac{30}{1000}$	$\frac{42}{1000}$	$\frac{65}{1000}$	$\frac{70}{1000}$	$\frac{85}{1000}$
Vitamin A, IU	12 million	25 million	40 million	40 million	40 million
% CP	40	68.9	81.7	90.7	52.7
% Ca	1.9	2.2	3.0	4.8	9.3
% P	1.0	1.8	1.8	-	-
% Salt	3.0	4.2	7.9	7.0	9.0

Table 11-9
Willow Lane Feedlot Performance Simulations

	Feedlot Ration	Initial wt.	Final wt.	Days on feed	As fed feed requirements				Cost of gain, $/cwt.			Projected profit, $
					Corn Silage	Ear corn	Shelled corn	Spl.	Feed	Non Feed	Total	
Heifers to feedlot	2	550	850	154	3670	1570	-	108	31.5	10.4	41.9	9.0
Steers to feedlot	2	600	1100	214	5910	2530	-	175	30.4	8.7	39.0	13.2
Steers to grass, then	2	600	1100	246	2575	1104	-	82	14.2	10.0	24.2	86.5
Heifers to grass, then	2	550	850	190	4460	1910	-	138	5.0	12.9	17.8	58.9
Steer calves grass, then	1,2,3	450	1100	347	3908	1545	381	110	18.2	10.8	28.9	79.8
Yearling str. grass, then	2,3	600	1100	239	1365	995	365	62	13.9	9.7	23.6	89.8
Steers to feedlot	2,3	750	1100	151	3374	1856	365	116	33.4	8.7	42.1	1.2
Steers to feedlot	1,2,3	600	1100	217	6943	1678	365	137	30.8	8.8	39.6	10.4
Steers to feedlot	"	600	1100	228	9177	996	365	120	31.9	9.2	41.4	3.2
Steers to feedlot	1,2,4	600	1100	214	6696	1321	616	144	30.5	8.7	39.1	12.7

Prices, $/cwt.

	In	Out
450 # steers	60	49
600 # "	55	49
750 # "	52	49
550 # heifers	45	45

Non-feed = 20¢/hd/day

Table 11-10
Willow Lane Feedlot Bid Chart

Purchase wt., lb	Feeding system	Heifers - purchase price, $/cwt.							Steers - purchase price, $/cwt.						
		40	42	44	46	48	50	52	48	50	52	54	56	58	60
					Profits/hd., $							Profits/hd., $			
550	#2	37	26	15	4	- 8	- 19	- 30							
550	Grass, then lot, #2	86	75	64	53	42	31	26							
450	Grass, then 1,2,3								133	124	115	106	97	88	80
600	Grass, then 2,3								131	119	107	96	84	72	60
600	#2								55	43	31	19	7	- 5	- 11
600	1, 2, 3								45	33	21	9	- 3	- 15	- 27
750	2, 3								31	16	1	- 14	- 29	- 44	- 59

11-2 LONG-RANGE FEEDING SYSTEM PLANNING

Choice of feeding system and facilities to use for the calves from weaning to slaughter is dictated by many factors. The following are the primary factors that must be weighed in the choice:

1. As more grain is added to the ration, cattle gain more rapidly, reducing maintenance energy, nonfeed costs, and the amount of feed and manure that must be handled and stored per unit of weight gained.
2. The cost per unit of energy from forages is typically lower than that of concentrate since it has fewer alternative uses. Also, more energy can be harvested per acre of corn as silage than as grain; but one must be careful in these calculations since it is energy available for gain per acre and cost of gain that is important.
3. Calves fed forages during the post-weaning growth phase make compensatory growth when placed on a high concentrate ration, resulting in a partial compensation for the previous higher maintenance cost.
4. The energy in forages is utilized more efficiently when the diet contains little or no grain.

The gain simulation (Telplan 56) was used to compare expected performance and costs with various feeding systems at different prices of corn. The following assumptions were used:

1. Land requirements. These are based on harvested yields of 100 bu corn grain (85% DM) and 16.7 tons of corn silage (32% DM) per acre. Adjustments were made for storage and handling losses, reducing availability to cattle to 97.3 bu and 15 tons, respectively.

2. Costs used. (a) The price of corn silage/ton (32% DM) = 6 (net corn price, $/bu) + $3. Corn silage priced in this manner returns the same net income/acre of corn as if harvested and sold as shelled corn. The formula adjusts for differential fertilizer, harvesting, and storage costs including storage losses, between corn harvested as grain and as silage. (b) Protein-mineral supplement was priced at $5.64/100 lbs when corn is $1.50/bu and $11.27/100 lbs when corn is $3.00/bu. Price was varied in relationship with corn prices since the price of soybean meal is determined, in part, by the opportunity cost of raising corn. (c) Nonfeed costs for the continuously operating feedlot are $25/head plus daily charges ranging from 16.5¢ for the all silage ration to 15.0¢ for high concentrate systems to reflect differences in feed and manure handling costs. Nonfeed costs for the lots feeding one group/year are $58/head plus a daily charge ranging from 9.9¢ for the all silage ration to 8.4¢ for high concentrate rations. Costs included are: in-and-out marketing, veterinary, death loss, power and fuel, bedding, and labor. Interest on the feeder was charged at 5¢/day. These charges allow a reasonable return to labor and equity capital. Thus, nonfeed costs can be directly compared, with the system having the lowest feed plus nonfeed cost having an advantage.

Relative prices of concentrate compared to forages and nonfeed costs determine the feeding system that will be used. Table 11-11 compares the predicted cost of gain at $1.50/bu, or $3.00/bu for #2 shelled corn. Feed costs of gain are given first; they assume the use of soybean meal as a protein supplement. The nonfeed costs include an average return to equity capital observed in cattle feeding. Thus, the system can be compared directly on a total cost basis; those with the lowest feed plus nonfeed costs will have an advantage. The projected performance data agrees with a summary of 17 university experiments prepared by R. D. Goodrich at the University of Minnesota.

The corn prices at which high silage rations become least cost varies with whether or not the lot is kept full continuously or only one group is fed each year. For the continuously operating lot, high silage systems become least cost at about $3/bu for #2 shelled corn. For those feeding one group/year, high silage rations become least cost at less than $3/bushel.

At past prices, high silage systems were more competitive than this because cost of supplemental nitrogen has been less from NPN. This is very important, due to the high need for supplemental nitrogen in high silage rations. At $190/ton for urea, high silage rations become least cost at $2.11/bu for #2 corn if only one group is fed/year and $2.80/bu when the lot is operated continuously.

Overall, the two-phase system may be the best, considering risk and best use of the corn crop. It is difficult for most cattle feeders to change systems every time grain prices change, and this system is not likely to be far from the most profitable system as grain prices change. This system is near the average of that presently used over the postweaning growth period in the United States, indicating that overall the industry has chosen this system as being close to the most profitable on the average. Some shifts are made

Table 11-11
Profitability of Various Feeding Systems at Two Prices of Corn

	All silage	Two phase	25% corn	50% corn	70% corn
Ration from 450 to 750 lb feed as % of ration DM					
Corn	--	--	23	50	68
Protein supplement	11	11	11	12	12
Corn silage	89	89	66	38	20
Ration from 750 to 1050 lb feed as % of ration DM					
Corn	--	53	25	53	72
Protein supplement	7	5	5	5	5
Corn silage	93	42	70	42	23
Expected performance for the total feeding period, 450 to 1050 lb.[a]					
Daily gain	1.85	2.16	2.09	2.44	2.57
Feed/gain	8.77	7.41	7.98	6.80	6.35
NE required, Mcal	3229	2905	3020	2727	2728
Turnover rate, head/year/space	1.06	1.24	1.20	1.41	1.48
Land required, acres	.49	.54	.62	.64	.67
Turnover Rate	1.06	1.24	1.20	1.41	1.48
Feed cost, $/100 lb gain (supplement priced relative to corn)					
Corn @ $1.50/bu	20.1	19.6	20.7	19.5	20.0
Corn @ $3.00/bu	36.4	37.0	38.9	37.7	39.3
	--------Continuous turn lot--------				
Non-feed cost, $/100 lb gain	15.8	13.8	14.1	12.4	12.0
	------------One turn lot------------				
Non-feed cost, $/100 lb gain	15.9	13.7	14.8	13.6	13.1
	------------One turn lot------------				
Total cost, $/100 lb gain					
Corn @ $1.50/100 lb	36.0	33.3	35.5	33.1	33.1
Corn @ $3.00/100 lb	52.3	50.6	53.7	51.3	52.5
	--------Continuous turn lot--------				
Total cost, $/100 lb gain					
Corn @ $1.50/bu.	35.9	33.4	34.9	31.9	32.0
Corn @ $3.00/bu.	52.2	50.7	53.0	50.1	51.4

[a] Expected performance is on a pay weight to pay weight basis for carrying an average frame steer, started in average condition, using a growth stimulant and fed in a nonstress environment from 450 to 1050 lb. A 14 to 17 day period with no gain, intake at 80% normal, was assumed at the start of the feeding period. A .5% empty body weight outshrink was used. Weight gains are adjusted for differences in fill.

in the length of the grain feeding period as the price of grain changes.

Many other factors enter into the postweaning decision on choice of feeding system. Cattle types available and their relative market value are important considerations. Although little published data is available, some cattle types may utilize high silage diets relatively better than others. Further studies are underway to explore possible cattle type-ration interactions.

Machinery needs for different systems, limits on the amount of corn silage that can be harvested and handled, the impact of continuous corn silage system on soil structure, and ability to get various crops planted and harvested on time must also be considered. Different crops have different harvest periods, which often means that from a total system perspective, timeliness and labor utilization favor a combination of silage and grain. Further, using a combination of silage and grain allows the use of the poorest corn for silage. Often harvesting much of the crop as silage becomes the only means of salvaging much of the corn crop, due to the late planting, dry or wet weather, and/or early frost. Many feeders prefer a two-phase system for its flexibility; they can alter the length of time on the all silage ration after the cattle are on feed, depending on the silage and grain available and market prices.

11-3   BEEF HERD FEEDING SYSTEM

The foundation underlying a least-cost feeding program is built on two factors: (1) matching the available feed supply with the cow's nutrient requirements. This involves having the best quality feeds available when the cow's nutrient requirements are at a peak during early lactation and rebreeding, and then utilizing feeds that have little alternative use such as the poorest quality hay or crop residues when the cow is just being maintained. (2) Minimizing the use of harvested feeds and maximum use of otherwise wasted feeds. This suggests maximum use of pasture and crop residues that have limited alternative uses, to minimize equipment, fuel, labor, and housing costs. This is of particular importance in the North because of the long winter feeding period and relatively short growing season.

In most areas, there are two major situations that determine the approach to selecting the best feed production system: Where more land is available than can be fully utilized, the factors limiting the number of cattle that can be kept are the labor or capital available and/or desires of the operator; or where land is limited, the goal is to most profitably utilize the land available. These determine the extent of pasture and hay land improvement and breeding plan that is best.

The basic decisions that must be made in either situation are:

1. How much should I improve my pastures?
2. What do I use for supplemental feeding in the winter?
3. Does it matter if I cut my hay once, twice, or three times?
4. How much machinery can I justify?

Unlimited Land

The conclusions that can be reached from most of our calculations under these conditions are as follows:

1. Even though land is unlimited it pays to improve pasture, and any money invested in agronomic practices should be invested first in reseeding and fertilizing practices. The main reason for this is when pastures are unimproved, there is a flush of growth early in the spring but the amount and quality of feed from the native grass declines rapidly in late summer and early fall at a time when the cow is still lactating, and supplemental feeding must begin earlier.
2. The least-cost plan for producing winter feed is usually to make one or two cuttings from unimproved native grass and/or maximize use of crop residues. The reason for this is unimproved native hay can be harvested at the peak of growth and at that point it contains enough nutrients for wintering beef cows that calve in the spring, with no cost inputs other than harvesting costs. Corn silage has not been profitable as a winter feed where land is unlimited.

Limited Land

In nearly every case, maximizing production by reseeding to the most productive pasture and hay crops and fertilizing and liming results in maximum profit when maximum return per acre of land is the goal. Returns depend on the price of beef; for example, beef production may be maximized but not dollar returns. More cows are supported and the feed production is better matched to the cows' requirements at various times of the year when the most productive forages are developed.

It is usually more profitable to improve the pastures than to improve the hay. The factors involved are an increased carrying capacity and a longer grazing season.

The use of corn silage supports more cows per acre, but is often not as profitable as the best hay production system unless 12 tons or more of silage per acre could be consistently produced in combination with using the best pasture system.

Thus, even with current production costs at an all-time high, it still pays to maximize pasture and hay production under these conditions.

Fall Calving vs Spring Calving

If calves are to be sold at 7–8 months of age, fall calving is feasible if maximum use is to be made of the land and high yielding corn silage is used in combination with the best pasture system. A higher price can normally be obtained for the calves in the spring than in the fall because there are fewer of them in the spring. Another reason for fall calves might be labor availability. Otherwise, fall calving does not appear to be feasible in the north under most conditions because high quality supplemental feeds must be used in the winter in order to support a lactating cow, whose nutrient requirements are nearly twice that of a dry, gestating cow.

In a spring calving system, pasture production is at a peak when the cow's nutrient requirements are at a peak, and is usually adequate to meet her requirements without supplemental feeding. Where high milking cows are used and the calves are not weaned and sold until 10 months of age or the calves are fed to yearling or slaughter weights, however, fall calving may be feasible.

Determining the Best System for Use of Grass: Cow/Calf, Cow Yearling, or Stocker Program

Land owners have several options for use of grass. They can use beef cows and sell the calves at weaning, keep the calves to yearling weights, or purchase stocker cattle. They also can consider retaining ownership through the feedlot phase.

The first consideration in choosing one of these alternatives is whether the calves should be placed in the feedlot at weaning or kept on grass and hay to short or long yearling weights. It must be determined whether the energy available as grass or hay can be used most profitably if fed to more cows with the calves removed at weaning or fewer cows and the calves kept to short or long

Table 11-12

Energy Requirements of a Beef Herd with Calves Removed at Weaning or After 180 or 360 Days on an All Forage Program

	Beef Herd	Calves removed 180 days after weaning, or purchased stockers or grazed 180 days		Calves removed 360 days after weaning, stockers wintered over, then grazed	
		Steers	Heifers	Steers	Heifers
TDN required, lb[1]	5000	1640	1350	3680	3150
lb TDN per beef cow[2] unit plus yearlings kept from this unit	5000	738	405	1656	1045

[1] Beef herd includes 1000 lb beef cow, 15% replacement heifers and 4% bulls. TDN requirements for steers and heifers are on a per head basis and assumes weaning weights of 450 and 420 lb, respectively and an overall gain of 1 lb per head daily on hay or pasture.

[2] The total herd requirements to carry the calf crop 180 days after weaning would be 5000 + (1640 x % steer feeder calves) + (1350 x % + (3680 x % steer feeder calves) + (3150 x % heifer feeder calves).

yearling weights. Table 11-12 gives the energy requirements of a beef herd with the calves removed at weaning or at 180 or 360 days after weaning. These values can be used to estimate the cows or cows plus yearlings that can be kept on a given amount of energy. Table 11-13 gives equivalent sale prices for calves removed at weaning or at 180 or 360 days after weaning. The calculations made to develop this table are given in the footnotes.

The calculations were based on 100 cows kept when calves are removed at weaning, allowing the values to be easily adjusted to various other herd sizes. One can use the requirements given in Table 11-12 and your own percent calf crop, however, to develop this comparison for your own herd.

One should adjust the equivalent sale prices for any changes in costs that occur if fewer cows and retained calves make up the herd as compared to more cows, and the calves sold at weaning. This value was adjusted when the calves were removed after 180 days for the expected higher labor requirement due to more total hours required for this number of cattle. All other differential costs seemed to nearly cancel out. For example, less cull cows being sold was nearly canceled out by the lower cow depreciation cost when fewer cows were kept.

These calculations show that when grain prices are low or feeder cattle are in short supply, calf prices will be high, favoring

Table 11-13

Estimating Equivalent Sale Prices for Calves Removed at Weaning or 180 or 360 Days After Weaning

Item	At weaning	180 Days after weaning	360 Days after weaning	
			Cow/calf	Stocker steers
TDN available, tons	250	250	250	250
No. cows[1]	100	81	66	136
Calves to remove[2]				
Steers	45	–	–	–
Heifers	30	–	–	–
Yearlings to remove[3]				
Steers		36	30	
Heifers		24	20	
Weight to sell				
Steers	45 x 450#	36 x 630#	30 x 810#	136 x 810#
Heifers	30 x 420#	24 x 590#	19 x 750#	
Cull cows	15 x 900#	12 x 900#	10 x 900#	
Expected sale values, $/cwt. – low grain prices				
Steers	$50.00	$46.00	$43.00	$43.00
Heifers	40.00	39.00	39.00	
Cull cows	32.00	32.00	32.00	
Gross returns[4]	$19,485.00	$19,411.00	$18,886.00	$16,768.00
Expected sale value, $/cwt. – high grain prices				
Steers	40.00	40.00	40.00	40.00
Heifers	32.00	32.00	32.00	
Cull cows	24.00	24.00	24.00	
Gross returns[4]	15,372.00	16,195.00	16,440.00	19,584.00

[1,3] Number of cows and yearlings that can be carried on the TDN available estimated by dividing the TDN available by the TDN required per beef cow unit on the various programs, as outlined in table 11-12. For example, where the calves are kept 180 days after weaning and 250 tons of TDN are available, the cows that can be kept are

$$\frac{250 \times 2000}{5000 + 738 + 405} = 81,$$

the number of steer feeders is 81 x % steer feeders and the number of heifer feeders is 81 x % heifer feeders produced.

[2] Based on a 90% calf crop and 15 heifers saved for replacements per 100 cows.

[4] Gross returns for stockers = sale value – purchase cost.

sale at weaning. When grain prices are low, cost of gain is usually lower in the feedlot than on grass. There are times when grain prices are moderately high but feeder cattle are in short supply, due to the phase of the cattle cycle. The higher cost of feeder cattle will be offset by higher prices for finished cattle under these conditions. When grain prices are high, however, gains are lower cost on grass than in the feedlot. Under these conditions, heavier cattle sell for as much or more than calves, encouraging cow/calf producers to keep their calves to yearling weights.

In the long run, one must be careful in shifting back and forth between calf and yearling programs. One could end up selling off cows when they are lowest and buying them back or replacing them when cattle prices are at a peak. At the bottom of the cattle cycle, stocker cattle are high priced, and most cow-calf operators who switched to a yearling program wish they had kept their cow herd.

Obviously, someone has to produce the feeder calves. Overall management system and type of forage available should be the main factors determining whether a cow-calf or stocker program is best for an individual operator. Many farmer-feeders purchase feeder cattle in the fall, graze them on corn stalks or milo residue until winter, then place them in their feedlot. They do not have grassland available, so beef cows are not feasible. Many wheat producers in the plains use stockers to graze wheat pasture, then place the stocker cattle in the feedlot in the spring. Many do not have range pasture or are not set up to handle a cow-calf enterprise. Cow/calf producers in dry areas will keep the number of cows their grass will support in the dry years, then will purchase stockers to utilize surplus grass in the years when moisture is more abundant.

Those producers who have large amounts of forage that has little alternative use, however, should keep a base cow herd, making shifts between cow/calf and cow/yearling programs as grain and feeder cattle prices change. There is a trend toward more calves being kept to yearling weights, to give more marketing options rather than having to be dependent on the feeder cattle price in the fall. Also, the increase in feedlot size has encouraged this shift. As feedlots become larger, it becomes more difficult to handle calves. They are not able to procure and handle large amounts of roughage to grow calves. The main factor involved, however, is the health problems experienced when large numbers of calves are assembled in the same lot.

Over a period of years more profit should be realized by retaining ownership until the cattle are finished. Over the long range the profits in beef production must be divided between producing feeder calves and finishing cattle if both types of operations are to survive. Thus, owning the calves through finishing should stabilize income over a period of time. The calf producer should have several advantages such as reduced marketing, transportation, and death loss costs. This also allows producers to get feedlot performance and carcass data on their calf crop that can be extremely valuable in making herd improvements. More capital is required for retaining ownership. One of the problems is that many cow/calf producers do not have enough intermediate to long-term capital available to support both a cow/calf operation and that needed for retaining ownership until finished for slaughter.

Suggestions for Placing Cattle in a Feedlot

Select a feedlot with a reputation for honesty and ability to feed cattle efficiently and at a low total cost of gain. Check with extension specialists in the area where you intend to have the cattle fed and with others who have placed cattle in that area for suggestions in this regard.

Put cattle together in uniform groups of 100 to 200 head. Most feedlots have pens that require at least 100 head and as many as 200 head. Many producers do not have this many calves, especially of one sex or type. It may be desirable to pool your cattle with other ranchers into uniform groups of similar sex, type, and weight. This will help in minimizing costs and maximizing sale price, as the cattle will likely all be sold at the same time.

Wean the calves 30 to 60 days before shipping to minimize stress.

Load the cattle with a moderate fill of grass hay and move and handle them as quietly as possible during sorting and loading. Instruct the trucker to move them to the feedlot as quickly as possible to minimize stress and shrink. Be sure the diesel stacks extend above the trailer. You can expect the calves to shrink about 3 percent plus about 0.6 percent for each 100 miles they are shipped, and all of these steps will help reduce shrink, death loss, and sickness. Normally, it takes 10 to 15 days to recover shrink losses, and death losses average 2 to 3 percent on calves and 1/2 percent to 1 percent on yearlings. However, it can take more than 30 days to recover shrink, and death losses can exceed 10 percent with improper handling prior to shipment, during transit, and on arrival at the feedlot.

Be prepared to pay feed and overhead bills at least monthly as you can expect to be billed either once or twice a month for these

Table 11-14
Developing a Budget for Feeding Raised Cattle or Stockers to Finished Weights

Estimating requirements	At weaning		180 Days after weaning		360 Days after weaning	
	Steers	Heifers	Steers	Heifers	Steers	Heifers
Weight at farm, lb	450	420	630	590	810	750
Expected sale weight, lb	1000	850	1050	900	1100	950
Lb gain	550	430	420	360	290	200
Lb feed/lb gain[1]	7.3	8.2	8.2	8.6	9.0	9.5
Total lb feed required[1]	4015	3526	3444	3096	2610	1900
Estimated days in feedlot[1]	220	187	162	153	109	87
Estimating costs						
Expected feed costs[2]	214.80	188.64	184.25	165.63	139.64	101.65
Transportation to feedlot[3]	11.25	10.50	15.75	14.75	20.25	18.75
Veterinary and medical	3.00	3.00	2.00	2.00	2.00	2.00
Yardage[4]	33.00	28.05	24.30	22.95	16.35	13.05
Interest on cattle[5]	13.56	8.71	13.36	9.60	11.51	8.00
Interest on feed[6]	6.47	4.83	4.09	3.47	2.09	1.17
Taxes, insurance, misc.	4.00	4.00	3.00	3.00	2.00	2.00
Death loss[7]	6.75	5.12	3.01	2.29	3.85	3.35
Market value of feeder calf[8]	224.80	170.45	301.04	229.01	385.41	335.33
Cost of calf and feedlot costs	517.83	423.30	550.72	452.70	583.10	485.30
Breakeven sale price	51.78	49.82	52.51	50.29	53.10	51.23
Projected sale price[9]	54.00	52.00	54.00	52.00	54.00	52.00
Profit per head for feedlot phase[10]	22.00	18.70	16.20	15.30	10.90	8.70

[1] Based on average commercial feedlot gains and feed conversions.

[2] Based on a ration cost of $5.35 per cwt., 90% DM basis on an 80% concentrate ration with feed grain at $5.70/cwt., roughage at $60.00/ton and supplement at $180.00/ton.

[3] Based on trucking rates of $1.00/loaded mile for a 40,000 lb haul and an expected average haul of 1000 miles.

[4] Estimated to be 15¢ per head daily. Feedlot yardage charges to cover the lot overhead cost of feeding cattle plus profits to the feedlot owners may vary from 10 to 20¢ per head daily plus 10% markup on the feed.

[5] Based on 10% per annum interest cost.

[6] Based on 10% per annum interest on 1/2 of feed cost.

[7] Based on death losses of 3% of calves and 1% of yearlings. The feed consumed by cattle that died was accounted for in the feed conversions.

[8] Based on price that could be paid to give a net return of 10¢/hd/day. This value will depend on cost of gain relative to sale value of finished cattle.

[9] Based on present outlook for finished cattle when the cattle would be ready for market.

[10] Profit per head = (sale price/cwt. − break even price/cwt.) × $\frac{\text{expected sale weight}}{100}$

items. Capital needs are outlined in Table 11-14, and similar budgets should be prepared for obtaining a line of credit.

Winter Feeding Systems

The greatest out-of-pocket cost for a beef herd is for winter feed. A number of methods are being used for winter feeding. Included are combinations of grazing or feeding harvested crop residues, grazing fescue or native range, feeding round or square bales, utilizing large hay packages (stackers and large round balers) or hay crop, corn, or sorghum silage. The factors involved in determining the best winter feeding system include labor, nutrient needs of the cattle and expected wastage, and cost of harvesting, storing, and feeding.

The least cost way to winter a beef herd is to graze crop residues or native range or fescue, supplementing as needed with hay, grain, range cubes, or silage. In most winter grazing systems, the cattle receive only the pasture during the first 60 to 90 days, then supplemented hay, grain, or range cubes will be given as needed, with the amount depending on the quantity and quality of the forage remaining. In a five-year Iowa study, it was found that grazing crop residues was lower cost than feeding harvested crop residues, due to elimination of equipment, manure handling, and feeding costs. In a South Dakota study, it was determined that wintering costs were lowest when maximum use was made of crop residues. Many now utilize harvested crop residues, however, due to needing to fall plow. Thus, the whole farm system must be considered. Many cannot afford to leave crop residues for grazing, due to the impact on yields and labor if the land is not fall plowed. The best hay harvesting system will vary among beef operations.

Using small square bales is still probably the most economical method for small beef herds in areas where used haying equipment is available. Large package hay systems are becoming widely used, due to a reduction in labor and storage requirements. In an analysis of some data from Ohio State, used conventional balers and wagons were found to cost least for 50 tons or less hay/year, but a round bale system costs least between 50-100 tons of hay. This indicates that for herds of 30-50 cows or less, conventional square bale systems are the least cost; but above this size, a large round bale system can be considered. In some areas, large balers can be rented or are available on a custom basis.

If not carefully managed, however, round bale systems can be more expensive than the data indicates. When square bales are hand fed, intake is controlled so that nutrients are not fed in excess of their needs. It is a more flexible system. If hay is short, grain can be substituted for part of the hay. For example, 8 lbs hay and 8 lbs ear corn could be fed in place of 20 lbs hay. When hay is limit fed in good hay racks, waste can be almost negligible. When round bales are left in the field for unrestricted winter grazing, storage nutrient and wastage losses can run as high as 40 percent, compared to 5-10 percent in a well-managed square bale system.

Further, beef cows may consume 20 to 30 percent more than they need under unrestricted grazing conditions. Strip grazing or feeding large bales in racks or behind panels is necessary under most conditions to control total losses to 10-20 percent.

12

Marketing Finished Cattle

12-1 WHEN TO MARKET SLAUGHTER CATTLE AND METHODS OF MARKETING

As cattle increase in weight, cost of gain increases on a given ration. Thus, selecting the optimum time to market cattle involves a consideration of the rate of increase in cost of gain in relation to the increase in the value of the carcass. Once the final weight is determined, the feeder must decide what method of selling to use such as the central public markets or direct selling. If sold direct, he also has to decide whether to sell on a live or carcass weight basis and if the selling price should be based on a flat price for all cattle or on carcass quality and cutability grade. He then must decide whether to top out his cattle or sell the whole pen at once.

When Cattle Should Be Marketed for Optimum Returns

Figure 12-1 shows the increase in cost of gain of steers and heifers as they increase in weight on typical feeding programs at average current prices for feedstuffs. The first reason for this increase in cost of gain is that more energy is required per pound of gain as the proportion of fat in the gain increases. Secondly, as cattle near the fatness of the low choice grade, they tend to consume less energy above their maintenance requirement, leaving less energy available for gain. Figure 12-1 is based on average feed intakes, and cost of gain may increase at an even greater rate than shown here as it becomes increasingly difficult to keep cattle on feed as they increase in fatness.

Based on the estimates given in Figure 12-1, Figure 12-2 shows average projected profits per head for average size cattle. Figure 12-2 indicates that at higher prices for finished cattle, maximum profits per head are obtained when cattle are fed beyond their optimum slaughter weight because the sale price is still above the cost of gain at heavier weights. If these profits are adjusted for cutability differences, however, maximum profits are still achieved when cattle are slaughtered at their optimum slaughter weight, as indicated by the dashed lines. Furthermore, Figure 12-3 shows that when lots are kept full year around, maximum profits per head of feedlot capacity are obtained when cattle are sold as they reach their optimum slaughter weight. Feed costs and sale prices will vary considerably from those used in Figure 12-1 through 12-3, but the relative effect of slaughter weight on profits should still apply.

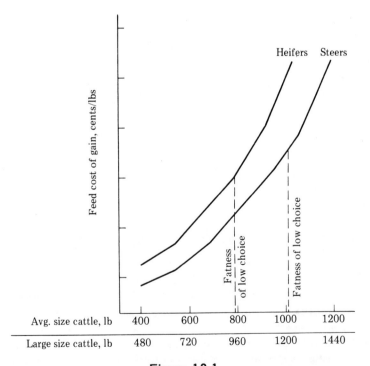

Figure 12-1
Relationship of Cost of Gain to Weight

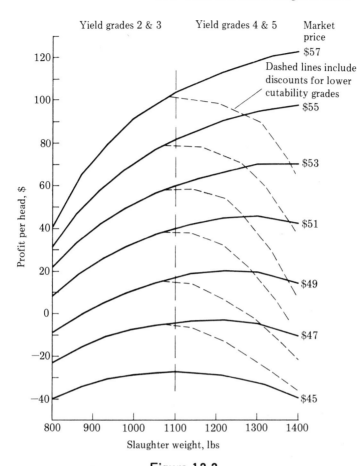

Figure 12-2
Effect of Grade and Market Price on Optimum Slaughter Weight

Therefore, even though at times cost of gain is still below sale price when cattle reach the optimum slaughter weight, more efficient use of the feed will be made by utilizing it in younger, lighter cattle. Also, when cattle are marketed continuously about the same net financial position is usually achieved if feeder cattle are purchased on the same market as the fat cattle were sold. In other words, if the fat cattle are sold on a low market, they can be replaced by feeders on the same low market. Furthermore, as consumers become increasingly fat conscious, beef is processed at the plant where slaughtered, and more cattle are slaughtered at a young age, discounts on excessively fat cattle will increase and price spreads between good and choice cattle will decrease or more

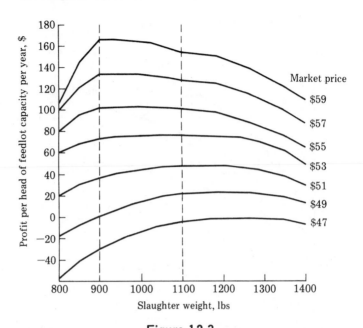

Figure 12-3
Optimum Slaughter Weights for Continuously Full Feedlots

cattle will be allowed into the choice grade at a lower fat content. Cattle feeders must become increasingly aware of being able to evaluate cutability factors in live cattle such as backfat thickness and rib eye area, in addition to quality grade. This requires getting slaughter data on cattle sold and relating it to the live animal.

Methods of Marketing

The methods used in marketing cattle depend on several conditions. First of all one has to decide whether or not the whole pen should be sold at once or topped out (removing cattle from the pen as they reach their optimum slaughter weight).

Topping out is advisable where the cattle in a pen were different in weight at the start, and the difference in weight represents a difference in age. Under these conditions it is probably advisable to top out lots once or twice and then sell the remainder. When cattle are of similar sex and weights at the start, however, then it is probably advisable to sell the whole pen at once when they average 80 to 90 percent choice, assuming the cattle are capable of reaching the choice grade. If the cattle are topped out under these conditions, the cattle left after topping are likely to be unable to make economical gains beyond that point.

The selling methods to use depend on several conditions, and they will be discussed separately.

Selling direct. There are several things a feeder should know if he intends to sell his cattle direct to a packer. First of all, he needs to know the price of carcass beef and the strength of the dressed beef market and the potential quality and cutability grades of the cattle he has. The price of beef can be obtained from *Cattle Fax*, USDA *Market News*, and the *River Provisioner* as they are commonly used as a basis for pricing cattle. Secondly, he needs to know the needs of packers in his area for numbers of cattle to keep the slaughter plant in operation. Thirdly, he needs to know current kill costs and offal credits to get an idea of current packer margins, which can be obtained from the USDA market news service. If he intends to sell the cattle on a live weight basis, he needs to know how to calculate live prices from carcass beef prices, or vice versa. Live price is determined by multiplying the dressed beef price × the expected dressing per cent or dressed beef price is determined by dividing the live price by the average dressing per cent, then the answer is multiplied by 100. For example, a 1000-lb steer at about the fatness of low choice can be expected to have a carcass weight of about 600 lbs, or a dressing percentage of 60. If the price quoted is 42 cents per lb alive, then the carcass value is 42¢/.60 = 70¢/lb. Also, he needs to know the expected amount of shrink under different weighing conditions and how to adjust the price offered for the expected shrink (Tables 12-1 to 12-3).

Table 12-1

Effect of Ration and Grade on Dressing Percentage of Beef Breed Cattle[1]

	Standard	Good	Choice	Prime
	- - - - - - - Dressing % - - - - - - - - -			
High silage	54.5	56.5	58.5	60.5
High grain	56	58	60	62

Source: Research trials at Michigan State University, 1974-1977 and the Ohio State University, 1968-1970.

[1] Beef breed steers will have a two to three higher dressing percentage than Holstein steers.

Table 12-2
Estimated Effect of Time Off Feed and Water on Dressing Percentage

	None	8 hrs	16 hrs	24 hrs
No haul	58	59	60	62
50 - 100 mile haul	60	61	62	64

Based on experimental data from Michigan State, Ohio State and Kansas State Universities.

Table 12-3
Dressing Percentage of Mature Cows

	Commercial	Utility	Cutter	Canner
Range	52-54	47-55	43-52	38-46
Average	53	51	47	43

Dressing percent varies with the degree of fatness, types of ration, and handling procedures. The following tables summarize these effects, assuming cattle are loaded out of the lot without withholding feed and water and are weighed at an auction or packing plant within 50-100 miles of the feedlot.

When comparing prices, these effects must be taken into account. For example, a price of $42.00/cwt on cattle out of the lot with no shrink is as good as $43.45 after 16 hr without feed and water ($42 \times 0.60/0.58$).

If the feeder is aware of these factors and becomes skilled in marketing, there are probably some advantages to selling direct such as less transportation and shrink, no unnecessary handling costs and fewer bruises, sale price is determined before cattle leave the lot, and being able to obtain carcass information. There are several ways to sell cattle direct, and the method to use may depend on existing conditions and the confidence you have in the potential buyer. It may be advisable to sell based on live price and weight if you are concerned about the ability of the cattle to grade or when the market is strong and competition is good. Weighing conditions that are fair to all concerned should be carefully agreed upon, however. If the feeder is not satisfied with the weighing conditions offered, but still wants to use a live price, then it might be best to sell on live price but with a guaranteed dressing percent. If cattle are muddy or the dressing percent is in doubt for other reasons but you still want to know the price of the cattle before they leave the lot, then it might be best to sell based on a flat overall carcass beef price. If you feel that the cattle are better quality than the price offered, however, then it might be best to sell them based on carcass weight and a price schedule for different quality grades.

In addition, if the carcass pricing schedule recognizes the value of high cutability and you market the type of cattle that will have a high cutability then it might be advantageous to sell based on carcass weights and prices for quality and cutability grades.

Several factors should be understood when cattle are sold based on carcass weights such as who stands the condemnation and bruises, and if standardized slaughtering and trimming procedures are followed for all cattle slaughtered, how soon the cattle will be slaughtered (tissue shrink probably starts after 12-14 hours off feed) and graded, how much slaughter information can be obtained, and how soon payment is made.

Above all, in selling direct you must have confidence in those you are dealing with and treat them fairly and honestly at all times. A wise feeder knows that he needs the buyer as much as the buyer needs him. Eliminate those that are not honest and fair and then deal justly with those you trust.

Selling on central public markets. If you do not sell often enough or do not have the information necessary to become skilled at marketing, then selling on the central public markets may be the best way to market fat cattle. Additional reasons for using terminal markets might be when the seller does not have the kind of cattle in demand by local packing plants, if the terminal market is relatively close, or as a means of periodically checking out the ability to sell cattle direct. It would be a good idea to let your local auction know when you have cattle ready for sale, and they can advise you when there will be buyers present for your type of cattle.

Marketing systems should be developed and used that will minimize the time and handling between the time the cattle leave the feedlot and the beef is in the consumer's hands to minimize handling and shrinkage costs. In addition, pricing systems are needed that maximize pricing based on edible beef produced and the current market value of edible beef. As more of the carcasses are processed at the slaughter plant, it will become increasingly possible to develop pricing systems based on both quality and cutability grade. This will improve the ability of a cattle feeder to market the cattle he produces for what they are worth by enabling him to know what quality he produces. In addition, by taking advantage of improved sources of market information such as *Cattle Fax* and other markets' news services, he can determine what this quality is worth.

Addresses for these services are:

1. *Market News*, U.S. Dept. of Agriculture, 609 Livestock Exchange Building, Omaha, Nebraska 68107
2. *Market News*, U.S. Dept. of Agriculture, 536 S. Clark St., Chicago, Ill. 60605
3. *Cattle Fax*, 1001 Lincoln St., Denver, Colorado 80203
4. *Drovers Journal*, One Gateway Center, 5th at State Ave., Kansas City, Kansas 66101

Selling for freezer trade. When selling direct to consumers, most prefer to quote a price on a carcass weight basis plus slaughtering and processing costs. The consumer should be aware, however, of the loss in cutting a carcass to trimmed retail cuts, as discussed in the next section.

	Yield Grades				
	1	*2*	*3*	*4*	*5*
Fat Trim	7.6	12.7	17.8	22.9	28.0
Bone and shrink	10.4	9.9	9.4	8.9	8.4
Trimmed retail cuts	82.0	77.4	72.8	68.2	63.6

Some other factors important in selling for the freezer trade are as follows:

1. You will likely want to have cattle at various stages of growth on hand so customers do not have to wait too long for beef. Also, encourage them to place their orders in advance so you can plan your feeding program.
2. Feed the size and type of cattle that will consistently provide the size of carcass and quality preferred by your customers, and avoid feeding the cattle beyond their optimum slaughter weight to avoid excessive fat trim and beef with a high fat content. If small carcasses are preferred, feed earlier maturing

cattle or heifers. If very lean, moderate-sized carcasses are preferred, consider feeding larger type cattle such as Holsteins or crosses with the larger breeds and slaughter them at 1000 to 1100 lbs.
3. Feed cattle that will reach the desired slaughter weight by 15 to 17 months of age or less, to ensure obtaining tender beef.
4. Work with a slaughter and processing plant that is clean and conscientious. Nothing will "turn off" a customer faster than picking up his meat in a dirty, smelly slaughter plant that appears to be operated by dishonest people.

There are excellent profit opportunities in developing a freezer trade, if properly conducted. People will pay a premium for beef they know the origin of and know it will consistently be good. Also, you can price your product more on a retail basis where the price does not drop as rapidly as it does at the producer level.

12-2 FACTORS AFFECTING CARCASS GRADE

The market value of a beef carcass at the present time is primarily determined by two factors: (1) the *quality* of the meat (palatability) and (2) the *quantity* or amount of lean meat available.

The USDA has established grades to represent the differences in both the quality and quantity of edible meat in a beef carcass. The differences in quality of the meat are represented by the USDA quality grades and differences in quantity or amount of salable lean are represented by the USDA yield grades. The USDA grader must now determine both of the USDA grades if the carcasses are to be graded. On certain types of beef the packer may choose to use his own "house grade" or "packer brand" in place of the USDA grades. Carcasses that meet requirements for the prime and choice quality grades are usually stamped with the USDA grades and those carcasses that fail to grade USDA choice often receive one of the packer's house grades.

Beef quality grades are important in determining carcass value because they serve as guides to the eating characteristics of the final product. The eating characteristics of beef are measured by the palatability of the cooked product—its tenderness, juiciness, and flavor. Yield grades identify the proportion of trimmed, retail cuts that can be obtained from the carcass.

Development of Present Grade Standards

The USDA beef quality grades—Prime, Choice, Good, Standard, Commercial, Utility, Cutter, and Canner—have been used since 1927 to identify differences in the palatability of beef. The major factors used to determine quality grades have been (1) marbling, (2) maturity and (3) conformation.

Since 1965, USDA yield grades have been available for identifying differences in proportion of trimmed retail cuts. Until recently, a packer could choose to have the USDA grader determine one or both of these grades.

On February 23, 1976 several revisions in the grading standards went into effect. The changes are:

1. All beef graded will be graded for both quality and yield, not just one or the other. This was implemented to properly identify differences in trimmed retail cuts and to discourage production of overfat cattle. This should result in premiums for those cattle with a minimum of fat trim and a discount for those with excess fat trim. This should encourage the industry to produce leaner cattle, reducing the cost of producing beef.

2. The marbling requirements have been reduced for cattle between 9 and 30 months of age. The old standards assumed that as an animal increased in age, more marbling was required to insure palatability. According to the USDA, recent research shows that there are no significant differences in the eating quality of beef from cattle ranging in age from 9 to 30 months ("A" maturity group). Further, they felt that since a much higher proportion of fed beef cattle now reach market weight at less than 24 months of age, the previous increase in marbling requirement was considered to be unnecessary and wasteful.

This change will result in some of the older cattle that didn't grade choice (because of marbling) now grading choice.

For example, a 2-year-old steer with a small amount of marbling previously would have graded good rather than choice because it needed a modest amount of marbling for choice. However, a 12-month-old steer with a small amount of marbling would have graded choice. Both would grade choice under the new standards.

3. Conformation (shape of carcass) has been eliminated as a factor in determining carcass grade. Research has shown that shape of the carcass has little influence on the proportion of high priced cuts or the quality of the meat.

USDA compared how our steer and heifer beef would grade under the old and new systems, assuming no changes in feeding practices:

	Old Standards	New Standards
	---------------- % of total ------------------	
Prime	4.5	6.5
Choice	54	68.5
Good	40	21
Standard	1.5	4

The reason for the dramatic drop in the proportion grading good is the ability of older cattle with a small amount of marbling and more angular cattle with a small or higher amount of marbling to grade choice under the new standards. Also, some cattle that had inadequate marbling for the good grade but had choice conformation could grade good because of their conformation. Those types will grade standard. Thus, the good grade is now more uniform.

Determination of Quality Grades

Marbling, the amount and distribution of small flecks of fat within a muscle system, is the most important single factor in determining quality grades. The evaluation of marbling is made on the cut surface of the rib eye by partially separating the hind from the forequarter between the twelfth and thirteenth ribs. Marbling contributes to the overall juiciness and flavor of beef. Several degrees of marbling have been established and are used as guides in grading beef carcasses. Figure 12-4 illustrates the lower limits of eight of the nine degrees of marbling (practically devoid not shown).

Maturity is also an important factor in determining beef quality grades. The primary indicators of maturity are color, size and shape of the rib bones, ossification of cartilage, particularly the "buttons" on the vertebrae, and the color and texture of the lean. Advanced maturity is often associated with decreased tenderness. Five maturity groups, A through E, have been established for ease of reference with group A indicating carcasses from very young animals and group E indicating carcasses from animals with evidences of advanced maturity or old age. The approximate age ranges of these maturity groups are as follows:

A—9 to 30 months
B—30 to 48 months
C—48 to 60 months
D—over 60 months
E—over 60 months

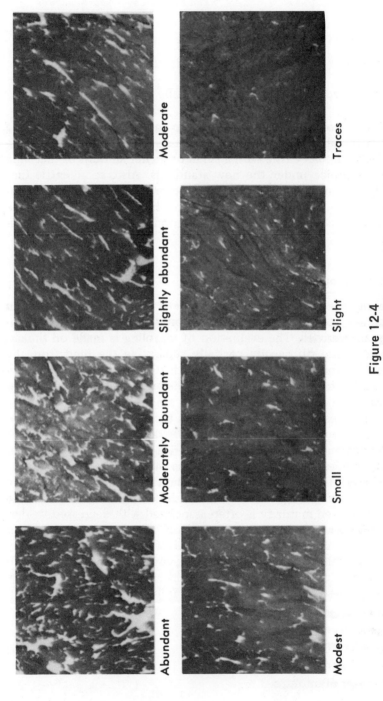

Figure 12-4
Lower Limits of Typical Degrees of Marbling Referred to in Grading Carcass Beef

12-2 Factors Affecting Carcass Grade

After the maturity group and degree of marbling have been evaluated, the two values are combined, with the use of the chart in Figure 12-5, into a single quality evaluation. The chart shows the minimum amount of marbling permitted for each of the quality grades and indicates that within each grade with progressive increases in maturity there is a progressive increase in the marbling requirement. For example, the minimum marbling requirement for choice varies from a small amount for the young carcasses to a modest amount for carcasses having the maximum maturity permitted in the choice grade. The chart also shows that cattle in the C, D, and E maturity groups are not eligible for the prime, choice, good and standard grades. For example, a steer with a typical slight amount of marbling and a typical A maturity would fall into the average good grade. The majority of market steers and heifers that fail to make the choice grade lack the degree of marbling necessary.

Each grade is associated with a specific degree of quality, thus enabling consumers to utilize the meat most efficiently by preparing it in the manner for which it is best suited.

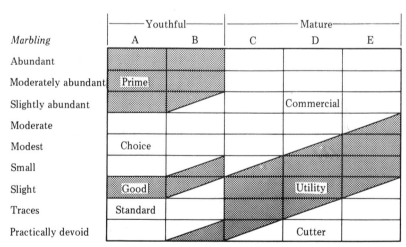

Figure 12-5
Relationship Between Marbling, Maturity, and Quality

Beef Yield Grades

Since 1965, USDA yield grades (also referred to as cutability grades) have provided an additional marketing tool for use by all who buy or sell cattle and beef carcasses. Yield grades are a means of identifying the most important value-determining characteristic—the amount of trimmed retail cuts that can be obtained from a beef carcass. Specifically, yield grades are based on the percentage of boneless, closely trimmed retail cuts from the round, loin, rib, and chuck. These four wholesale cuts account for more than 80 percent of the carcass value. There are five USDA yield grades numbered 1 through 5. Carcasses with yield grade 1 have the highest yield of retail cuts, while carcasses with a yield grade of 5 have the lowest yield of retail cuts. Yield grades for beef carcasses are applied without regard to sex or quality grade. Table 12-4 shows the percent of boneless, closely trimmed retail cuts that can be cut from the round, loin, rib, and chuck for each of the five yield grades.

Yield grades are determined by using the following four factors: (1) fat thickness, in.; (2) rib eye area, sq. in.; (3) percent kidney, heart, and pelvic fat; and (4) carcass weight, lb.

The amount of fat over the outside of a carcass is the most important factor in determining yield grade because it is a good indication of the amount of fat that is trimmed in making retail cuts. The measurement is made between the twelfth and thirteenth ribs over the rib eye at a point three-fourths of the length of the rib eye from its chin bone end. This measurement may be adjusted to reflect unusual amounts of fat on other parts of the carcass (Figure 12-6). Retail yield is reduced approximately 2 per-

Table 12-4
Percent of Boneless Retail Cuts from Round, Loin, Rib, and Chuck

Yield grades	
1	52.4% and above
2	52.3% to 50.1%
3	50.0% to 47.8%
4	47.7% to 45.5%
5	45.4% and below

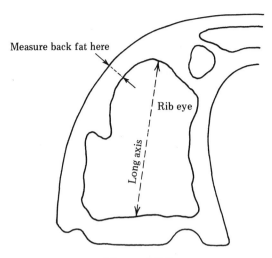

Figure 12-6
Measuring Fat Thickness

cent for each 0.075 inches additional fat thickness. However, with present chilling and handling procedures, it is desirable to have approximately 0.2–0.3 inches of outside carcass fat to protect the carcass from drying and discoloration.

The rib eye is the largest muscle in the carcass, lying on each side of the backbone running the full length of the back. When the carcass is separated into a fore and hindquarter between the twelfth and thirteenth ribs, a cross section of the rib eye is exposed. The area of the rib eye is used in determining the yield grade because it is an indicator of the total amount of muscle in a carcass. Among carcasses of the same fatness and weight, an increase in the rib eye area indicates an increase in the yield of retail cuts. A desirable rib eye area is a minimum of 2 square inches/100 lb of carcass.

The amount of kidney, heart, and pelvic fat around the kidneys and in the pelvic and heart areas also affects carcass yields. Because all of this fat is removed in trimming, increases in these fats decrease the yields of retail cuts. The amount of kidney, pelvic, and heart fat is estimated and expressed as a percent of the carcass weight. The average amount of kidney, heart, and pelvic fat usually present at various quality grades is shown in Table 12-5.

Table 12-5
Average Internal Fat Related to Carcass Grade

Grade	Percent
Prime	4.5
Choice	3.0
Good	2.5
Standard	2.0
Commercial	4.0
Utility	2.0
Cutter and Canner	1.5

The following method is used to determine yield grade in beef carcasses. These factors can also be estimated in live cattle to determine their potential yield or cutability grade.

I. Determine a preliminary yield grade from the following schedule:

Thickness of fat over rib eye	Preliminary yield grade
.2 inch	2.5
.4	3.0
.6	3.5
.8	4.0
1.0	4.5

II. Determine the final yield grade (1 to 5) by adjusting the preliminary yield grade as necessary for variations in kidney fat from 3.5 percent and for variations in area of rib eye. These adjustments are made as follows:

A. Rate of adjustment for area of rib eye in relation to warm carcass weight.

1. For each square inch of rib eye area *more* than the area indicated in Table 12-6, subtract 0.3 from the preliminary yield grade.

Table 12-6
Minimum Rib Eye Area Needed for Various Carcass Weights

Warm carcass wt.	Minimum area of rib eye
500	9.8
525	10.1
550	10.4
575	10.7
600	11.0
625	11.3
650	11.6
675	11.9
700	12.2
725	12.5
750	12.8

    2. For each square inch of rib eye area *less* than the area indicated in Table 12-6, add 0.3 to the preliminary yield grade.

  B. Rate of adjustment for percent of kidney, pelvic, and heart fat.

    1. For each percent of kidney, pelvic, and heart fat *more* than 3.5 percent add 0.2 to the preliminary yield grade.

    2. For each percent of kidney, pelvic, and heart fat *less* than 3.5 percent subtract 0.2 from the preliminary yield grade.

Example: Determination of yield grade given the following data: 0.6 in. fat thickness, 12.5 sq. in. of rib eye area, 625 lb carcass, and 2.0 percent estimated kidney, heart, and pelvic fat.

I. Preliminary yield grade: 0.6 in. fat = 3.5 preliminary yield grade.

II. Rate of adjustment for rib eye area, percent kidney, heart and pelvic fat.

    A. 625 lb. carcass should have 11.3 square inches (from Table 12-6)

       12.5 (sq. in.) actual data
      −11.3 (sq. in.) from Table 12-6
        1.2 sq. in. rib eye area more than indicated in Table 12-6.

1.2 × .3 (rate of adjustment) = 0.36 yield grade adjustment
3.5 preliminary yield grade −0.36 = 3.14 after adjustment for rib eye area

B. 3.5% kidney, heart, and pelvic fat is considered normal
    3.5% normal percent kidney, heart, and pelvic fat
    −2.0% actual data
    ─────
    1.5% difference

1.5 × 0.2 (rate of adjustment) = 0.3
3.14 adjusted yield grade − 0.3 = 2.84, or 2.8 final yield grade

Note: When used in the meat trade, fractional parts of the final yield grade are dropped. For the above example, the yield grade is a 2.

Table 12-7 shows the expected pounds of fat trim, bone, and trimmed retail cuts per hundredweight of carcass for each of the five yield grades.

The values indicate that as pounds of fat increase from yield grades 1 to 5, the pounds of trimmed retail cuts decrease.

Figure 12-7 presents a typical breakdown of a choice steer carcass and various wholesale cuts that are processed. The hind quarter represents 48 percent of the weight of the carcass but approximately 60 percent of the value. The fore quarter represents 52 percent of the carcass weight, but approximately 40 percent of the dollar value. It is anatomically impossible through selection to change the relationship of high priced cuts to low priced cuts, or the percentage of hind quarter to fore quarter through selection. The correlation of one major muscle is so high to that of another in the beef carcass that increasing the muscle mass of the round, as an example, proportionately increases the muscle mass of the chuck in the same carcass. This leaves the breeder and producer with the task of altering the total fat and fat distribution in cattle in order to change carcass composition.

Table 12-7

Pounds of Fat Trim, Bone, and Trimmed Retail Cuts per Hundredweight for Each of Five Yield Grades

	Yield Grades				
	1	2	3	4	5
Fat trim	7.6	12.7	17.8	22.9	28.0
Bone and shrink	10.4	9.9	9.4	8.9	8.4
Trimmed retail cuts	82.0	77.4	72.8	68.2	63.6

	Saleable Beef—lbs	Other lbs

- **CHUCK** 164.8 lbs (26.8% of total carcass)

Blade pot-roast	59.3	
Stew or ground beef	32.1	
Arm pot-roast	22.3	
Cross rib pot-roast	10.7	
Boston cut	9.9	
Fat and bone		30.5
TOTAL	134.3 lbs	30.5 lbs

- **BRISKET** 23.4 lbs (3.8% of total carcass)

Boneless	9.4	
Fat and bone		14.0
TOTAL	9.4 lbs	14.0 lbs

- **RIB** 59.0 lbs (9.6% of total carcass)

Standing rib roast	24.2	
Rib steak	12.4	
Short ribs	4.7	
Braising beef	2.7	
Ground beef	3.5	
Fat and bone		11.5
TOTAL	47.5 lbs	11.5 lbs

- **LOIN** 105.8 lbs (17.2% of total carcass)

Porterhouse steak	18.7	
T-bone steak	9.5	
Club steak	5.2	
Sirloin steak	41.4	
Ground beef	2.9	
Fat and bone		28.1
TOTAL	77.7 lbs	28.1 lbs

Chuck 164.8 lbs; Rib 59.0 lbs; Loin 105.8 lbs; Round 137.8 lbs; Brisket 23.4 lbs; Short plate 51.0 lbs; Flank 32.0 lbs; Shank 19.1 lbs

- **SHANK** 19.1 lbs (3.1% of total carcass)

- **SHORT PLATE** 51.0 lbs (8.3% of total carcass)

Plate, stew, short ribs	40.8	
Fat and bone		10.2
TOTAL	40.8 lbs	10.2 lbs

- **FLANK** 32;

- **FLANK** 32.0 lbs (5.2% of total carcass)

Flank	3.2	
Ground beef	12.6	
Fat		16.2
TOTAL	15.8 lbs	16.2 lbs

MISC. 22.1 lbs (3.6% of total carcass)

Kidney, hanging tender	3.6	
Fat, suet, cutting losses		18.5
TOTAL	3.6 lbs	18.5 lbs

- **ROUND** 137.8 lbs (22.4% of total carcass)

Top round (inside)	21.0	
Bottom round (outside)	20.3	
Tip	13.1	
Stew	8.3	
Rump	4.8	
Kabobs or cubes	2.1	
Ground beef	14.2	
Fat and bone		54.0
TOTAL	83.8 lbs	54.0 lbs

SUMMARY (1000 lbs choice steer)

Dresses out 61.5%	615 lbs
Less fat, bone and loss	183 lbs
Saleable beef	432 lbs

Figure 12-7
Retail Cuts from a Typical Choice Yield Grade 3 Steer

Value differences of $5 to $6 per hundredweight between adjacent yield grades (2 and 3, 3 and 4, etc.) are quite common. With the advent of yield grades, retailers can order beef of a specific yield grade and carcass weight, knowing approximately how many pounds of edible lean will be available for sale.

Estimating Yield Grades in Live Cattle

Evaluating live cattle as well as their carcasses in terms of their yield grade is very useful in appraising their value. Cattle with a desirable yield grade (high yield of retail cuts; yield grade 1 or 2) will be heavily muscled and have little outside fat cover. Cattle that are fat, wasty, and poorly muscled will have a less desirable yield of retail cuts (yield grades 4 or 5). Differences in both fat thickness and muscling affect the appearance of the live animal, and, because fatness and muscling have opposite effects on yields of retail cuts, evaluating live animals for yield grade requires an ability to make separate and accurate evaluations of these two factors.

Cattle can vary a great deal in external fat thickness at slaughter time. Therefore, the ability to estimate fat thickness correctly is very important in determining yield grade. Differences in fat thickness can be best estimated by observing areas where fat is deposited most rapidly—the brisket, flanks, twist, and over the back and around the tailhead. As cattle increase in fatness, these areas become progressively fuller, thicker, and deeper in appearance. In general, the deeper the animal, relative to its length, the more fat it will carry.

The muscular development of an animal can best be evaluated by observing those body parts that are the least affected by fatness—the round and forearm area. The thickness and fullness of the round and forearm are largely due to thickness of muscling.

It is necessary to know how to accurately estimate quality and yield grades to do a good job of marketing cattle. In order to become more skillful in estimating yield grades in live cattle, it is helpful to evaluate a group of cattle individually and then observe their carcasses in the cooler. In the cooler it is important to compare the visual estimates with the final quality and yield grades as well as the actual degree of marbling, fat thickness, rib eye area, etc.

Yield grades provide an indirect means for reflecting consumer preferences for beef with a high ratio of lean to fat. Thus, they can be effective in bringing about changes which will eliminate much of the waste now present in the production and marketing of beef.

When used in conjunction with quality grades, yield grades will provide a means of identifying strains of cattle and production methods that will produce high quality beef with a minimum of waste fat, which should lead to better values for consumers and greater returns for producers. The heritability coefficient for marbling is high (0.6). Thus, breeders can identify superior lines for this trait through progeny testing.

Management Factors Affecting Carcass Grade

Research has shown that shape of the carcass has little influence on the proportion of high priced cuts or the quality of the meat. *Thus, the problem is to produce cattle with small amounts of backfat and at the same time have at least a small amount of marbling.* Research has shown that large amounts of marbling are not needed to obtain high quality beef, but it does appear that at least a small amount of marbling is needed to obtain a desirable palatability. This section will discuss the influence of various factors thought to influence both backfat thickness and marbling, weight, sex, age, feeding program, and time on feed. The comparisons are based on the minimum amount of fat necessary to give a small amount of marbling, which is the amount needed to grade low choice.

Weight, frame size, and sex. Weight for the frame size is the most important factor influencing the proportion of muscle and fat in the carcass and where the fat is deposited. Research has shown that fat and muscle develop at the same time as part of the normal growth process. Fat is deposited in the viscera and kidneys and between the muscles early in life. As cattle continue to grow and approach mature weight an increasing amount of fat is deposited over the outside of the body. Then at some point, noticeable after there is at least 0.3 inch of backfat, fat is deposited in the muscle, and beyond this point the more backfat an animal has, the more marbling it is likely to have. Some backfat is desirable. Probably 0.2–0.3 inch of backfat is needed to prevent rapid chilling of the carcass and drying and discoloration of the meat. However, amounts above this are wasteful unless the trim fat is needed to blend with nonfat beef. Most cattle will have at least 0.3–0.4 inch of fat before their carcasses will contain rib eyes with a small amount of marbling.

Because of this normal growth pattern, the biggest help in determining when an animal will grade choice is the weight at which it can be expected to have 0.3–0.5 inch of backfat. The weight at which this occurs will vary, depending on the sex and frame size. Heifers mature at lighter weights than steers and steers mature at lighter weights than bulls. Further, long-bodied, long-legged cattle mature at heavier weights than short-bodied, short-legged cattle. Thus, the earlier maturing cattle will have 0.3 to 0.5 inch of backfat at lighter weights than later maturing cattle. Table 12-8 gives weights at which cattle of various breeds and types can be expected to have 0.3 to 0.5 inch of backfat and a small amount of marbling, giving a yield grade of 2 to 3 and a low choice quality grade. The estimates in Table 12-8 are based on recent research at Cornell University, Ohio State University, Michigan State University, University of Wisconsin, and the U.S. Meat Animal Research Center at Clay Center, Nebraska.

Visual appraisal and these weights used in combination give a good estimate of the point where the carcass is likely to have a yield grade of 2 or 3 and a small amount of marbling. Animals beginning to show signs of fullness in the brisket and in the flanks and patches around the tailhead are likely near this point. If the brisket and flanks become completely full, heavy patches appear around the tailhead, and the underline sags, the animal is likely to have a yield grade of 4 or 5 and more than a small amount of marbling.

Age. Most cattle feeders feel that yearlings must be fed to heavier weights than calves to grade choice. Research results have not provided a clear answer to the weight at which a yearling placed in the feedlot off grass will grade choice (have a small

Table 12-8
Slaughter Weights at Which Cattle of Various Frame Sizes, Sexes, and Breeds Can Be Expected to Have 0.3 to 0.5 Inch of Backfat and a Small Amount of Marbling

Breed and Type	Slaughter weight, lb.	
	Steers	Heifers
Small type Angus or Shorthorn	900	700
Large type Angus or Shorthorn	1100	900
Small type Hereford	950	750
Large type Hereford	1150	950
Brahman, Charolais and Limousin crosses	1150	950
Maine-Anjou and Simmental crosses	1200	1000
Holstein	1300	1100
Chianina crosses	1350	1150

amount of marbling) as compared to a calf of similar body type placed in the feedlot at weaning. It probably depends on their age or length of time they have been on a low level of nutrition before being placed in the feedlot. A few research trials and observation suggest that most yearlings need to be about 50 to 75 lbs heavier than comparable calves to have 0.3 to 0.5 inch of backfat and a small amount of marbling.

Feeding programs and time on feed. Nine experiments at Michigan State University were completed recently in which calves were fed the lowest energy rations (all corn silage) or the highest energy rations normally used in feedlots. The results were as follows:

Quality grade		Yield grade		Carcass fat	
All silage	High grain	All silage	High grain	All silage	High grain
9.3	9.9	2.7	3.2	28.7	33.2

9 = high good; 10 = low choice

Eight different frame sizes were used in these experiments. All data were adjusted to the same carcass weight within a cattle type. Carcass grade was improved 6.5 percent by high grain feeding. Yield grade, however, was 18.5 percent poorer, and carcass fat was increased 15.7 percent by high grain feeding. These data suggest that cattle fed high silage rations continuously need to be fed to 50-75 lbs heavier weights to obtain the same carcass grade. In experiments at other stations, differences in carcass grade due to feeding program have been small. The Michigan data, however, is likely representative of what occurs on the average on corn-corn silage rations, as the feed intakes and rates of gain were similar to those normally obtained on these rations by cattle feeders.

Many cattle feeders use time the cattle have been on their ration as an indication of readiness for slaughter. This is useful when a certain weight and size of cattle are always fed and the environment and ration composition do not vary. However, weight for the frame size and visual appraisal are the best guides under most conditions.

Breed. On the average, most breeds of cattle will tend to have the desirable amount of outside fat and marbling when they reach the appropriate weight for their frame size and sex as shown in

Table 12-8. However, there are some differences between breeds in ability to deposit intramuscular (marbling) fat. Some observations can be made, based on experience and some research. Angus tend to deposit more intramuscular and less backfat than most other breeds at the same stage of growth. Holsteins tend to have less backfat and more internal fat (kidney, heart, and pelvic fat) than British breeds (Angus, Herefords, and Shorthorns) at the same stage of growth. Charolais, Simmental, and Chianina breeds appear to be intermediate between the British breeds and Holsteins in fat distribution.

12-3 PLACING SLAUGHTER CATTLE IN A SHOW

Market steers should possess the functional and economic traits that all phases of our beef industry can produce and market in large uniform numbers. Completeness and normality should be two terms emphasized. The steers should possess the best combination of gainability, feed conversion, structural soundness, cutability, and quality grade. Do not accept extremes in either direction for those respective carcass traits and select normal appearing steers that are not camouflaged by artificial feeding or management. More specifically, top steers should be mentally indexed with the best combination of the following characteristics:

1. Live weight: 1,000-1,200 lbs
2. Carcass weight: 600-750 lbs
3. Fat thickness: 0.3-0.5 inches
4. Rib eye area: 12.0-15.0 inches
5. Yield grade: 2.0-2.9
6. Marbling: small +, or above preferably
7. Quality grade: choice or above
8. Structurally sound
9. Adequate body capacity, width of chest, and good middle for extra feed intake
10. Frame size of 4-5 on a 1-7 rating to provide an adequate frame for fast gain and optimum slaughter weight
11. A total body shape that would visually indicate that the steer calved with ease

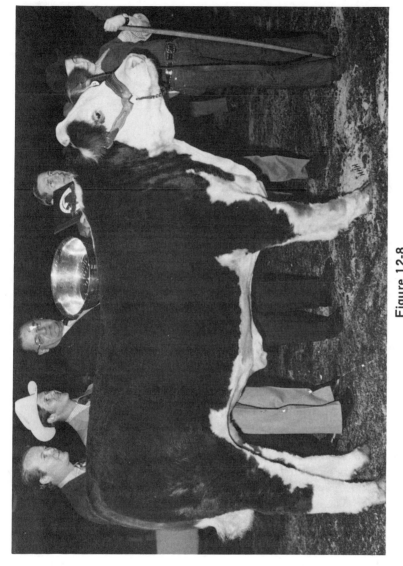

Figure 12-8
International Livestock Exposition Champion
Hereford-Simmental Steer

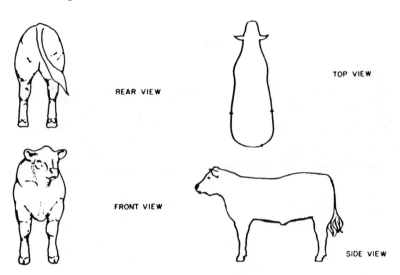

Figure 12-9
Dimensional Views of a Well Muscled Steer

Top market steers should look as good on pasture, in the feedlot, packing house, and supermarket as they do if selected as champions at our major livestock shows (Figure 12-8).

Figure 12-9 displays the four dimensional views of an extremely heavy muscled trim steer. The best reference points for muscling in order of priority to use visually are: **(1)** the wide stance in front and behind; **(2)** the extreme width through the point of shoulder and center of the quarter as viewed from in front and rear, respectively; **(3)** the prominent bulging forearm, stifle, and jump muscle viewed from all four angles; **(4)** the butterfly-shaped rump and loin muscle that protrude from the vertebrae column; and **(5)** the creases outlining the prominent muscles throughout the steer's body. These same diagrams also display the reference points of an extremely trim steer. The reference points listed in order of priority for determining fat are: **(1)** a clean trim brisket and twist; **(2)** an extremely trim and angular shape and definition to the fore and rear flank areas; and **(3)** the freedom from fat at the point of the shoulder and down the spinal column. These four dimensional views of a steer easily give him the physical appearance to yield grade 1.0 or less on the cutability scale. However, a yield grade steer with a 2.5-3.0 is a more realistic goal in selection to avoid the extremes that lead to double muscling and other negative production responses realized in recent years.

Figure 12-10 represents the USDA pictures and descriptions of the 5 yield grades.

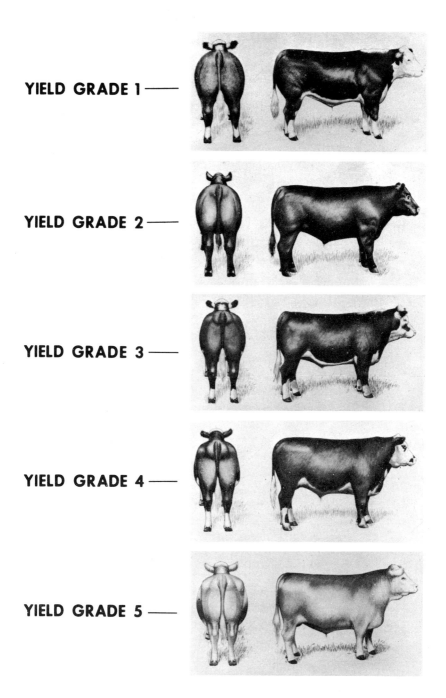

Figure 12-10
Visual Appearance of Cattle with Different Degrees of Condition

13

Facilities and Feed Storage for Beef Cattle

Beef cattle are able to withstand environmental stress better than dairy cattle, swine, or poultry. The reasons for this include heavy hides, tendency to deposit fat on the outside of the carcass, and heat produced during fermentation. There are excellent references on specific recommendations for space requirements, building and lot layout, feed storage and handling, feeding equipment, and cattle handling equipment. Large feedlots use consultants to design and construct their facilities. Farmer-feeders and cow-calf producers usually build their own. Professional help is available to them through their local county extension service. When the local extension agent is contacted, he will schedule a farm visit to discuss the needs of the specific operation. If needed, he will schedule another visit, bringing in the extension agricultural engineer from the state university. This service is free. The extension personnel can be very valuable, as they have the opportunity to visit many farms, and can utilize experiences in working with a wide variety of situations to solve specific individual problems.

This chapter will briefly discuss the general approach to facilities and equipment for beef cattle. Suggested references for specific recommendations include *Midwest Plans Service Beef Housing and Equipment Handbook*, *Great Plains Cattle Feeders Manual*, and your state university agricultural engineering extension service fact sheets. These references are all available through the local county extension service office, usually located in the county seat town.

Tables 13-1 and 13-2 summarize space requirements for cattle; a brief discussion will be given that outlines facilities needed for beef cattle.

13-1 BEEF COW CALF

Shelter

Most beef herds in the United States exist without housing. Actually, fewer problems are encountered when they do not have access to housing. Respiratory and scours problems are increased, and manure must be handled when beef herds have access to a

Table 13-1
Space Requirements for Beef Cattle

Type of facility	Square ft/head
Conventional	
Lot surfaced with free access to shelter	20 in barn, 30 in lot
Lot surfaced, no shelter	50
Dirt lot; surfaced 10-12 ft behind feedbank and around waterers	
Sloping well drained, with mounds (Mounds are 6-8 ft high, 6 ft wide on top)	150 lot, 25 mound
Nearly level, with mounds	400 lot, 25 mound
Nearly level, no mounds	1000 +
Confinement	
Solid floor (bedded or cleaned approximately weekly)	30
Totally or partly slotted	17-20
Calving pen	100; 1 pen/12 cows
Sick pens	50 (enough to hold 2-5% of herd)
Feeders	**Inches feeder/head**
All eat at once	
Calves to 600 lb	18 - 22
600 lb to market	22 - 26
Mature cows	26 - 30
Pre-weaning calves	14 - 18
Feed always available	
Hay or silage ration	6 in. or more (limited by bunk capacity)
High grain ration	3 - 4 in.
Calf creep	1 space/5 calves
Waterers	
1 space/40 head. Capacity:	Feeder cattle, 15 gal/hd/day
	Cows, 20 gal/day

Table 13-2
Dimensions for Beef Cattle Corrals and Working Facilities

1. Holding area	20 sq. ft/head
2. Working chute	
Length	20 ft. or more
Width	26 in. or 16 in. at bottom and 26 in. at top
3. Corral and chute fence	
Minimum height	50 - 60 in.
Depth of post in ground	30 in.
4. Loading chute	
Width	26 - 30 in.
Length	12 ft.
Rise	3 1/2 in./ft.
Ramp height	stock trailer - 15 in.
	pickup truck - 28 in.
	straight truck - 40 in.
	semi-truck - 48 in.

barn. They should have access to an area protected from the wind, however. The information given in Chapter 8 on the effects of cold stress show that wind speed greatly affects energy requirements for maintenance.

Access to woods, valleys, or constructed windbreaks will usually provide adequate protection from wind. Many consider the best wintering area to be a woodlot near a stream that flows year-round that is accessible by truck for winter feeding. With the advent of large round bale hay systems, many store large bales in a fenced-off area near the wintering area.

The greatest need for protection is during winter or early spring calving. Many calve outside even in the winter, as long as the cows can be observed closely and the calves can be dried off immediately after birth. Most that calve in the winter or early spring in northern climates, however, have a simple housed calving area. Usually existing buildings are utilized. This allows easier observation, protection for the calves, and availability to catch pens where individuals can be confined if nursing or health problems are encountered.

Most also utilize existing buildings for hay storage. However, the use of large round bales, which can be stored outside, is increasing due to reduction in labor and elimination of the need for a building to store hay for winter feeding.

Due to increases in building costs, the movement is away from utilizing buildings for beef herds. Returns from beef cows usually are not high enough to justify much investment in housing. Further, the chances of scours and pneumonia problems are increased. Many that have encountered scours problems have a well-drained, protected sodded field close to the house that is saved just for calving.

Fencing

Boundary fences are typically four or five strand barbed wire or woven wire fence. The use of woven wire is decreasing except for drive alleys and lots due to its high cost. The use of electric fences is increasing due to the development of dependable electric chargers, ease of installing and moving electric fence, and its relative low cost. One wire about two feet from the ground is usually adequate. Many will use two wires when small calves are in fields near roads or other areas where calves must be kept in.

Electric fences are highly effective, and can actually be better for restraining animals than barbed wire. However, only the best chargers should be used, and any possibility of a "short" in the line should be eliminated.

Weeds and grass can be a problem with electric fences, as they reduce the charge when they grow up around the fence, especially after a rain. Some spray a grass and weed killer or soil sterilant under the wire to prevent this problem.

Equipment

Other than farming tools and hay and silage storage, the main equipment needed for beef herds includes feed bunks, waterers, and corrals for treating sick animals, restraining cows so a new, timid calf can nurse, or for routine procedures such as tattooing, AI breeding, implanting calves, grub and lice treatment, pregnancy checking, vaccination, dehorning and castrating, and worming. Most corrals for beef herds are simple and low cost, due to the small number of cattle present in most beef herds. The larger the herd and the more the cattle are handled, the more important the corral and squeeze chute becomes.

Corrals. Designs and dimensions can be obtained from the references given previously. The following are general guidelines in developing handling facilities.

1. Locate them in an area easily accessible from all of the pastures and wintering area, so they can be easily utilized at all times of the year.
2. Have a catch pen where the cattle can be driven in by one or two persons or will go in on their own.
3. Have a narrow alley at least 20 feet long along one fence within the catch pen, into which several cattle at a time can be driven and confined.
4. Have a good headgate. Check with producers or your extension agent in your area to see which seem to be working the best.

Water. Water must be available at all times. Year around streams are one of the most commonly utilized sources of water. However, in some areas they cannot be utilized due to environmental control regulations.

Springs and ponds are widely utilized. They should be fenced out, and freezeproof tanks located below them. They can be kept open by heaters or by allowing the water to flow through them continuously. Many use automatic waterers, when cattle have continuous access to a barn or lot near the farmstead. Tanks are better if the cattle all come to the barn at the same time for water, however.

Hay storage. Must be adequate to maintain quality. Square bales must be protected from the rain. One ton of baled hay will occupy 300 to 400 cubic feet. Therefore, if 2 tons of hay are needed/beef cow, 600 to 800 cubic feet of hay storage are needed/beef cow maintained. Existing structures should be utilized when possible to reduce costs. Many use a simple roof for hay storage.

Small round bales are usually left in the field for winter grazing, using electric fence to divide these pastures into strips. Nutrient losses due to weather are 14-20 percent. Feeding losses will be minimal if strip grazed. However, unrestricted access will result in losses of 20-50 percent.

Large round bales should be stored at a well-drained, easily accessible site. The sides should be left facing east and west for rapid drying after a snow or rain, and the bales should be about 18 inches apart for air movement. Nutrient losses are 10 percent or less during storage. Feeding losses can be held to 5-10 percent by using feeder panels, rings, or electric fence to limit access.

In the Northern Plains, much of the hay is stored in large loose stacks. Storage nutrient losses are 10-20 percent. Stacks are placed with the front towards prevailing winds. They should be a minimum of 2 feet apart, and stored in a well-drained area. A grapple fork on a front-end loader is used to remove the hay from the stack and place the hay into a feeder.

13-2 FACILITIES FOR GROWING AND FINISHING CATTLE

More attention must be given to facilities for growing and finishing cattle, due to their close confinement and the effect of environment on cost of gain. Table 13-3 shows the impact of feedlot conditions on cost of gain at high and low corn prices, for a 450-lb steer calf fed to low choice grade on a two-phase, corn-corn silage feeding program.

It is clear that the higher the feed requirement and the more costly the feed, the more valuable a no stress environment becomes. The degree of feedlot improvement that can profitably be made depends primarily on the current environment, the price of feed and nonfeed variable costs, and the annual use cost of facilities, which is determined by the interest rate, taxes, repairs, and the potential life of the facility. Most good outside feedlots are environment "5" with short periods of environments "1" or "3".

Table 13-3
Effect of Environment on Cost of Gain*

	Code 1 Outside lot, with frequent deep mud and cold rain	Code 5 Outside lot, well mounded, bedded during adverse weather	Code 7 No mud, wind chill or other stress. With shade and good ventilation
Daily gain, lb	1.93	2.21	2.34
Feed/gain, lb	8.45	7.37	6.94
Increase in cost per cwt. of gain, $/cwt. gain			
Growth stimulant fed	7.69	2.20	--
No growth stimulant	13.16	7.01	--
Corn @ $2.20/bu.	7.69	2.20	--
Corn @ $3.30/bu.	10.57	3.03	--

*Environment codes 1-7 and their effect on energy requirements are described in Table 8-13.

The maximum amount that can be invested to improve facilities can be estimated by dividing the expected reduction in feed and overhead costs (with appropriate adjustments for change in manure value and improvement in sale price of cattle due to improved dressing percentage) divided by the percentage annual use cost. For example, most producers wanting to improve facilities would have an environment 4 or 5 lot.

Assume calves are typically fed from 450 to 1050 lbs, and the lot is kept full year around; at 600-lb gain/head and a 2.34 lbs daily gain, payweight to payweight, 256 days are required per head. Assume 280 days are required, as most are not able to keep their lots completely full. The producer can feed 365/280, or 1.3/head/feedlot space/year. As shown in Table 13-3, the expected savings in feed and nonfeed variable costs are $2.20/cwt gain, or $17.16/head capacity/year (6 cwt × $2.20 × 1.3). If the annual use cost is 17.4 percent (15-year life, 12 percent opportunity cost on capital, 3 percent property taxes and maintenance), he could pay up to $17.16/.174, or $98.62/head of capacity to improve his lot. If his lot was an environment code 1, he could afford to pay up to (6 × $7.69 × 1.3)/.174, or $344.72 per head capacity to improve his lot. (Most lots are not this bad on a year around basis, however.) Additional adjustments would be needed for any improvement in selling price due to improved cleanliness, and thus improved dressing percentage of the cattle, increases in manure value, and capital risk.

An excellent summary of various types of housing costs and capital requirements for various types of facilities has been prepared by Dr. David Petritz, Department of Agricultural Economics, Purdue University. Changes in performance, labor, manure value, and dressing percentage for various types of housing have been determined by Ralph Smith, Superintendent of the University of Minnesota Experiment Station, Morris, Minnesota. Those interested in more detail should contact these sources for summaries of their data. In this section a brief description of the types of housing commonly used for growing and finishing cattle will be given. Space requirements for the various types were given in Table 13-1.

Dirt lot, no shelter. Most feedlots in the western Corn Belt, Great Plains and far west are of this type. Due to the low rainfall in these areas, the environment is equivalent to a type 5 or better most of the year. In the northern parts of these regions the lots usually contain windbreaks. In the areas with greater moisture, mounds are usually provided. Pens usually contain 100 to 200 head or more.

In the central and eastern Corn Belt and other regions of the United States with 25 to 30 inches or more annual rainfall, most feedlots contain shelter. There are outside lots in these areas, however, where the site has a gravel base and/or is well drained. These lots will not likely exist in the future in the eastern half of the United States, due to regulations on runoff and pollution, and the need to control odor in more populated regions.

Conventional. In this type, cattle have access to both an outside lot and a barn or shed. Most are cemented both inside and out. The inside area is usually bedded, and cleaned 2-4 times or more/year. The outside lot is scraped as needed, often weekly. Most small feedlots (less than 100 head) are of this type. The disadvantage of this type is the need for bedding and the difficulty of keeping bedding dry due to cattle bringing in moisture from the outside lot. Also, cost may be as high as for some confinement systems, due to the larger area needed/head. These lots will likely decrease in numbers, due to these effects and the need to control lot runoff.

Solid floor confinement. Cattle are totally confined to a building, with 100 to 200 head/pen. There are several types in this category.

1. Manure scrape/manure pack: A cement slab 10-15 feet behind the feed bunk is scraped approximately every 7-10 days. The rest of the barn has a dirt floor, and is bedded. A manure pack is allowed to accumulate and is then removed usually two times/year (before spring planting and after fall harvest). This system works well, except bedding is needed.
2. Solid floor scrape: The total floor is cemented. The cattle are removed every 7-10 days, and the floor is cleaned. Little or no bedding is used. According to the summary by Petritz, this is the lowest cost confinement housing that can be built.
3. Sloping floor with gutter cleaner: This system is relatively new. The cattle work the manure down a sloping floor to a gutter cleaner, which carries the manure to a storage area. This system is currently being tested at the University of Nebraska. The manure may be refed to beef cows after fermentation in the storage area.

Slotted floor confinement. Cattle are confined on a slotted floor in pens of 100 to 200 head. The manure is worked through the slots by the hooves of the cattle. Provision for handling the manure is usually one of the following.

1. Deep pit: The manure accumulates in a pit, usually 8-10 feet deep. At 17-20 square feet/head, the pit will store the manure produced in approximately 6 months. This allows cleaning before spring planting and after fall harvest. If high silage rations are fed, they may require cleaning 3 times/year. Table 13-4 summarizes expected manure accumulation and value on different feeding systems. This type of facility is popular with farmer-feeders, due to the low labor and maintenance requirements, preservation of the manure value, and ability to synchronize cleaning with farming operations.
2. Slotted floor scrape: A scraper is pulled with a cable through a channel under the slotted floor several times daily. The manure is either accumulated in a storage area or is hauled daily. This system is lower cost to build than the deep pit. However, maintenance costs are higher, and freezing can be a problem in the Corn Belt and Northern Plains. It is considered by many to be a high risk system; if the cable breaks or the scraper freezes, there is no provision for handling the manure.
3. Flush flume: The floor consists of tubelike sections with slots in the top. The manure is flushed from these flumes several times daily into a lagoon. The advantage of this system is a lower construction cost of the facility and elimination of manure storage under the cattle. However, when the cost of building the lagoon and irrigation equipment needed to place the manure on the fields is included, this system is usually as expensive as a deep pit system. These systems are typically being built next to large corn fields in less populated areas so irrigation equipment can be used to remove the accumulated manure from the lagoons.

13-3 MAKING AND STORING HIGH QUALITY SILAGE AND HIGH MOISTURE GRAIN

Storage needs for feed must be part of the planning for any efficient beef management system. Details on bin sizes and design of feed storage systems can be obtained from the *Midwest Plans Service Beef Housing and Equipment Handbook*. This section will deal with storing high-moisture grain and whole plant silage, due to the increase in their usage and the need to understand procedures needed to properly preserve them.

In order to successfully preserve any high-moisture feed, the forage and grain must ferment in an environment that is as free of

Table 13-4

Manure Production and Net Cost of Feeding, Manure Handling, and Storage on Various Feeding Systems[a]

	All silage			Silage plus added grain				High concentrate	
Measure	\-\-\-\-\-\-\-\-\-\-\-\-\-\- Percent grain in ration dry matter \-\-\-\-\-\-\-\-\-\-\-\-\-\-								
	30	40	50	60	70	80	90	100
Urine (1/day)	6.4	6.4	6.4	6.4	6.4	6.4	6.4	6.4
Feces (lb/day)	30.3	27.4	24.4	21.4	18.4	15.5	12.5	9.5
Total volume (gal/day)	5.34	5.02	4.65	4.28	3.91	3.54	3.17	2.80
Nutrients available (lb/day)[b]								
Nitrogen (N)	.13	.13	.13	.13	.13	.13	.11	.11
Phosphorus (P_2O_5)	.10	.10	.09	.08	.08	.07	.07	.07
Potassium (K_2O)	.12	.10	.10	.09	.08	.08	.07	.06
Fertilizer value (¢/day)[c]	4.06	3.90	3.74	3.50	3.42	3.18	2.95	2.87
Feeding, manure storage and handling cost (¢/day)	22.15	21.62	21.01	19.04	18.36	17.68	16.89	16.25
Manure credit (¢/day)	4.06	3.90	3.74	3.50	3.42	3.18	2.95	2.87
Net cost/day (¢)	18.09	17.72	17.27	15.54	14.94	14.50	13.95	13.38

[a] Woody, H.D., J.R. Black and D.G. Fox. 1978. Michigan State University. Cost of feeding, manure handling and storage and manure credit for all silage versus all grain ration.

[b] % of nutrients available to corn plant: N 55%; P_2O_5 75%; K_2O 90%.

[c] Cost/lb: N 11.5¢; P_2O_5 16¢; K_2O 8¢.

air as possible. Preservation is achieved by the formation of organic acids during fermentation by anaerobic bacteria as soon as plant cell respiration ceases. The pH of the ensiled feed drops to 4.0 to 4.5. This prevents the growth of undesirable organisms whose end products cause reduced palatability, nutrient loss, or toxicity. The most notorious of these are the aflatoxins. The method used to exclude air includes some combination of grain or forage moisture content and processing before storage, rapid filling and packing to exclude air internally, and sealing the silo to prevent spoilage at exposed surfaces. Various minimum rates of feeding are necessary to prevent spoilage at the exposed surface, requiring silo sizes to be compatible with the amount to be removed daily, based on the number of cattle fed and ration used.

Complete discussions on harvesting, storing, and feeding whole plant silage and high-moisture grain can be obtained from Extension Bulletin E1139, Michigan State University or Bulletin PM 417, Iowa State University. This discussion will outline key management factors in making high quality silage.

Silage Additives

Several types of products are being added to silage to either increase preservation or to add supplemental nutrients.

Fermentable carbohydrates (molasses, ground grain, whey). These may be beneficial under certain conditions with hay crop silage. However, corn silage normally contains a large quantity of sugars and starches, which allow rapid fermentation and production of the acids that preserve silage.

Biological additives. Research results with adding bacterial cultures to silage have been highly variable. Many factors likely influence the response to these types of products, including bacteria numbers and types present on the crop at harvest, type and variability of cultures used, and available fermentable carbohydrate in the forage and moisture content of the silage. Under some, as yet undefined, conditions, the addition of bacterial cultures may increase lactic acid content, which in turn may help preserve the silage. Nebraska studies with adding two of the most commonly used cultures, *Aspergillus oxyzae* and *Bacillus subtilis* to corn silage resulted in improved preservation in only one of three trials. In a feeding trial, however, the treated silage gave a 9.3 percent better feed efficiency.

Improvements in cattle performance by using these products have also been obtained by Georgia and Florida Experiment Stations.

Intensive thorough experiments with these types of products are now underway at two midwestern experiment stations, which should result in definitive recommendations on the use of these products in the near future.

Urea, ammonia, and limestone addition. Research at several experiment stations indicates that the addition of these products has the common effect of buffering acids produced, resulting in a longer production of lactobacilli and a 40 to over 100 percent increase in lactic acid production. Less energy is lost during fermentation when lactic rather than when acetic, propionic, and butyric acids are produced. Calculations by Oklahoma researchers indicate that fermentation in the silo can conserve more energy than a corresponding fermentation in the rumen. About 10 percent of the digestible energy is lost as methane during fermentation in the rumen. Less rumen methane is produced when the silage contains high levels of lactic acid. This means that theoretically prefermenting the carbohydrates to lactic acid in the silo can lower energy losses in the rumen, leaving more net energy for the animal. Thus, under ideal ensiling conditions, overall utilization of the corn plant energy can be improved if the silage is made at the proper moisture content (30–35 percent DM) and a buffer is added.

The addition of ammonia to corn silage has also been shown to reduce mold growth; it is thought that proteolysis of the original plant protein may be reduced by ammonia addition.

However, these benefits are often more than offset by high requirements for protein of lightweight beef cattle or high-producing dairy cattle, whose needs cannot be met by entirely substituting NPN for a natural protein supplement. The whole corn plant is low in amount of protein. Further, protein quality is reduced during fermentation, being reduced to 60 to 70 percent of the value of that in the fresh corn plant. Thus, proper supplementation is critical to ensure maximum utilization of the silage energy. NPN addition can greatly reduce the costs of providing supplemental protein, but the degree of the amount of substitution of NPN treatment for supplemental natural protein that can be done profitably will depend on the stage of growth or lactation and cost of natural supplemental protein. The proper use of NPN treatment of silage in various feeding programs is discussed in Chapter 10.

When to Harvest Corn or Sorghum Silage

The following are some important factors in determining the best time and practices to use in harvesting corn silage:

- Maximum dry matter (DM) per acre
- Maximum digestibility of nutrients in the silage
- Maximum DM stored per cubic foot of silo capacity
- Minimum of seepage loss from the silo

To best meet these goals, harvesting should start when the kernels are in the early dent to late dent stage of maturity, and be completed as soon as possible. At this time, the dry matter of the corn plant is from 30 to 40 percent. Dry matter in the kernels will vary from 50 to 65 percent. Calendarizing hybrids will help to meet these objectives.

If corn silage is harvested when the dry matter is 30 percent or less, extensive seepage will occur, especially with silos 60 ft or taller. This results in a loss of nutrients (seepage is about 8 percent DM) and severe erosion of the walls and hoops of tower silos. There is also danger of the silo collapsing due to extreme pressures generated from the added weight of the material. To eliminate excessive seepage in tower silos the DM content of the silage must be above 35 percent.

In bunker silos seepage will not normally occur nor will extreme pressures be a problem with 30 percent DM silage. With bunkers, harvest should begin earlier than the maximum yield stage because extra moisture in the silage is necessary to ensure exclusion of air, good packing, and proper fermentation.

To determine a more exact amount of dry matter or moisture in the silage before starting to ensile, a relatively accurate moisture tester to weigh, dry, and reweigh the dried material is available. It requires about 30 minutes to run. The forage harvester must be run in the corn field to obtain a good sample of silage for the moisture test.

Another method of telling when corn is ready for harvest is to shell kernels from several ears and take them to the elevator for a moisture test.

Effect of maturity on dry matter yield per acre. Research data from Michigan State University and other experiment stations relating stage of maturity of corn silage to dry matter yield per acre are summarized in Figure 13-1.

These data show that yield per acre increases until the plant

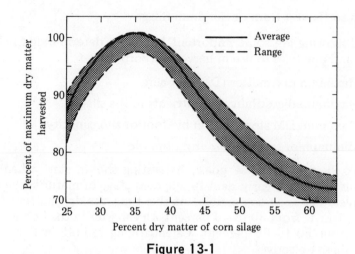

Figure 13-1
Effect of Stage of Maturity of Corn Silage on Dry Matter Harvested per Acre
(Summary of Research Conducted at Michigan, Indiana and USDA)

reaches approximately 35 percent DM or until the first killing frost. It will then level off for 5 to 10 days (depending upon the extent of frost, wind, and rain) and then begin declining at a rapid rate. This is due to the loss of leaves and tassels from standing stalks and the loss of the entire stalk from lodged plants.

Figure 13-2 shows DM accumulation in the corn plant for typical 120-day corn, assuming no frost during the 120-day growing season and no loss of leaves, tassels, and whole plants due to lodging.

Both Figures 13-1 and 13-2 show that there is an advantage in delaying corn silage harvest after it has reached 35 percent dry matter or after all kernels are in the hard dent stage of maturity.

Factors Affecting Nutrient Content

Drought damaged corn. If the corn has been stressed all summer, has no ears and is short, the energy value will be about 70 percent of normal corn silage. This is an estimate based on experience, since no controlled research has been conducted under these conditions. Nebraska studies suggest corn which has not been as severely stressed (perhaps 10 to 20 bu grain/acre) will have 80 percent of the energy of normal corn silage.

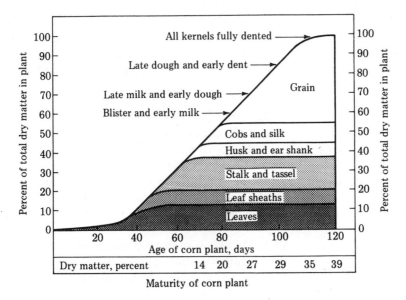

Figure 13-2
Effect of Stage of Maturity of Corn Silage
on Total Dry Matter Accumulation

Michigan studies suggest corn which was stressed by heat and drought during pollination (has normal stalk development, but relatively barren stalks and ears; approximate 3½ bu grain/ton silage or 40 to 60 bu/acre) has about 90 to 95 percent the energy value of normal corn silage.

Droughty silage typically has a higher protein content than normal silage. However, most of this protein is found in the plant rather than in the grain under these conditions, making it more easily degradable in the rumen. As a result, NPN supplementation does not appear to work as well as in normal silage, according to Nebraska studies. Thus, it becomes especially important to supplement drought corn with a natural protein source for calves up to 600 to 700 lbs.

The DM content of droughty silage must be in the normal range (30 to 35 percent) to make good silage. If the corn did not set ears and is green, or if the ears are all brown and the stalk is green, the moisture content will be too high; but hot, dry weather can cause rapid moisture drop, so careful observation of changes in moisture content to determine when to harvest is essential.

Although nitrate levels in drought-stricken corn may be high, ensiling will reduce more than half the nitrates to ammonia, which can be utilized. For this reason, nitrate toxicity rarely occurs with feeding ensiled drought-damaged corn. However, if the drought damage is extreme, and high levels of nitrogen were applied to the soil, a nitrate test on the silage should be conducted.

High sugar corn. This type of corn concentrates sugars in the stalk rather than starch in the ear and will contain about the same amount of energy as average corn silage, even though it does not have ears. Thus, its use should be based on how it compares with other varieties in dry matter yield. Keep in mind that it has little alternative use other than silage and tends to remain high in moisture and water-soluble dry matter.

Brown midrib corn. This type of corn silage will have a somewhat lower grain content, but the stalk is more digestible due to a lower lignin content. It will have a somewhat higher net energy value than average corn silage. However, stalk breakage can be high if a high wind or an early snow occurs before harvest. In Michigan studies, it was somewhat more difficult to harvest as grain, due to stalk breakage, and silage yield per acre was lower than normal corn.

Waxy corn. The protein in waxy corn is less degradable in the rumen, which should increase corn protein bypass to the small intestine. Also, the starch is in a form that is more easily degraded in the rumen. An average of experiments with waxy corn showed improvements of approximately 2 percent in average daily gains and feed efficiency from feeding waxy corn grain. A Michigan study did not show any advantage for silage made from waxy corn compared with normal corn. It would appear that yields per acre of waxy corn need to be near that of normal corn to be considered.

Opaque-2 (high lysine) corn. This type of corn contains higher levels of the amino acid lysine, which is low in normal corn. It has been shown to improve performance in swine; they must obtain all needed amino acids from the daily ration. However, results with cattle have been mixed. The likely reason is that cattle apparently can synthesize enough of this amino acid in the rumen; 1/2 to 3/4 of the corn protein is degraded in the rumen and then resynthesized into bacterial protein.

Fineness of chop and packing. Fine chopping (1/4 to 3/8-inch cut) will increase packing. Tight packing will increase storage capacity and will reduce oxidation losses due to entrapped air or air entering silage that is not well packed.

Fine chopping silage will reduce sorting by cattle, and mixing of the total ration is facilitated. Some Michigan studies indicate that fine chopping silage increases daily gain and feed efficiency. However, extremely fine silage may not be beneficial in high-grain beef cattle rations.

Small grain whole plant silage. Nutrients harvested/acre can usually be at least doubled by harvesting small grain crops (barley, oats, rye, and wheat) as whole plant silage rather than grain. It is usually best to harvest oats at the early to medium dough stage for the best silage. It can be direct harvested at this stage, as the moisture will be down to about 70 percent; it should be 60 to 70 percent moisture for storage in a bunker or cement upright silo to allow packing for elimination of air and to avoid improper fermentation. Careful packing and sealing is important, particularly as oats have a hollow stem which makes packing difficult. Fine chopping (1/4 to 3/8 inch) is important for good packing.

Oats silage can also be made at the late boot or early heading stage. At this point the protein content will be higher, but the total protein yield will be reduced because of greatly reduced yields/acre (Table 13-5). If cut at this stage, the oats should be cut and allowed to dry down to 65-70 percent moisture before harvesting for silage.

Wheat, barley, or rye can also be made for silage. Yields will be similar to those in Table 13-5. Protein content will be lower in most other small grain silages, however (7-9 percent CP). Energy value of other small grain silages will be similar to oats silage (50-60 percent TDN in the dry matter). Wheat and barley should be harvested by the milk stage and rye by the late boot stage, just before the heads show. If harvested later than this, they will be fluffy and hard to keep, and palatability is reduced. Proper moisture content, fine chopping, and packing recommendations are the same as those given for oat silage.

Hay crop silage. Almost any hay or pasture crop can be harvested as haylage or silage, and has the advantage of getting it harvested more rapidly to avoid weather damage and leaf loss. Also, it is more adaptable to mechanized harvesting, storing, and

Table 13-5
Effect of Stage of Maturity on Oats Silage Yield per Acre

Stage	Moisture when cut, %	Protein in dry matter, %	Relative yields/acre at 68% moisture, tons/acre
Boot	85	22	3
Heading	82	18	4.5
Milk	78	15	6.5
Early dough	71	13	9.0
Late dough	60	11	10.0

feeding. In addition, more dry matter is harvested/acre as field losses are 10 to 20 percent less than with hay.

Regardless of how the hay crop is harvested, the time of harvesting should be based on the stage of maturity. The hay crop should be harvested at the late bud to early bloom or head stage to obtain maximum yields of digestible protein and energy. The hay crop should be wilted down to 60 to 70 percent moisture before harvesting as silage to avoid seepage and improper fermentation. It should be chopped fine (1/4 to 3/8 inch) to make packing and feeding easier. Water should be added to increase the moisture content if it becomes too dry (about 4 gallons/ton are required for each 1 percent increase in moisture content). The hay crop can be made as low-moisture silage (haylage, 45-55 percent moisture) if it is to be stored in a gas-tight or tight cement upright silo. Unless tightly packed, however, excessive heating will occur, causing extensive heat damage to the protein and loss of energy. The silage will appear brown, dark brown, or black, depending on the degree of damage. In Michigan and Minnesota studies, 1/3 of the samples taken from farms were heat damaged and averaged 79 percent of the expected protein value, regardless of silo type. Heat damaged protein can be determined by having an acid detergent fiber-nitrogen (ADF-N) test.

High moisture grain. Excessive fermentation occurs at a high grain moisture content, resulting in increased energy losses during storage and reduced dry matter intake. If the moisture content is too low, exclusion of air becomes more difficult and mold growth and excessive heating can occur, resulting in energy losses and heat-damaged protein. An acceptable range in grain moisture content is 25 to 30 percent, with an upper limit of 33 percent. This is a logical operating range, as harvesting and handling of the grain become easier below 30 percent moisture. However, field losses will increase as the season progresses, due to increased harvest losses as the grain becomes drier and risk of wind, snow, and stalk rot causing ear droppage and down stalks. To avoid these problems, harvest should begin when the grain is about 30 percent moisture and should be completed when the grain is about 25 percent moisture.

When these recommendations for grain moisture content are followed, ear corn will be 28 to 32 percent moisture content, with 25 to 40 percent moisture of the ensiled ground ear corn being acceptable limits.

Better utilization of sorghum grain is obtained when it is stored whole at 28 to 32 percent moisture, then rolled before feeding.

It is possible to obtain acceptable preservation at moisture contents as low as 21 percent in oxygen-limiting structures. However, caramelization due to excessive heating has been reported at a moisture content of 24 percent.

The effect of processing on utilization of high-moisture grain was discussed in Chapter 9. However, various degrees of processing are used to allow compaction as a means of excluding air, unless an oxygen-limiting structure is used. Most grinding into storage is done at the silo, using a hammermill, burr mill, or a blower with a recutter attachment. A rollermill is commonly used to process high-moisture grain before feeding.

Field grinding of ear corn is done with a mill behind gathering rolls or the cylinder of a combine. Special heads are available for forage harvestors that snap the ears; then they are chopped with the knives in the harvester, and then blown into a wagon or truck.

Propionic Acid Treatment of High Moisture Corn

Treating high-moisture corn with propionic acid is an alternative to drying corn to prevent spoilage, providing the corn is to be fed to livestock. This organic acid will preserve high-moisture corn stored in conventional storage when applied correctly and at the proper rate. The acid will kill mold spores and prevents the production of mycotoxins. This allows high-moisture corn to be stored in bins, or concrete floors, in corn cribs, and on the ground when the corn is protected from moisture by plastic sheeting. The feeding value of the acid-treated corn appears to be similar to untreated high-moisture corn on a dry basis. Also, treated high-moisture corn can be fed in a self-feeder without spoilage and small amounts can be fed from a silo without danger of spoilage. The biggest disadvantage is the cost of treatment. The cost per bushel depends on the moisture content of the grain and the length of storage.

For corn that is 25 to 30 percent moisture the cost is about 15 to 18¢ or more per wet bushel for materials. In addition, the applicator and bin liners must be purchased. The acid will cause corrosion in a silo or steel bin unless a liner is used.

Presently, the place for acid treatment of corn appears to be where the amount of high-moisture corn harvested or purchased exceeds the present storage capacity, or if the feeder doesn't want to invest in silos or bunkers for storage of high-moisture corn. Also, existing unused structures such as corn cribs or steel bins can be used with a small investment in liners.

To ensure proper preservation the following practices should be carefully followed in treating high moisture corn with propionic acid.

1. Accurately measure the moisture content. Do this in each load if possible. Then calibrate the applicator to apply the correct amount for the moisture content of the corn and the desired storage period. Failure to apply enough acid or in getting it evenly distributed will usually result in spoilage.
2. Coat metal and concrete bins or silos with a sealer or cover with a plastic liner.
3. Apply in a well-ventilated area; avoid breathing the vapors; and avoid contact with skin. Wear rubber gloves and aprons.

Corn plant residue. This is the material left in the field after the grain has been harvested. Thus, it contains the leaves, stalks, cobs, and husks. About 1/3 of the TDN in the corn plant is left in the field after the ear is removed. Some beef producers make silage of the residue, for wintering beef cows. If the corn grain is harvested at 25 to 28 percent moisture, there may not be adequate moisture in the residue for proper ensiling. The stalks will contain about 60 percent moisture, but the leaves and husks will only contain 15 to 20 percent moisture. Chopped refuse will usually not contain more than 45 to 50 percent moisture. Thus, water should be added to ensure satisfactory ensiling and preservation. About 6.2 gallons are required to increase the moisture content one percentage unit. Thus, 62 gallons/ton would be required to raise it from 50 percent moisture to 60 percent moisture.

System Planning

While most farm operators do some planning before making a major purchase, often their purchases are made in a hurry. Time is not taken to do enough planning to ensure a relative uniform silage flow of sufficient volume and with components that get the job done with reasonable labor, energy, and silage loss.

Good planning takes some effort, as it requires digging out reliable information and utilizing a planning technique that relates silage flow to components of the system. Two planning tools that provide an organized planning approach are the *System Planning Guide* and the *flow diagram*.

This information was prepared by R. L. Maddex and R. G. White, Department of Agricultural Engineering, Michigan State University.

A system planning guide is shown in Figure 13-3. An example is worked out based on harvesting 1,500 tons of corn silage and information in Tables 13-6 and 13-7 using an allowable harvest season of 3 weeks (21 days) and moderate weather which would permit harvest 1 day out of 2. Acres required are based on an estimated yield of 12 tons per acre of 35 percent DM silage. Flow rates are determined by dividing the total tons by estimated harvesting days and hours. No allowance is made for Sundays, which could reduce the total possible harvest days by two or three for some farm operators. Three transport wagons are to be used in order to reduce lost field time for the harvesting unit. The size of the tower silo needed to store 1,500 tons of silage is selected from Tables 13-7 and 13-8. (The amounts needed can be determined in Table 10-11.)

A planning guide is a good tool for estimating flow rates. Each farm operator must put in his own figures for total tons, working days and hours, transport vehicles, and feeding rates. The more realistic the figures used in planning, the more accurate will be the results. This planning technique is applicable to any silage volume.

Silo Structures and Selection

High-quality silage can be made in all types of structures. Quality will vary, depending on the condition of the silo, moisture level, stage of maturity when harvested, fineness of chop, and other management factors. Important factors in selection of silos are: (1) size, (2) feeding arrangements, (3) materials other than corn

Harvesting

	Example	Your Farm
Total tons to be harvested (estimated)	1,500	
Acres to be harvested (at 12 tons per hour)	125	
Harvest days per season (1 day out of 2)	11.5	
Hours of harvest per day	6	
Total harvest hours per season	63	

Flow Rate

	Example	Your Farm
Tons per day	137	
Tons per hour	24	

Transport Vehicles

	Example	Your Farm
Tons per vehicles	8	
Number of vehicles	3	

Unloading into Storage

	Example	Your Farm
Minutes to unload transport vehicle	15	
Tons per minute	.5	
Tons per hour	30	

Unloading from Storage

	Example	Your Farm
Total pounds or tons fed (50 lbs x 60 animal)	3,000	
Pounds per feeding	1,500	
Pounds or tons per minute removal from storage	200	
Removal time	7½	

Storage

	Example	Your Farm
Tower silos required — size	24 ft x 60 ft / 24 ft x 70 ft	
OR		
Horizontal silo (50 cu. ft per ton) — size	60 ft x 100 ft	

Figure 13-3
Silage System Planning Guide

Table 13-6

Capacity Ranges for Forage Harvesters*

Harvester size	Tons/silage harvested per hour	Tractor HP
Small	9 to 18	60 to 100
Medium	15 to 28	100 to 150
Large	20 to 40	125 to 200
Self-propelled	30 to 60	

*Based on a well-managed operation. Includes about 35 percent lost time for adjustments, repairs, changing wagons, etc. Under less favorable conditions, and/or with older machines, these figures should be discounted by 10 to 25 percent.

13-3 Making and Storing High Quality Silage

Table 13-7

Capacity of Tower Silos

Size (ft)	Tons		
	DM	40% DM	32% DM
14 x 50	62	148	195
16 x 50	81	190	252
18 x 50	103	245	320
18 x 60	124	296	392
20 x 50	125	300	394
20 x 60	155	370	483
20 x 70	184	456	574
24 x 50	182	435	570
24 x 60	223	530	697
24 x 70	265	628	827
30 x 60	348	673	886
30 x 70	413	825	1,087

Based on refill of silo with final silage level 3 to 5 ft below top of silo walls.

silage that may be stored in silo, and (4) investment required for total storage and feeding systems.

Losses in storage. Tight structures, good distribution and packing, and use of a plastic cover properly weighted down keep losses low. Losses in bunker silos are also influenced by depth and width of material stored. Less surface is exposed for the deeper silos. In a well-managed operation, silage losses are estimated at 5 to 10 percent for concrete tower silos and 10 to 15 percent for bunker silos.

Table 13-8

Number of Animals Fed per 2-Inch Silage Layer for Various Size Tower Silos and Feeding Rates

Silo diameter ft	Approximate lb silage in 2-in. layer lb	Lb of silage per day per animal						
		20	30	40	50	60	70	80
		Number animals to consume 2-in. silage layer at above rates						
16	1,675	84	56	42	34	28	24	21
18	2,120	106	71	53	42	35	30	28
20	2,615	131	87	65	52	43	37	33
22	3,165	158	105	79	63	53	45	39
24	3,770	188	126	94	75	63	54	47
26	4,430	222	144	111	88	74	63	56
28	5,130	256	171	128	103	86	73	64
30	5,890	295	196	147	118	98	84	74
36	8,480	424	283	212	169	141	121	106

*To determine height of silo, multiply 2 inches by feeding days and divide by 12.
Example: To feed 98 animals at a rate of 60 pounds per day would require a 30 foot diameter silo. To feed 98 animals at this rate for 365 days =

$$\frac{2 \text{ inches} \times 365}{12 \text{ inches}} = 61 \text{ feet of silage}$$

Thus, 61 feet of silage plus 5 feet of unused silo from settling requires a 66-foot high silo.

Tower silo. The design of the tower silo is determined by the manufacturer. Silage juice is the worst enemy of tower silos. A maximum silage moisture content of 68 percent is recommended. Continual seepage from silage high in moisture (70 percent or higher) will cause deterioration of all types of tower silos. An adequate foundation for silos, on well-drained soil, becomes increasingly important as silos increase in size and height. In addition, it is necessary to provide drainage for seepage away from the silos to protect the foundations.

The capacity of tower silos is shown in Table 13-7. The tonnage shown for the various size silos can vary by a plus or minus 10 percent for any tower silo. Time of harvest, length of cut, rate of filling and unloading, which influences oxidation losses in the silo, and even varieties will contribute to variation of total tons in both tower and horizontal silos. The total dry matter tons in a silo remain relatively constant for a particular harvest operation even though the actual moisture content of the silage may vary considerably.

Table 13-7 is based on research data and has also checked out closely with some large farm silos where corn silage was weighed in and out.

Bunker horizontal silo. Bunker silos can be completely aboveground, partially in the ground, or completely in a bank. The first consideration, however, is for drainage out of and away from the silo. Other considerations are access to the silo and orientation to prevent snow accumulation in the silo.

The density of the packed silage has a direct bearing on oxidation losses, which are not seen by the farm operator, as well as visible spoilages. The deeper the silage the less the loss in weight and quality. From a practical standpoint, 12 to 16 ft of settled silage is recommended. Putting the silage in at a slightly higher moisture (68 percent to 72 percent) and chopping it short will increase the density. Good packing of the silage is necessary. A wheel tractor will provide more packing pressure than a track-type tractor.

Several factors should be considered in the length and width of a horizontal silo. The best method of filling a bunker silo is to unload the silage on a concrete floor, then push it upon the silage silo with a front blade and tractor. A width of approximately 50 feet is needed to turn and maneuver vehicles in the silo for unloading. Narrow silos require more backing of vehicles or unloading of silage on the front apron, resulting in a longer push to get materials in the back of the silo. With settled silage depth of 12 ft or more, the density of the silage is great enough to prevent any amount of spoilage on the face of the silage pile so that it isn't necessary to remove several inches per day. If silage depths are less than 10 ft, 2 to 4 in. should be removed daily from the face of the silage.

The capacity shown for bunker silos in Table 13-9 is based on 40 lbs per cu ft, which is 50 cu ft of storage space per ton of silage. A close estimate of the capacity for bunker silos can be made by determining the tons per foot of length, then multiplying by a given length. This also provides a method for determining the length of a silo. For example, a silo 60 ft wide with an average depth of 12 ft would have 720 cu ft (60 × 12 ft) per foot of length. The capacity per foot of length would be 50 cu ft/ton = 14.5 tons per ft of length.

If the silo was 100 ft long the total storage capacity would be 14.5 tons/ft × 100 ft = 1,450 tons, minus approximately 10 tons lost by the sloping front of the silage pile.

Table 13-9 also shows the amount of silage per slice of thickness for 1-in. and 12-in. slices. This information can also be used to determine the size of silos or the feeding days for a given length. For example, if 200 cattle were fed 30 lbs of silage per day for 300 days, it would require 6000 lbs or 3 tons. Table 13-9 shows 1 ton per inch for a 50-ft wide silo, 12 ft deep, so it would require approximately 3 in. per day to feed the 200 cattle. The length required to feed 365 days from the 50-ft wide silo would be 3 in. per day × 365 days = 1095 in., or 91 feet.

Filling Silos

Rapid filling is desirable for both tower and bunker silos to minimize oxidation losses and spoilage. Distribution of silage in silos is also desirable for all tower silos and necessary for large diameter

Table 13-9
Capacity of Bunker Silos (12-Ft Deep) and Amount of Silage per Slice

Length (ft)							Width	Amount silage per slice	
								Thickness	
60	80	100	120	140	160	200		1 in.	12 in.
Tons							Feet	Tons	Tons
288	384	480	576	672	768	960	20	.4	4.8
432	576	720	864	1,008	1,152	1,440	30	.6	7.2
576	768	960	1,152	1,344	1,536	1,920	40	.8	9.6
720	960	1,200	1,440	1,680	1,920	2,400	50	1.0	12.0
864	1,152	1,440	1,728	2,016	2,304	2,880	60	1.2	14.4
1,152	1,536	1,944	2,292	2,688	3,072	3,840	80	1.6	19.2

Settled depth

silos. The ideal distribution of silage would maintain the silage along the side walls slightly higher than the center during filling with only a slight crowning at the center to top out the silo. Directing all silage at the center of the silo and letting it roll out can result in fluffy spots along the walls; this can result in the slipping of unloader drive wheels, the tipping of unloaders, and sometimes pockets of spoilage.

The filling of horizontal silos will vary with the construction or location of the silo. Most permanent horizontal silos are constructed above ground. The best method of filling horizontal silos is to dump the silage on the concrete floor, then put it up on the pile to the maximum depth (Figure 13-4), keeping the slope on the pile as steep as possible. This reduces the exposure of the silage to the air and minimizes oxidation losses and spoilage. A 50-ft wide silo permits turning inside the silo and unloading silage close to the silage pile. Narrower silos may require unloading silage on the apron in front of the silo and pushing silage farther with tractor and blade. The least desirable method of filling an *above ground*, horizontal silo is by driving over the silage pile, because this results in layer filling, causing more silage loss and usually requires either light loading of the transport vehicles or a second tractor to pull the load onto the silage. Narrower silos and temporary silos partially below ground or in a bank may be filled by driving off a bank. A 50-ft wide by 80-ft long silo permits better maneuvering of vehicles for unloading and thus easier filling than a 40 × 100-ft silo. If the silo is filled to a settled depth of 12 ft to 16 ft, the increased spoilage on the face of the silage in a wider silo is minimal.

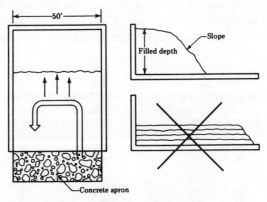

Figure 13-4
Placement of Silage in a Horizontal Silo

Index

A

Accuracy of selection, 25-26
Acute indigestion, 171
Allergic reactions, 172
American dairy breeds, 55-57
— Ayrshire, 56
— Brown Swiss, 56
— Guernsey, 56
— Holstein Friesian, 55
— Jersey, 56
Animal breeding fundamentals, 19-22
Antibiotics, 202
Artificial insemination management, 116-122
— cow fertility, 116-117
— heat detection, 117-119
— insemination, 119-120
— semen quality, 116

B

Backfat measurement, 367
Beef consumption, 147
Beef-cow unit, 3, 6
Birth weight, 70
Blackleg and malignant edema, 126, 166
Bloat, 171
Body capacity, 80
Bovine virus diarrhea (BVD), 126, 166
Brahman and Brahman crosses, 53-55
— American Brahman, 53
— Barzona, 55
— Beefmaster, 55
— Braford, 54
— Brahmental, 55
— Brangus, 54
— Charbray, 55
— Sahiwal, 55
— Santa Gertrudis, 54
Breeding season management, 138-139
British breeds, 37-44
— Angus, 38
— Belted Galloway, 39
— Devon, 39

British breeds *(Contd.)*
— Galloway, 39
— Hereford, 39
— Lincoln Red, 40
— Luing, 40
— Milking Shorthorn, 41
— Murray Grey, 41
— Polled Hereford, 41
— Red Angus, 42
— Red Poll, 42
— Scotch Highland, 42
— Shorthorn, 42
— South Devon, 43
— Sussex, 44
— Welsh Black, 44
Brucellosis (Bangs), 125, 140
By-products of beef, 2, 9, 11

C

Calving difficulty, 70-71
Calving season management, 136-138
Carcass grade factors, 373-376
— feeding programs and time on feed, 375-376
— weight, frame size, and sex, 373-375
Carcass quality grade, 363-365
— effect of marbling, 363-365
— effect of maturity, 363-365
Carcass yield grades, 360, 366-370, 379
— factors affecting, 366-367
— method for estimating, 368-370
— percent boneless trimmed, round, rib, loin, and chuck, 366
— percent fat, bone, and retail cuts, 360
Castration, 124
Cattle cycle, 145-146
Choice of feeding systems, 340-351

Choice of feeding systems *(Contd.)*
— beef herd, 343-351
— calves from weaning to slaughter, 340-343
— winter feeding systems, 351
Choosing a breeding system, 59-66
Classification of breeds, 57-60
— mature size, 57-58
— milk production, 57-58
— muscle type, 57-58
— sire, dam or two-way, 59-60
Colostrum, 124, 138
Computer comparison of different feeding systems, 340-343
Computerized ration balancing, 323-340
— actual bid chart, 339
— actual rations and supplements, 335-337
— case example, 324-327
— computer input form, 325-336
— computer programs available, 323
— economic projection, 333-335
— feedsheet formulation, 330-331
— gain and feed use projection, 331-332
— grade and sale price projections, 332-333
— ration formulation results, 327-328
— simulated feed requirements and projected profits, 338
Consumption of beef, 1
Continental European breeds, 44-53
— Aubrac, 44

Continental European breeds *(Contd.)*
— Beef Friesian, 46
— Blonde d' Aquitaine, 46
— European Brown Swiss, 46
— Charolais, 46-47
— Chianina, 47
— Gasconne, 47
— Gelbvieh, 48
— Limousin, 48
— Maine-Anjou, 48
— Marchigiana, 48
— Meuse-Rhine Issel (MRI), 49
— Montbeliard, 50
— Normande, 50
— Norwegian Red, 51
— Parthenaise, 51
— Piedmont (Piemontese), 51
— Romagnola, 52
— Salers, 52
— Simmental (Fleckvieh, Pie Rouge), 52-53
— Tarentaise, 53
Corn-corn silage rations
— expected performance, 268-271, 321
Corn plant residue, 400-401
Corn silage additives, 391-392
— biological additives, 391-392
— fermentable carbohydrates, 391
— urea, ammonia, and limestone, 392
Corn silage feeding systems, 134-136
Corn silage quality and quantity, 321, 391-397
— brown mid-rib corn, 396
— drought damaged corn, 394-395
— factors affecting, 391
— fineness of chop and packing, 397
— high sugar corn 396
— maturity, 393-395

Corn silage quality and quantity *(Contd.)*
— opaque-2 (hi-lysine) corn, 396
— silage additives, 391-392
— when to harvest, 393
Costs of feeding (weaning to slaughter), 5-8
— breakeven, 7
— feed, 7
— non-feed, 7
— production, 7
Costs of producing beef, 13-18
— management errors, 16-17
— practices that affect cost, 15
Costs of producing calves, 3-5
— breakeven, 4
— feed and operating, 4
— land, 5
— machinery and equipment, 4
— total annual, 3, 131
Cow-calf *vs* cow-yearling *vs* stocker program, 345-349
Cow culling criteria, 98-99, 105
Cow herd health calendar, 127-128
Creep feeding calves, 140, 304-305
Crop residues, 135-136, 143
Crossbreeding, 61-65
— three breed terminal cross, 62
— three breed rotational cross, 62
— two breed crisscross, 63
Crude fiber, 205-206, 246
Cutability (*see* Yield grade)

D

Dehorning, 124
Detector pads, 118
Diarrhea, 173
Diseases of cow herd, 96-97, 126

Diseases of cow herd *(Contd.)*
— Anaplasmosis, 97
— Bangs (*see* Brucellosis)
— Blackleg and Malignant Edema, 123, 126
— Brucellosis, 96, 124-125
— BVD, 123, 126
— Grass Tetany, 126-127
— IBR, 123, 126
— Leptospirosis, 96-97, 124-125
— PI_3, 123, 126
— Scours (Clostridium), 124
— Trichomoniasis, 97
— Tuberculosis (TB), 96
— Vibriosis, 97, 126
Double muscling, 78-79
Dressing percentage, 357-358
— effect of ration and grade, 357
— effect of time off feed and water, 358
— mature cows, 358
Dry matter intake, 265-273
— adjustment for feeding rumensin, 269-270
— factors causing poor intake, 271-273
— for growing and finishing cattle, 265-273

E

Edible products of beef, 2
Energetic efficiency, 8
— related to grain feeding, 9-14
— related to slaughter weight, 8
Energy utilization, 203-218
— crude fiber, 205-206

Energy utilization *(Contd.)*
— digestion and metabolism, 203-204
— digestive disturbances, 203-205
— fat, 206
Energy values of feeds, 206-210
— digestible energy (DE), 208
— gross energy (GE), 207
— metabolizable energy (ME), 208
— net energy (NE), 209-210
— total digestible nutrients (TDN), 208-209
Estimated breeding value (EBV), 29, 80-82
Estimating live yield grade, 372-373
Estrous synchronization, 120-122
— progestogen and beef A.I., 121
— prostaglandins and beef A.I., 121
Estrus (*see* Heat detection)
Ether extract (fat), 246
Expected progeny differences (EPD), 80-82
External parasites, 123, 125

F

Facilities for cow and calf, 382-386
— corals, 384-385
— equipment, 384
— fencing, 384
— hay storage, 385
— shelter, 382-384
— water, 385
Facilities for growing and finishing cattle, 386-389

Facilities for growing and
  finishing cattle *(Contd.)*
— conventional, 388
— dirt lot, no shelter, 387-388
— effect of environment, 386
— slotted floor confinement,
  388-389
— solid floor confinement,
  388
Fall calving *vs* spring calving,
  243-244
Fat as energy source, 206
Feed additives and mineral
  supplements for all
  growing and finishing
  rations, 310
Feed analysis, 245-255
Feed composition values,
  248-253
— adjustments, 257-260
Feed processing, 260-265
— economic advantage, 264
— expected improvement
  in performance, 263
— exploding, 262
— extruding, 262
— grinding, 261
— high moisture storage,
  262-263
— micronizing, 262
— pelleting, 261
— popping, 262
— roasting, 262
— rolling, 261
Feeder cattle management
  (newly arrived cattle),
  162-165
— energy level and source,
  163
— guideline rations, 165
— minimizing stress, 162
— nutrition, 163-165
— protein requirement, 164

Feeder cattle management
  (newly arrived cattle)
  (Contd.)
— vitamin and medicated
  supplements, 164
— water, 164
Feeder cattle selection,
  149-154
— body condition, 151
— breed effects, 150
— economics, 153
— feeder grade, 153-154
— frame, 150
— health, 152
— heifers *vs* steers, 149-150
— yearlings *vs* calves, 151
Feeding cow herd for 12
  month reproductive
  cycle, 240-242
Feeding grain on pasture,
  305-306
Feeding replacement heifers,
  242-244
Female selection factors,
  96-103
— age, 102
— health, 96-97
— performance standards,
  103, 142
— reproductive efficiency,
  97-99, 101, 138
— type, 99
Femininity, 101
Fertility,
— bulls, 72
— cows, 97-99, 116
Filling silos, 405-406
Forage testing, 136
Forage management systems, 131-136
Formulating rations, applying energy values, 210
Founder, 171

412 Index

Frame size, 74-75
— impact on nutrient requirements, 178-180
Free-choice mineral mixtures, 297-298

G

Generation interval, 24-25
Gomer bulls, 118
Grass tetany, 126
Growth stimulants and feed additives, 200-201
— recommended products, 201
Guideline rations for beef herd, 290
— dry 1100-lb mature cow, middle third of pregnancy, 291
— dry 1100-lb mature cow, last third of pregnancy, 291
— lactating 1100-lb cow (average milker), 292
— lactating 1100-lb cow (heavy milker), 292
— mature herd sires, 293-297
— two-year-old pregnant heifers, 290
— weaned heifer calves, 290
Guideline rations for bulls on 140-day feed test, 299-303
Guideline rations for newly arrived feeder cattle, 165
Guidelines for feeding bull calves, 299
Guidelines for growing and finishing beef, 306-310
— high grain rations, 307-309
— high roughage rations, 306-307

Guidelines for growing and finishing beef *(Contd.)*
— protein-mineral adjustments for frame, sex, and age, 308
— rations, other than corn silage, 308
— supplements for corn silage rations, 309-310

H

Handling cattle prior to market, 159
Hay crop silage, 397-398
Hay feeding systems, 134-135
Health program for feeder cattle, 166-176
— commercial treatments, 173-176
— common signs of illness, 168
— handling sick cattle, 167-168
— summary of health management, 170-171
Heat detection (estrus), 117-119
— aids, 118
— factors affecting heat, 117
— recognizing a cow in heat, 117
Herd health program for cow herds, 122-128
— basic principles, 123
— breeding herd, 123-124
— calves, 124-125
— herd health calendar, 127-128
Heritability,
— defined, 22
— estimates, 23
Heterosis, 61, 64-65
High moisture corn, 398-399
— proprionic acid treatment, 399

I

IBR, 126, 138, 141
Identification systems, 110-116
— ear tags, 114
— freezebrand, 112-113
— hotbrand, 113-114
— nose print, 111-112
— numbering systems, 114-116
— tattoo, 110-111
— use of numbers in registered name, 115
Inbreeding, 65-66
Independent culling levels, 28
Infectious Bovine Rhinotracheitis (IBR), 166
Internal parasites, 123, 125

J

Judging show steers, 376-377

L

Leptospirosis, 125
Linebreeding, 65-66
Lump jaw, 172

M

Magnesium deficiency (*see* Grass tetany)
Management calendar (beef cow herd), 136-143
Manure handling on slotted floor, 338-390
— accumulation and value, 390
— deep pit, 389

Manure handling on slotted floor *(Contd.)*
— flush flume, 389
— slotted floor scrape, 389
Market information, 159
Marketing feeder cattle, 156-161
— auction sales, 157
— direct sales, 157
— special feeder sales, 157
Mating decisions, 83-91
Mature weight, 70
Methods of marketing, 356-361
— selling direct, 357
— selling for freezer trade, 360
— selling on central public markets, 359-360
Milk production, 57-60, 105-107
Mineral requirements, 187-195
— calcium, 188
— magnesium, 191
— phosphorus, 189-190
— potassium, 189-190
— salt, 191
— sulphur, 191
— trace minerals, 192-194
Minerals, balancing a ration for, 194-199
Moisture content of feeds, 274-280
— correcting for, 274-279
— testing, 279-280
Most probable producing ability (MPPA), 33-34, 107
Muscle patterns, 76-77

N

Net energy (NE), 209-218
— adjusting for frame, breed,

Net energy (NE) *(Contd.)*
    environment, and growth stimulants, 213-218
— requirements for growing and finishing cattle, 211-218
— using requirement tables for steers, heifers, bulls, 210-216
Nitrogen Free Extract (NFE), 246
NPN and mineral treatment levels, 318
— for corn silage, 318
— high moisture snapped ear corn, 319
NPN silage treatment, 225-226
NPN treatment of corn silage, 311-315
— economics of, 315
— performance of cattle, 311-313
— recommendations, 313-315
Nutrient requirements of breeding cattle, 233-244
— dry pregnant cows, 236
— effect of cold weather, 238-239
— lactating cows, 237
— two-year-old heifers in last third of pregnancy, 235
— weaned heifer calves, 234
Nutrient requirements of growing and finishing cattle, 226-233
— level of grain and protein requirements, 229
— other factors affecting DM intake and gain projections, 230-231
— typical ration formulations for varying weight and energy levels, 231-233
— weight groupings, 226-229

O

Optimum slaughter weight, 353-356
— effect of grade and market price, 355
— for continually full feedlots, 356
— relationship to cost of gain, 354
Other breeds and crosses, 57
— Beefalo, 57
— Hays Converter, 57
— Longhorn, 57
Outcrossing, 65

P

Parainfluenza 3 (PI_3), 166
Pasteurella, 126, 141
Pasture improvement, 132-133
Pasture renovation, 133-134
Performance pedigree, 69
Performance records, 30-35, 103
— most probable producing ability (MPPA), 33-34, 107
— postweaning, 34-35
— reproduction, 35
— weaning, 31-32
Pinkeye (Keratoconjunctivitis), 125
— control, 140
Polioencephalomalacia, 172
Possible change (PC), 82
Pregnancy test, 123, 140-141
Protein requirements, 219-222
— for gain, 219
— for maintenance, 219
Protein supplement guidelines, 223-224
Protein utilization, 222-223
Purebreeding, 65-67

Purebreeding *(Contd.)*
— inbreeding, 65-66
— linebreeding, 65-66
— outcrossing, 65

R

Ration formulation, 280-288
— balancing methods, 281-288
Replacement heifer selection, 98-99, 104
Reproduction,
— performance data, 35
Resource requirements for cow herd, 128-131
— annual production costs, 131
— buildings, equipment and machinery, 129
— investment in cattle, 129
— labor, 130
— land for hay and pasture, 128-129
Retail and wholesale cuts, 370-371
Rumensin, 202-203

S

Sampling feeds of analysis, 255-256
— dry grain, 256
— hay, 256
— high moisture corn, corn silage, and haylage, 255-256
Selecting a breed, 37-57
Selecting a feedlot, 349-350
Selection differential, 22
Selection index, 28
Selection response, 22-26
Selection to improve overall merit, 27-28
Semen quality, 116

Shrink, 160-161
Silage system planning, 401-402
Silo structures and selection, 401-405
— bunker horizontal silo, 404-405
— tower silo, 403-404
Sire selection factors, 69-82
— pedigree data, 68-69, 83-91
— performance data, 70-71, 83-91
— physical traits, 72-78, 83-91
— National Sire Evaluation, 80-81, 83-91
— precautions, 91-92
— price, 93
— progeny data, 79-80, 83-91
— summary, 92-93
Skeletal size (*see* Frame)
Small grain whole plant silage, 397
Space requirements for, 382-384
— confinement housing, 382
— conventional shelter, 382
— corals and working facilities, 383
— feeders, 382
— water, 382
Straightbreeding, 59-61
Structure problems, 76
Synchronization (*see* Estrus synchronization)
Systems of selection, 26-27
— family selection, 27
— mass selection, 26
— pedigree estimation, 27
— progeny testing, 26-27, 79-80
— use of correlated traits, 27

T

Tandem culling, 28
Testicle size and shape, 73-74
Tools for cattle improvement, 29-30
Total digestible nutrients (TDN), 208-209, 246

U

Udder and teat evaluation, 102-103
Urea and other ammonia compounds, use of, 224-225
Urinary calculi, 173
U.S., cattle numbers, 147
Uterine infections, 172

V

Vibriosis, 126, 138
Vitamins
— A, 124, 136-137, 258, 195, 198

Vitamins *(Contd.)*
— B vitamins, 200
— D, 124, 136, 198-199

W

Water analysis, guide to use, 184-187
Water requirements and quality, 136, 164, 180-187
Weaning management, 140-141
Weaning weight, 70, 103
Winter feed budgets, 293-297
— using corn silage, 295
— using hay, 294
— using hay and corn silage, 296
Winter feeding management, 136, 142-143

Y

Yearling weight, 70, 103
Yield grade, 8, 13